THE *Bar U*

THE Bar U

& CANADIAN RANCHING HISTORY

SIMON M. EVANS

Library and Archives of Canada Cataloguing in Publication

Evans, Simon M.
The Bar U and Canadian ranching history / Simon M. Evans.

(Parks and heritage series, ISSN 1494-0426 ; 8)
Includes bibliographical references and index.
ISBN 1-55238-134-X

1. Bar U Ranch National Historic Site (Alta.) 2. Ranching—Canada,
Western—History. I. Title. II. Series.

FC3670.R3E92 2004 971.23'4 C2004-903075-2

Second Printing 2005
Third Printing 2007

Published by the University of Calgary Press
2500 University Drive NW, Calgary, Alberta, Canada T2N 1N4
www.uofcpress.com

This book has been published with the help of a grant from the Canadian Federation for the Humanities and Social Sciences, through the Aid to Scholarly Publications Programme, using funds provided by the Social Sciences and Humanities Research Council of Canada.

We acknowledge the financial support of the Government of Canada through the Book Publishing Industry Development Program (BPIDP) for our publishing activities. We acknowledge the support of the Alberta Foundation for the Arts for this published work.

Printed and bound in Canada by Houghton Boston
♾This book is printed on acid free paper
Cover design, page design and typesetting by Mieka West.

Canada Council for the Arts Conseil des Arts du Canada

Front cover photograph: Charlie Millar, cowboy at the Bar U. Glenbow, NC-17-1.
Back cover image: O.N. Grandmaison, "July" (detail). By permission of Ranchmen's Club, Calgary.

PARKS AND HERITAGE SERIES
ISSN 1494-0426

The Parks and Heritage series focuses on topics related to national parks and historic sites in North America. Both historical and contemporary, these books raise our awareness about the many facets of national parks, including the warden service, religion, ethnohistory, and environmental studies.

Where the Mountains Meet the Prairies: A History of Waterton Country *Graham A. MacDonald* · No. 1

Guardians of the Wild: A History of the Warden Service of Canada's National Parks *Robert J. Burns with Mike Schintz* · No. 2

The Road to the Rapids: Nineteenth-Century Church and Society at St. Andrew's Parish, Red River *Robert J. Coutts* · No. 3

Chilkoot: An Adventure in Ecotourism *Allan Ingelson, Michael Mahony, and Robert Scace* · Copublished with University of Alaska Press · No. 4

Muskox Land: Ellesmere Island in the Age of Contact *Lyle Dick* · No. 5

Wolf Mountains: A History of Wolves along the Great Divide *Karen R. Jones* · No. 6

Protected Areas and the Regional Planning Imperative in North America Edited by *Gordon Nelson, J.C. Day, Lucy M. Sportza, James Loucky, and Carlos Vasquez* · Copublished with Michigan Sate University Press · No. 7

The Bar U and Canadian Ranching History · *Simon Evans* · No. 8

Heritage Covenants and Preservation: The Calgary Civic Trust *Edited by E. Anne English, Kimberly E. Haskell, Sara Jennings, Michael McMordie, Frits Pannekoek* · No. 9

To those who brought the Bar U Ranch National Historic Site into being, and all those who have guided it through its first decade.

CONTENTS

MAPS

* MAPS INCLUDED IN THE COLOUR SECTION

Acknowledgments

This book is dedicated to all those who helped to bring the Bar U Ranch National Historic Site into being: those working for Parks Canada, the members of the Friends of the Bar U Historic Ranch, and the ranching community in general. The historical research carried out on the ranch has been a team effort. I owe an enormous debt to Alan McCullough.* His meticulous research in the National Archives helped to build a picture of the relationships between the ranch and the federal government. He also pieced together much about the Stimson family. Alan's analysis of contemporary diaries and his work on the "seasonal round" in ranching country was also very helpful. In Calgary, Barbara Holliday was indefatigable in her pursuit of oral history respondents. The transcripts of her interviews, and those carried out earlier by David Finch, add a personal and human flavour to the text. Barbara organized a wonderful collection of photographs, generously donated by individuals and families with Bar U connections. She has been a generous "cheerleader" for this book since its inception, and she read and commented on the whole manuscript.

I would like to thank Jim Taylor of Parks Canada for drawing me into the Bar U project in 1991, and for his ongoing support for my work which culminated in his reading the manuscript with a discerning eye. Bill Yeo also helped to create a productive and informal working environment at Parks Canada and put his immense breadth of knowledge at my disposal. He also evaluated the book and helped to smooth out some rough edges. In a more formal way, I acknowledge with gratitude the financial support of Parks Canada in funding several new maps which were vital to the book and judged to be useful for the interpretation of the historic site.

I remember with respect and gratitude the enthusiasm and drive of Stan Wilson and Omar Bradley, early "power houses" of the Friends group, who have died. I also enjoyed my meetings with Ann Clifford and I am so glad she finished her own book before her death. Her niece Jeannette and her husband Corky Rousseau have been helpful and supportive over the years and were kind enough to let me use the Clifford/Alwood papers over an extended period. Willa Gordon was always hospitable and made available the Elsie Gordon papers which contained some crucial material. While I was writing, I became reacquainted with Frank Jacobs, long-time rancher, author and editor of Canadian Cattlemen. It was great to be able to run ideas by him and to draw on his experience. I miss him since his death in November 2002.

Joy Oetelaar did some valuable research for Parks Canada using local newspapers, and through that work she came to know and respect George Lane so that she could write about him with authority. She was kind enough to vet my chapters on Lane. Bruce Roy, editor of *Fetlock and Feather* and Percheron enthusiast, helped with information and insights on all horse matters. Those presently guiding the fortunes of the Bar U Ranch National Historic Site have been unfailingly

* Alan McCullough was for a long time a historian with Parks Canada. He retired recently. The same applies to Barbara Holliday and Bill Yeo. All of these are still researching and writing, but I was fortunate to have them as colleagues from 1991 to 1995.

helpful: Ken Pigeon and Jean Gallup of Parks Canada, and Ralph Nelson, Don Moore, Hank Pallister, and Doug Nelson from the Friends.

Like all those writing about western Canada, I have relied heavily on the wonderful library and archival collection at the Glenbow. Doug Cass, Lindsay Moir, and Jennifer Hamblin have, as usual, gone well beyond the call of duty in appraising me of new acquisitions, tracing references, and maintaining their interest in the project over several years. Ric Lalonde drew the final copies of the maps and it is always a pleasure to work with a consummate professional.

More recently, I have enjoyed working with the team that Walter Hildebrandt has put together at the University of Calgary Press: John King, Mieka West, and copy editor Peter Enman.

My wife Stephanie has been living with the Bar U Ranch since 1991, when we returned to the west to work for Parks Canada. She probably knows Stimson, Lane and Burns just about as well as many of our relations! She has been a tremendous help by being an independent thinker, a thoughtful reader, and a constant support.

Preface

"... for according to what the wise say, the art and beauty of historical composition is, to write the truth ..."[1]
– Captain Bernal Diaz Del Castillo

The Prince of Wales displayed exemplary ability to grasp essentials and to express them in a pithy manner when he wrote "Some Ranch!" in the visitors' book of the Bar U Ranch in the fall of 1919.[2] During his visit he had been impressed with the size and diversity of the operations at the ranch, by the beauty and expansive openness of the foothills, and by the warmth of his welcome. He visited the ranch in the fourth decade of its existence and relished rubbing shoulders with some of the men who had contributed to the establishment of the ranch during the 1880s. But his remark has remained appropriate through succeeding decades and resonates even today. For much of its 122-year history the Bar U Ranch can claim to have been one of the most famous ranches in Canada. Its reputation is firmly based on the historic role which the ranch played, on its size and longevity, and by its association with some of the remarkable people who helped to develop the cattle business and to build the Canadian west. My aim in writing this book is to tell the story of the Bar U. However, it is impossible to separate the history of the ranch from the context of the times which shaped it. Moreover, the long history of the ranch illuminates the more general history of ranching in western Canada. The text which follows contains both narrative and analysis, and seeks to tell what happened and to explain why it happened.

The history of the Bar U Ranch falls into three historic periods, each dominated by an important figure: the era of the North West Cattle Company and manager Fred Stimson, 1881 to 1902; the George Lane era from 1902 to 1927; and the Pat Burns era, 1927 to 1950.[3] However, it is more than half a century since Burns and Company sold the heart of the Bar U to an enterprising young rancher, and an attempt is made in a brief concluding chapter to sketch the ongoing history of the ranch until the present.

The Bar U Ranch, or to give it its formal name, the North West Cattle Company, was one of the first major corporate ranches to be established in the Canadian prairie west. It came into being when Sir John A. Macdonald's Conservative government was persuaded "for the purposes of the Dominion" to modify the Dominion Lands Act to allow cattle companies and individuals to lease large blocks of pasture land on which to graze their herds. The ranch was born during a frantic period when investors in eastern Canada and Great Britain vied with each other to "get in on a sure thing." From the perspective of the twenty-first century, we have to make a real effort to try to imagine how investors, from wealthy individuals to ordinary people depositing their savings in building societies, were one and all caught up in the hysteria of the "beef bonanza" on the great plains of the United States. For a time in the 1870s and 1880s, stock in cattle companies and ranches attracted the same kind of interest, and paid the same extraordinary dividends, as did the "high-tech" stocks of the 1990s. In Canada, much of the interest proved to be speculative, and many paper cattle companies never got beyond the stage of publishing a glowing prospectus. However, four large corporate ranches were established. Two, the Cochrane Ranch Company and the North West Cattle Company, represented direct investment of Canadian capital, while the Oxley and the Walrond ranches were established by Canadian entrepreneurs who sought financial backing in Great Britain. Because of their size and their close connections with the government in Ottawa, these four ranches made up a "cattle compact" which dominated the Canadian ranching industry during its first decade.[4]

The 1880s was a revolutionary period of rapid change in the Canadian west. Risk capital was deployed on a scale not previously witnessed in the area. Ranches were established and carefully selected cattle herds were driven in from Montana. Soon, reserves and the NWMP were supplied with beef from the Canadian range, and the completion of the transcontinental railway linked to the ports of the St. Lawrence was eagerly awaited so that surplus cattle from western Canada could be sold in Britain. The Bar U played a key role in these developments. It suffered none of the early calamities experienced by the Cochrane herds, nor the financial problems faced by John Craig at the Oxley. Fred Stimson, the manager of the North West Cattle Company, had considerable experience in practical stock rearing and he – unlike the manager of the Walrond – was a full-time resident of the west with no competing responsibilities. Stimson was able to acquire key government supply contracts which ensured that his backers received some early returns on their investments. He also gained the respect of other ranchers and played an important part in the establishment of the first stockmen's association. The success of the North West Cattle Company owed much to the backing it received from the Allan family of Montreal. During the 1870s, the ships of the Allan Line began to carry more and more live cattle, from the United States and from Ontario, eastward across the Atlantic. As the cattle trade developed into a multi-million dollar export business, so Sir Hugh Allan determined to invest in a ranch in western Canada. Over the next twenty years, capital earned in the web of companies owned by the Allans underwrote the establishment and development of the Bar U. Equally important was the fact that the connections and experience of the Montreal-based directors secured a privileged position for Bar U cattle on the long journey to Liverpool and London. This was the first time that unique links to transport and markets favoured the Bar U Ranch; it was not to be the last.

The fame of the Bar U rose to new heights under the ownership of the extraordinary entrepreneur George Lane. Cattle raised at the ranch were fattened on other ranches owned by Lane, who controlled large sections of the remaining open range along the Bow and Red Deer Rivers. A continuous process of reassessment and innovation took place as Lane experimented with running imported stocker cattle under open range conditions on the eastern mixed-grass prairies, while intensifying his cow-calf operation at the Bar U by introducing irrigation and by growing alfalfa. Lane enjoyed a close personal relationship with James T. Gordon, and through him an ongoing link with the Winnipeg cattle shipping company of Gordon, Ironside and Fares. Once again Bar U cattle would receive exceptional consideration on their way to foreign markets. In the decade before the First World War, Lane diversified by building up an internationally famous stud of Percheron horses. His personal stature as the confidant of governments, a respected cattleman and horse breeder, founder of the Calgary Stampede and friend of the Prince of Wales, further enhanced the profile of the Bar U.

Pat Burns bought the Bar U in 1927. His purchase drew the ranch into an agribusiness empire which displayed a remarkable degree of vertical integration. Some of his ranches were cow-calf operations, while others specialized in fattening young cattle. Market-ready cattle were finished at his Calgary feedlot, and used as required in his meat-packing plant. The meat was then distributed through his network of retail outlets. The decade during which Burns ran the ranch personally was marked by the use of conservative, traditional methods. His life-long involvement in the cattle business meant that he had well established ways of organizing and running a multi-ranch operation. Moreover, the decade coincided with a period of unprecedented environmental and economic turmoil. Burns brought his ranches through the Depression by cutting production costs to a minimum and by absorbing consistent but manageable losses. The company which took over after Burns' death benefited from a period of rising prices and from the insatiable demands of a wartime economy. The last decade of the Bar U as a corporate identity was one of its most prosperous.

I have explored each major historical period of the Bar U's history by focusing my attention on four themes: the people who managed and worked the ranch; the economic context of the times and the markets the ranch depended upon; the methods used and the evolving technology of the cattle business; and, finally, the growth of the headquarters site and the gradual transformation of parts of the range as production was intensified.

In a recent discussion it was suggested that historians are under considerable pressure to "spice up" or glamorize history.[5] Characters like Stimson, Lane, and Burns tower larger than life over the oral and written record, and the story of their achievements needs no embellishment. I have gone as far as the information available has allowed me to outline some of the personal characteristics of these interesting and complex men, but their feelings, emotions, and inner motivations remain a mystery. I can only hope that in due course they may attract the attention of biographers, novelists or dramatists. The big ranch required a large and diverse group of people to make it run smoothly. I have described the Bar U community in each era and discussed the general nature of the labour force, before proceeding to tell the stories of a few of the men and women who lived and worked at the ranch. First Nations people have played an important part in the history of the ranch, and I have outlined their long association with the Bar U by concentrating on a period during which their contribution was particularly vital.

Broadly speaking, each historic period saw shifts in the relative importance of the factors of production. In Stimson's day almost all investment was in cattle, for land and labour were cheap. Lane bought deeded land, increased the cropped acreage, and developed irrigation; however, these investments in infrastructure never balanced his outlay on cattle and his herd of Percheron horses. The Second World War brought far-reaching changes. Labour costs rose dramatically, as did the value of the land and the "opportunity costs" large landholdings represented. The rapid changes and challenges of the 1950s were better handled by smaller, highly mechanized, family-operated units.

Access to changing markets spelled success or failure to cattlemen. Initially, contracts to supply the North West Mounted Police, the Indians on their reserves, and the workers building the railways, were of immense significance. The first ranches which could offer large lots of cattle on a reliable and consistent basis secured remunerative government contracts which virtually assured their profitability. As herds increased and more fat steers were produced than could be absorbed locally, so the surplus was moved eastward along the newly completed railway and across the North Atlantic to Great Britain. For two decades the British market absorbed the biggest and best grass-fed steers from ranches in the foothills. This important trade grew in response to particular circumstances and confounded basic economic principles. The development of sophisticated techniques to chill meat rather than to freeze it favoured producers in the southern hemisphere, particularly Argentina. The years leading up to the First World War saw the demise of the British market for Canadian cattle. For the past century, the fortunes of the Canadian cattle industry have been closely tied to the demand for cattle in the United States. Chicago has absorbed both fat cattle and feeders, and exports southward have improved domestic prices. But dependence on the huge market south of the border has meant that Canadian cattlemen have been subject to the roller coaster of successive "cattle cycles" as well as to the occasional imposition of devastating protective tariffs. For most of its history, because of its size and the knowledge and connections of its owners and managers, the Bar U Ranch was able to establish strategic relationships within these shifting markets.

The seasonal round of activities on a contemporary ranch in the foothills would be immediately familiar to men like Herb Millar and Charlie Mackinnon, who managed the Bar U herds around the turn of the century. Calving, roundup and branding, weaning and the drive to summer pasture, gathering for market, and the drudgery of winter feeding, still take place at their appointed times. Of course, the broad parameters of continuity mask a business completely transformed by the impact of science, mechanization, and public taste. It is hard to develop a coherent narrative which describes the constant innovations and refinements which have taken place in each seasonal operation from breeding to shipping. At the Bar U there was always a tension between what could be done to ensure the survival of calves or the weight gain of yearlings, and what the adoption of a new technique meant in terms of costs and returns. In general terms, the scale of operations at the Bar U militated against the early adoption of more intensive methods and kept alive some of the ethos of the open range. As long as cattle were cheap and profit margins substantial it did not make business sense to pamper cows on a large ranch. This was not a reflection of poor management or callous behaviour on the part of riders, but rather adherence to policies which, because of economies of scale, ensured acceptable and consistent returns.

The first maps and photographs of the headquarters site show a modest huddle of buildings and corrals fringing Pekisko Creek. The unfenced and unmodified range spread out in all directions around it. Over the next twenty years, more and more flat land was cut bi-annually for wild hay, and huge fields were fenced to keep out stock during the growing season. Under Lane, the site reached its maximum complexity as barns to house and show off the Percheron horses were built at the eastern end of the flood plain. An ambitious irrigation scheme intensified production of tame hay, and land was broken to grow oats and wheat. During the Burns era, several buildings were lost to fires and others, formerly used for stock, were converted to store grain or shelter machinery. Throughout the long history of the ranch, the native fescue grassland, which covered the greatest proportion of the ranch lands, was carefully managed as winter range. It remains a rare and valuable ecological asset, and an important legacy of the Bar U Ranch.

The historiography of the Canadian range has been dominated by a search for origins. More than a quarter of a century has passed since David Breen proposed his thesis that the ranching frontier evolved in very different ways to the north and south of the international border.[6] He argued convincingly that ranching in Canada developed within a political and legislative framework which had no parallel in the United States, and he showed that the social milieu which sprang up in the Alberta foothills during the 1880s differed fundamentally from ranching communities south of the 49th parallel. Much of the research done through the 1980s continued to use Breen's organizational model, and what had been proposed as a revisionist approach became the accepted paradigm. It is time to frame entirely new questions which reflect new visions of American wests and focus on the history of the cattle industry during the twentieth century.

Three important books published during the 1990s establish a new context for studies of ranching history. Terry Jordan's *North American Cattle Ranching Frontiers* stressed the variety of origins and the complex regional adaptations which occurred as ranching diffused over the plains and mountains. Richard Slatta compared and contrasted the cowboy experience throughout the hemisphere in *Cowboys of the Americas*, while Paul F. Starr's book *Let the Cowboy Ride* described the evolution to the present of five very different ranching systems from Texas and New Mexico to Wyoming, Nebraska, and Nevada.[7] Each of these books in its own way celebrates the diversity and complexity of ranching history. Together, they suggest that there are not two 'ranching frontiers' divided by a national boundary but rather any number of regional adaptations. The story of how Canada's range lands were occupied for ranching displays many distinctive characteristics, but may be studied as one of several regional systems within the grasslands of North America.

The "Cowboy Conference" that took place as part of the Glenbow Museum's exhibition on the Canadian Cowboy in 1997 explored some new perspectives on ranching history. Selected papers were published under the title of *Cowboys, Ranchers and the Cattle Business.*[8] While some participants looked at old questions in new ways, others broke new ground by examining the plight of ranchers during the Great Depression, and by following the fortunes of one ranch family into the modern period. Two of the contributors have gone on to publish books on ranching.[9] Warren Elofson, in *Cowboys, Gentlemen and Cattle Thieves*, challenges the thesis that Canadian ranching society was more cultured and law abiding than was the American version. He suggests that Breen's thesis "has been pushed too far" and needs to be reined in.[10] He makes the case that Canadian cowboys were more disorderly and violent, and less deferential to authority, than the followers of Breen and Thomas would have us believe. Elsewhere, Elofson stresses, in a general way, the importance of the physical environment in promoting technological change. In the United States, proponents of environmental history have taken this further, and interpretive models based on the ecological relationships between cattlemen and their ranges are beginning to emerge.[11] In *Trails and Trials*, Max Foran explores economic and institutional aspects of the cattle industry.[12] Ranchers fought with successive governments to obtain security of tenure, their effectiveness waxing and waning as the demand for beef fluctuated through two world wars and the Great Depression. He handles both the institutional framework and the economic side of ranching with detailed authority. Ian MacLachlan's focus in *Kill and Chill* is on the restructuring of beef production and the meat-packing industry in the modern period;[13] but he includes a perceptive introductory chapter on ranching history, and his conclusions concerning the close relationships between consumers, processors, and ranchers throw as much light on the past as they do on the present.

Studies of individual ranches are vastly outnumbered on the ranching historian's bookshelf by volumes dealing with particular themes, time periods, or lives lived on the range. Among the handful of exceptions are Harmon Ross Mothershead's book on the Swan Land and Cattle Company and William Pearce's work on the Matador Land and Cattle Company.[14] Both tell their particular stories while at the same time illuminating the history of ranching as a whole. In Canada, two huge spreads in British Columbia have been memorialized. Nina Woolliams has told the remarkable story of the Douglas Lake Cattle Company, while Dale Alsager has described life on the Gang Ranch.[15] Edward Brado sketched the early history of most of the major ranches in Alberta, and his book remains an excellent point of departure, as do journal articles on specific ranches.[16] In my own book on the E.P. Ranch, the ranch itself had to share the focus with the story of its charismatic owner.[17] This history of the Bar U Ranch is the first book-length study of one of the great corporate ranches of the Canadian prairie west.

Focus on a particular ranch over a long period of time has much to commend it. It allows one to probe at a level of detail which would be impossible to sustain if the study was more wide ranging. The utility of models can be put to the test using a detailed case study, and the long timeline means that it can throw light on a number of generalizations applicable to different time periods. At the end of the book I would like to be able to comment on what the Bar U tells us about the origins of ranching in western Canada; to weigh the relative significance of the physical environment versus economic and institutional factors as determinants in the development of ranching; to reflect on the reasons for the longevity of the ranch; and to review the role played by the Bar U as an ecological reserve of native range.

Let me conclude this introduction by returning to the title of the book. Can a study of a single ranch claim to throw light on Canadian ranching history? It is a tall order, for ranches in the interior valleys of British Columbia and those on the semi-arid mixed-grass prairies of southwestern Saskatchewan have very different stories from those in the Alberta foothills. However, I think it can. The Bar U Ranch came into being as a result of the National Policy and was the result of eastern Canadian entrepreneurial initiative and capital. During its first decades, the ranch supplied Indians and the North West Mounted Police with beef, and played a geopolitical role by helping to establish the economic independence and sovereignty of western Canada. Later, the ranch contributed to the flow of live cattle from the Dominion to the heart of the British Empire during a period when the cattle trade played a significant part in Canada's export-driven economy. Under George Lane, the fame of the Bar U transcended the newly established provincial boundaries, and the ranch became emblematic of a beloved and mythic theme in Canada's settlement history. During the Burns era, a fundamental shift occurred in Canada as emphasis on primary production gave way to secondary manufacturing and the growing importance of the tertiary service sector of the economy. Agricultural dominance was superseded by urban and industrial concerns, and the Bar U ceased to play a part on the national stage. Nevertheless, I think that one can claim that for much of its existence the Bar U Ranch played a role which far transcended its geographical location. The history of ranching in Canada is a theme of national historical importance, and the Bar U Ranch is the place designated to commemorate this history. The ranch headquarters was purchased by Parks Canada in 1991 and opened as a National Historic Site in 1995. The story of the ranch does contribute to the understanding of Canadian ranching history, and I hope the following chapters will demonstrate this.

Some 'housekeeping' matters: in order to maintain the flow of the story, all notes, references, and statistical data have been relegated to the back of the book. Place and space are very important in the history of the Canadian range. The twenty or so maps provide an atlas of ranching in western Canada and of the Bar U Ranch in particular. At the smallest scale, the end pieces provide an overview of important ranches in the Canadian prairie west; other maps show the routes to the west, the influx of cattle herds, and the major areas leased for grazing. At an intermediate scale, the complex ranching 'empires' of George Lane and Pat Burns are illustrated, while at a larger scale, the functional areas of the Bar U at different times are shown, and the gradual development of the headquarters site is charted. Finally, the photographs have been chosen to play an integral, rather than purely decorative, role in telling the story. Their usefulness and context should be obvious, and, with a few exceptions, the pictures and their captions are not referred to in the text.

1

A Western Journey: Checking out the Prospects

"I could imagine what it was like to be the first person to ride out onto our plains ... and could catch the tone of excitement in the diaries and letters I read describing the experience."[1] – Jill Ker Conway

The group of Canadians, standing on the levee of the Missouri River at Bismarck and watching the stern-wheeler "Red Cloud" unload, showed some signs of impatience. It was already mid-June, 1881, and they had left Bonaventure station in Montreal nearly two weeks before. The railway journey on the Northern Pacific, by way of Chicago and St. Paul, had only taken eight days, but they had been languishing in the railway's comfortable hotel, The Sheridan House, for a further week waiting for a boat to appear round the bend. Smoke had been seen early that morning and the vessel had secured alongside by noon. Now the gangways swarmed with 'roosters' unloading buffalo robes, furs, and wool. However, the huge piles of boxes and barrels piled along the makeshift quay, which represented the upstream freight, looked as if they would take at least another day to stow on board.

At first glance they were a disparate group. Matthew Cochrane and Duncan McEachran were well into middle age and were dressed in business suits and hats. Fred Stimson and his brother-in-law William Winder were in their thirties, and less formally attired. E.A. Baynes, Cochrane's son-in-law, was rather younger. Whatever their apparent differences, the men had much in common. They represented the small but vigorous business class of Montreal and the Eastern Townships of Quebec. An elite bound together by ties of birth and marriage, and which combined manufacturing, trade, and commercial interests with extensive landholdings and progressive agriculture. A combination which would have been anathema in the 'old world' of Victorian society.

The sense of urgency they shared was a product of their mission. They were on their way westward to the grasslands of the Canadian North-West. Finally, after two years of procrastination and negotiation, Sir John A. Macdonald and his Conservative government had been persuaded to revise the Dominion Lands Act and to provide, through grazing leases, an umbrella of security to investors who were prepared to establish large-scale cattle ranches. Moreover, the surveys for the transcontinental railway were well advanced and a Canadian syndicate had emerged to build it. Reports from the west suggested that the North West Mounted Police had been remarkably successful in imposing the "Queen's Peace" and had thereby taken some of the risk out of investment opportunities in this undeveloped region.

However, it was clear that there was no time to be lost. At present there were only a few hundred European settlers and a few thousand head of cattle in the Bow Valley district, but, as one observer pointed out, things were poised to change fast:

> … stock rearing is not like agriculture, requiring a long time for its development, and in less than ten years all will be changed, and the plains and hills over which one may now wander for days without meeting either game or human inhabitants, will be covered with cattle and dotted with ranches.[2]

Well before the details concerning the grazing leases were promulgated on 23 December 1881, the Department of the Interior started to receive requests for leases. Groups of business associates lobbied members of the Conservative government to press their claims; to insist on equal treatment; or to boast of what they could contribute. Those travelling to the west to evaluate prospects first hand were careful to leave powers of attorney with trusted friends so that they would not miss their chance should a sought-after lease be offered in their absence.[3] The newspapers reflected this sense of excitement for they were full of reports and speculation about the North-West. George Dawson was somewhat embarrassed to be adding to this 'media frenzy' and prefaced his report by saying, "… so much is written about the Northwest country at present, that the topic is likely to become tedious to those whose interests center chiefly on other matters, yet there is none of equal importance at this time in Canada."[4]

Unlike many British and American speculative investors who were lured to invest by the persuasive rhetoric of the "beef bonanza" and knew little of grass or cattle, the Canadians waiting at Bismarck were veteran stockmen who were already involved in the trans-atlantic cattle trade. The Hon. Matthew H. Cochrane was an acknowledged leader in breeding shorthorn cattle. In 1877 he had sold one cow, the Duchess of Hillhurst, and some of her offspring, at the Royal Show in London for record sums.[5] Duncan McEachran was Chief Veterinary Inspector for the Dominion of Canada, and had been responsible for reorganizing the quarantine stations and shipping points along the St. Lawrence River from which Canadian cattle were shipped to British ports. He was naturally conversant with every aspect of the cattle trade, conditions on ships, losses, freight rates, and insurance.[6] Frederick Smith Stimson had been running the substantial family farming enterprise in Compton County, Quebec, since he was eighteen years old, and had a thorough practical grasp of cattle breeding and feeding.[7] Finally, William Winder was the only one of the group who had experience in the western country.[8] He had served there for seven years with the North West Mounted Police. While still in the force he had shown an interest in the government farm and had already started to experiment with stock raising by buying cattle and running them on the open range. This interest resulted in a rebuke from Commissioner Irvine, and a mention by the Prime Minister in a debate in the House of Commons.[9] Winder had retired from the force the preceding April. He lived in Fort Macleod with his wife Julia and ran a store there, while his cattle were looked after by a foreman. Together these four men would establish major ranches and draw several millions of dollars in investment to the foothills country of Alberta.

Finally, the boats moored along the bank were loaded with freight and wood for fuel, and were ready to take on passengers. The Canadian party split up to take advantage of available cabin space. The Cochrane Ranch group, including Senator Cochrane, McEachran, Gibbs, Baynes, Thompson, and Wilson went aboard *The Red Cloud*, the pride of the I.G. Baker fleet and the fastest boat on the river. In her capacious holds, which could handle about three hundred tons of freight, were sixteen horses, a mule, and six purebred bulls, two dogs and a mass of packages destined for the ranch.[10] Fred Stimson and William Winder found their cabins on *The Benton* of the Block P Line owned by T.C. Power and Company. They had been joined by Charles Sharples, the son of a prominent family in Quebec City which had amassed a fortune in the timber trade. Sharples had been out west for three years and was to manage the ranch for Winder.[11] *The Benton*, too, was fully loaded, with goods in bond destined for Fort Walsh and Fort Macleod, and with building supplies, and mining equipment for Virginia City.

It took from ten to fourteen days for the sternwheelers to push their way up the winding Missouri. As the crow flies it was a little more than eight hundred miles to the head of navigation, but by river it was more like twelve hundred miles. In June, the level of the river was dropping and the current was slackening, but it was still some five feet above the low-water mark, and groundings on the constantly changing sandbars were infrequent. In most weather conditions navigation could be maintained at night, and the only stops were to take on more wood, or to drop off supplies at military depots or Indian agencies. The first four hundred miles up to the junction with the Yellowstone and Fort Buford were somewhat monotonous as the low muddy banks screened views of the gently rolling prairie. Thereafter, the river's

1.1 The Honourable Matthew Henry Cochrane, the Member of Parliament and stockman who did so much to persuade Sir John A. Macdonald, and his Conservative government, to establish a system by which range lands in western Canada could be leased in large units for ranching purposes. Glenbow, NA 239-25

1.2 The steamboat *Benton* under way on the upper Missouri River, 1879. She was owned by T.C. Power and Company. Fred Stimson travelled up the Missouri to Fort Benton aboard her in 1881 and thereafter favoured Thomas Power with his business over the rival I.G. Baker Company. Historical Society of Montana Archives, 955-104

1.3 Karl Bodmer, "View of the Passage Through the Stone Walls." The young Swiss artist was impressed with the scenery along the Missouri east of Fort Benton when he journeyed by on the steamer Yellowstone in 1833. Some thirty years before, Captain Meriwether Lewis had described this reach of the river as "beautiful in the extreme" in his journal. Joslyn Art Museum, Omaha, Nebraska

course was cut through sandstone gorges eroded into all manner of strange shapes, and progress was slowed by a number of rapids.[12]

By the end of June both vessels were safely docked at Fort Benton. This inland river port boasted of itself as "The Chicago of the Plains."[13] Situated at the head of navigation on the "Big Muddy," it did indeed seem that "all trails led to Benton" [Map 1.1]. Westward the famous Mullen Road wound through the mountains to Walla Walla, southward lay the territorial capital Helena and the gold fields, while the Whoop-Up Trail reached northward into the Dominion of Canada. The use of the Missouri route meant that goods could be transported from eastern centres to Fort Macleod for six to eight cents per pound, while rivals like the Hudson's Bay Company, using the northern rivers or the overland route from Winnipeg, were forced to charge three times as much.[14] The great merchant companies whose warehouses lined the levee, and whose cargoes filled the stern-wheelers had grown rich on supplying the fur trade and the gold fields. As Sharp explains:

> The great merchant princes of Fort Benton were the final agents for a vast system which provided Canadian and American ranchers, gold seekers, treaty Indians, and government forces with the products of the looms of Manchester and Hartford, of the forges and furnaces of Sheffield and Pittsburgh, and of the distilleries of Boston and Louisville.[15]

The wild frontier Benton of wolfers and whisky smugglers was gone, and the town tried to present a respectable and progressive face to visitors. Nevertheless, the editor of the newly established newspaper warned his readers that they could look forward to a wild time when four riverboats all arrived at the same time![16] Front Street paralleled the waterfront and boasted a growing number of solid brick buildings which housed merchants and stores. The prestigious Centennial Hotel offered a degree of comfort which was somewhat unexpected on the frontier, while the old Overlander catered to more humble tastes. The new Grand Union Hotel was under construction and due to open in 1882. Behind the long main thoroughfare a network of streets gave access to livery stables, blacksmiths, wheelwrights, harness makers, boot and shoe makers, tailors, barbers, grocers, and butchers.

The population of Fort Benton was estimated to be about fifteen hundred in 1881. It was made up of a polyglot company of people, most of whom were transient, following rumours and dreams from place to place. For the time being there were jobs to be found and money to be made in Fort Benton. Southerners re-fought the civil war with Yankees in the bars. Irish immigrants were prevalent and became a considerable political force. French Canadians and Métis frontiersman who had been involved for generations in the fur trade now turned their skills to transport, and were involved both on the river and with the overland bull trains.

The Canadian visitors arrived in Fort Benton during a period in which the town was experiencing remarkable growth, matching that of the hectic gold rush days of the 1860s. Thirty-six boats docked along the river bank that year, and they carried thirty-five million tons of freight and thirty-seven hundred passengers. The total value of 'river business' was estimated to have been over $200 million.[17] Much of this new prosperity was based on contracts with the government of Canada to supply the North

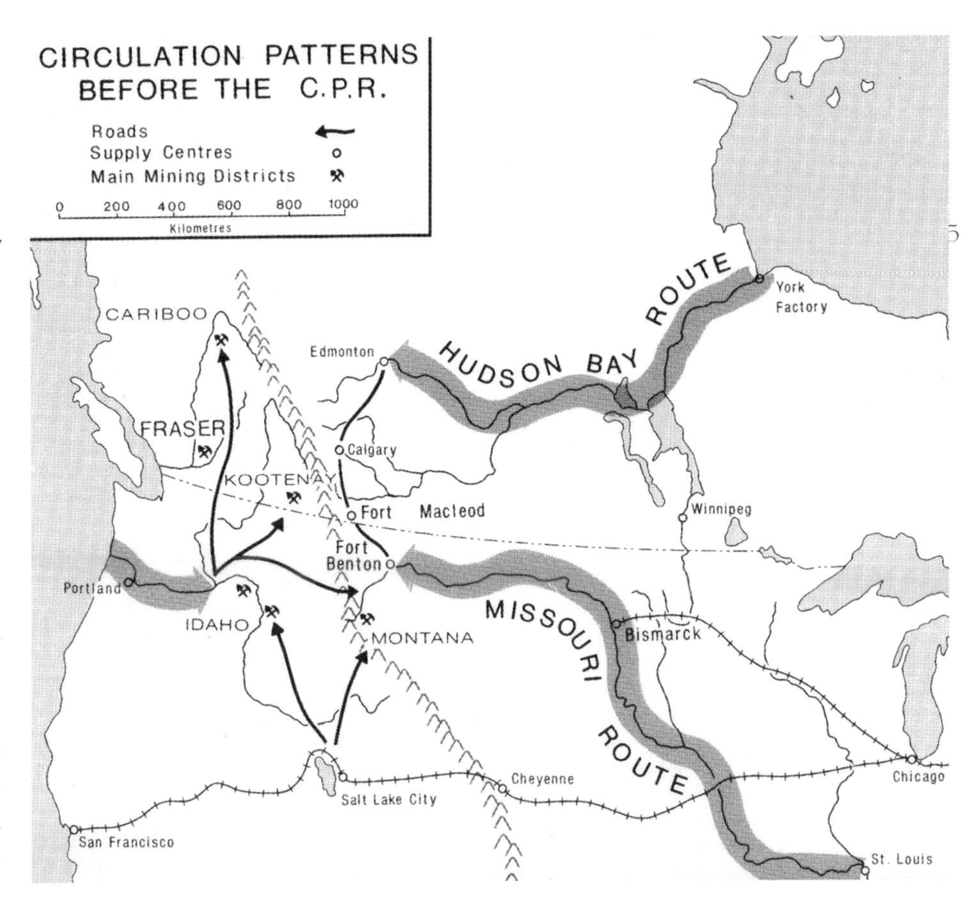

Map 1.1 The two major river routes which were used to reach the Canadian North West prior to the completion of the Canadian Pacific Railway.

1.4 Fort Benton, Montana, 1878. The river port at the head of navigation on the Missouri River, the town boasted of being "The Chicago of the Plains." It was indeed an important entrepot for the Montana mines and the Canadian west. The steamer *Benton* is moored on the extreme left of the picture. Glenbow, NA 2446-11

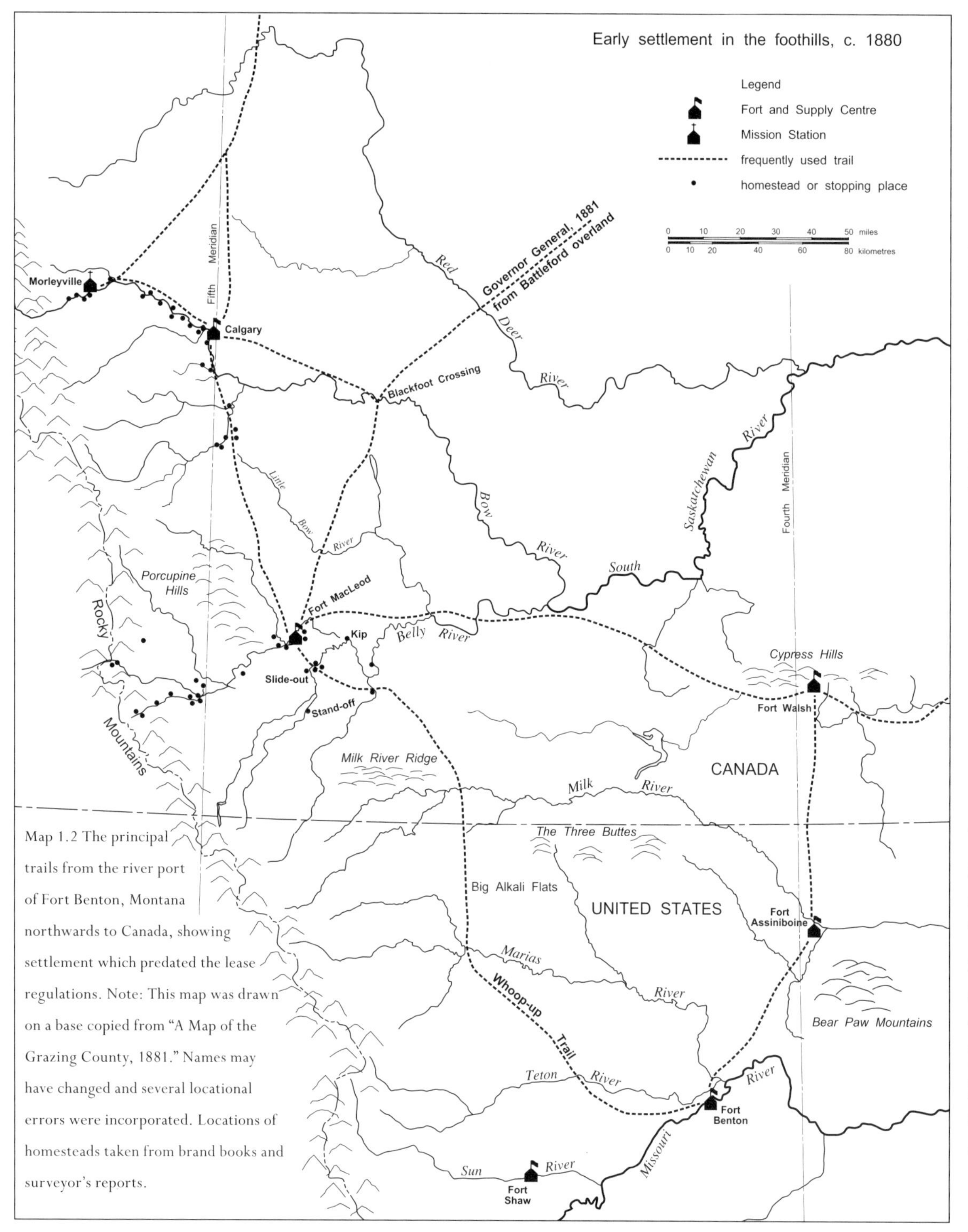

Map 1.2 The principal trails from the river port of Fort Benton, Montana northwards to Canada, showing settlement which predated the lease regulations. Note: This map was drawn on a base copied from "A Map of the Grazing County, 1881." Names may have changed and several locational errors were incorporated. Locations of homesteads taken from brand books and surveyor's reports.

West Mounted Police and to provide the Indians on the newly established reserves with meat and grain in accordance with the treaty provisions. Thus, while the northward extension of the Union Pacific railway to Dillon threatened to diminish the southern extent of Fort Benton's hinterland and the heyday of the gold mines and the fur trade was over, the dramatic expansion of legitimate trade with Canada prolonged and invigorated the commercial life of the river port. Colonel S.C. Ashby, who had been stationed for years on the Upper Missouri, commented:

> I must say that the great success that came to the Benton people was due largely to the fact that the Canadian government under Sir John A. Macdonald decided to send companies of troopers to what is known as the Belly River country.[18]

Thus, it was no surprise that the visiting Canadian entrepreneurs were welcomed and treated with some deference. The editor of the *Fort Benton Weekly Record* interviewed the principals of the party and wrote a column on "Cattle Raising in the North-West." He remarked that the Bow River country "is especially coming into notice just now, and capitalists at Ottawa, Montreal and throughout eastern Canada are looking to it as a favourable field for future harvests of money." He went on to emphasize the large scale of the planned operations, and the fact that Cochrane planned to breed from Hereford and Polled Angus bulls imported from the east and even from England. The editor felt that "in this respect we believe the Cochrane Ranch Company will eventually revolutionize the whole North West system of cattle raising."[19]

The Canadians spent a busy couple of days in Fort Benton readying themselves for the overland portion of their journey. Light wagons and teams were purchased or hired, guides and cooks were engaged, and stores loaded. They were impressed by the variety and quality of the goods available, but the casual chaos of the frontier town made them wonder whether their business would ever be concluded. Eventually, everything was ready, and early next morning the party climbed up the steep side of the coulee onto the level prairie and looked down on the river and the town for the last time. For the next three or four days, McEachran's journal chronicles the struggle the party had to adapt to travel on the Whoop-Up Trail. The heat, the alkali water, and the constant battles with mosquitoes and bulldog flies were compounded by broken nights guarding against imagined threats from Indians, and by savage prairie thunderstorms. Gradually, however, the rhythm of the trail asserted itself, and there was more time to appreciate the unique events of each day: the drama of a river crossing; the danger of riding over gopher holes; and the blessings of a good well at a stopping place[Map 1.2].

On the fifth day on the trail, the party gained a practical demonstration of how goods were moved along the Whoop-Up Trail. McEachran recorded the scene:

1.5 The levee of the Missouri at Fort Benton was used as a dock by the river boats. This photograph, taken in the early 1880s, shows piles of many different kinds of freight. Glenbow, NA 354-32

> While we ride on and admire the grand panorama, we decry a long white moving chain in the distance, which, on nearer approach, we find to be one of Baker and company's bullock teams, consisting of eight trains, of three wagons, each drawn by sixteen oxen, making 128 of as fine working cattle as it is possible to see. It was a grand sight. On the left front was the captain

> quietly walking by the side of his "cayuse" or saddle pony, he being the only mounted man in the party. Then, in one long line comes the slowly moving train, urged by the occasional crack of the huge bull-whip, as it twirls around the driver's head, while he shouts to the lagging beasts. The wagons are very large and strong, the front one being the largest, and usually carrying about six tons. The next is less in size and the third one still less. They are covered with white canvas spread over arches, which forms good protection from sun or rain. These teams make very long journeys, usually at the rate of about fifteen miles per day. The oxen have nothing but prairie grass to eat and often go twenty- four hours without water. The men are fed on pork, bread and tea, with preserved vegetables, corn and beans, and occasionally canned fruits, and sleep on the ground with a blanket. They have little comfort and very hard work, yet we found them contented and healthy.[20]

The land over which the party moved and the customs of the country were alien and exotic to the visitors from the east. After all, they had been travelling for nearly a month and were a long way from home. They naturally evaluated what they saw and experienced in terms of the humid and "civilized" east with which they were familiar. But the grasslands and foothills had been home to aboriginal people for generations, and were by no means a *terra incognita* to Europeans. It is true that the formidable "Blackfoot Confederacy" had proved something of a barrier to the westward expansion of the fur trade, and had prevented for a time the exploration and evaluation of the region. Nevertheless, the way of life of the plains Indians had already been altered beyond all recognition by the diffusion of the horse and the gun.[21] By the middle of the nineteenth century they were dependent on traders for supplies of powder, shot, and steel implements. A seasonal rendezvous at a trading post on the periphery of the plains became a characteristic feature of their yearly activities. Both missionaries and traders had criss-crossed the area for decades before the arrival of the North West Mounted Police in 1874.[22] Since that time, a regular system of patrols had been established and a constant stream of reports flowed back east. The boundary survey had been completed, and the survey and demarcation of the township and range grid was progressing. In 1881, the Department of the Interior published a map of the "Grazing Country," which showed in some detail the rivers and relief of the region and the progress of the survey.[23]

Although McEachran and his companions were awed by the scale of the country and the fact that it appeared almost deserted, a small but flourishing population of white settlers had regrouped themselves around the new nucleii of settlement represented by the police posts and along the trails. For example, surveyor Montague Aldous, employed on the standard survey in 1880, remarked that, "in the valleys of the Oldman, Belly and Kootenie rivers several farmers are settled and succeed in raising very good crops."[24] They were joined by policemen who had completed their terms of engagement, many of whom started small ranches. Superintendent William Winder commented that twenty-five of the thirty-nine men discharged had stayed in the region.[25] So the feasibility of ranching on a small scale in the foothills had been amply demonstrated. Ranches and farms clustered along the rivers, especially round Fort Macleod. Stock-rearing activities at this stage were closely integrated with arable agriculture. Wild hay was put up in considerable quantities, and the demand for oats and other fodder crops was rising.

The party of prospective cattlemen whose progress we have been following was but one of several groups traversing the prairie during the summer and fall of 1881. The most prestigious of these was the expedition of Canada's Governor General, the Marquess of Lorne. This 'royal progress' was orchestrated by the Canadian government to sweep away the negative image of Canada which was thought to be a major factor in slowing immigration from Britain. Every move and encounter of the Queen's son-in-law during his overland journey from Battleford to Fort Calgary, and on down the foothills to Fort Macleod, was reported in the British press.[26] Another very influential visitor was John Macoun, the Dominion Field Naturalist and Botanist, who was completing his book *Manitoba and the Great North-West*, which was to do so much to change people's perception of the agricultural potential of the Canadian prairies. Alexander Begg, his co-author, followed McEachran up the Whoop-Up trail during August, and probably met up with the Reverend A. Sutherland, who was also heading for the Bow Valley and the mission at Morley.[27] Geologist G.M. Dawson had also returned to the foothills for another field season of scientific inquiry.[28]

As one reviews the surviving writings of these men, more than a century after their adventures, one cannot but be impressed by their astute grasp of the special nature of the moment in history which they sought to record. Scientist and businessman, newcomer and veteran, all seem to have appreciated the drama of visiting a beautiful land in waiting. It seemed to them that the grasslands were enjoying a brief hiatus before a tidal wave of change transformed them, for the

1.6 A bull train en route on the Whoop-Up Trail between Fort Benton and Fort Macleod, like that described by McEachran. Four or five wagons would be hitched together to form a "train" and they were drawn by eight to twelve oxen. Glenbow, NA 98-11

1.8 The Marquess of Lorne was an enthusiastic artist. Here he sketched the Blackfoot encampment on the Bow River prepared for his visit. He entitled it "Pow-wow at Blackfoot Crossing." Glenbow, NA1269-13, printed in the Illustrated London News, December 10, 1881

1.7 The Marquess of Lorne, Governor General of Canada and son-in-law of Queen Victoria, visited the North West Territories during the summer of 1881. His visit was widely reported and drew attention to the productive potential and the attractive nature of the foothills and plains. Glenbow, NA 900-1

time of the buffalo and the plains Indians, who lived in symbiosis with the herds, had suddenly passed.[29] Railways had brought increasing numbers of sports hunters to the southern plains, while new technology had created an insatiable demand for buffalo hides.[30] Dawson and his companions on the boundary survey had looked down on a black sea of buffalo from the Sweetgrass Hills in 1874. Colonel Macleod, Major Crozier, and Mr. Fraser of the I.G. Baker store, all confirmed these reports and added their own eyewitness stories. But in his report for 1878, the commissioner for the North West Mounted Police stated that the best authorities were of the opinion that the bison as a means of support, even for the Indians in the southern district, would not last more than three years. The following year he commented that he had not expected that this prophecy would be so literally fulfilled. No observations of large numbers of bison are known for the southwestern Canadian plains after 1879, and very few observations of bison were made anywhere after 1880.[31] The Governor General's party came across a forlorn remnant of ten head lurking in the breaks of the Red Deer River, but McEachran reported no game of any kind.[32]

McEachran's harsh and xenophobic description of the parties of Indians with whom he came into contact is somewhat muted as he describes the total dependence of the nomadic tribes on the buffalo, and their consequent abject poverty now that the herds had disappeared.[33] He displays some sympathy for their semi-starved condition, although his major concern may have been that hungry Indians represented a risk to stock. Biologist Macoun suggested that cattle could occupy the now deserted feeding grounds of the buffalo. For example, he mentioned the Hand Hills area close to the Red Deer River:

> In former years it was noted for its rich pastures and for the enormous herds of buffalo wandering in its neighbourhood. At present the buffalo are all gone, the Indians having disappeared with them, the whole region is without inhabitants, and nothing is left but the waving grass on the hillsides and the wildfowl in the marshy flats.... Standing on a hilltop and looking over a wide area of grass covered hills and valleys – is it too much to say that here was room for millions of cattle to roam at will and get fat on the very richest of grasses?[34]

1.9 An embarrassing moment for an unknown member of the Governor General's party in the camp of the North Peigan Indians, close to Fort Macleod. Sketch by Sydney P. Hall, artist for *The Graphic*, London. Glenbow, NA1190-12

Both from their own observations, and from hearsay, the attributes of the land were evaluated and reported to the interested public in the settled part of the Dominion. G.M. Dawson articulated, perhaps for the first time, what was to become something of a mantra to cattlemen, and the foundation of their plea that farmers had no business intruding into grazing

1.10 Buffalo slaughtered for their hides which were made into drive belts for eastern factories. In 1876 alone, some 60,000 hides were shipped down the Missouri from Fort Benton. Frontier photographer L.A. Huffman did a series of studies of the last days of the great buffalo herds. See Mark H. Brown and W.R. Felton, *The Frontier Years* (New York: Bramhall House, 1955), pp. 69–79. Glenbow, NA 207-68

1.11 Duncan McEachran remarked on the bleached buffalo bones that his party encountered on the trail. Almost a decade later tons of the bones were collected and shipped east to be manufactured into fertilizer. Here, Buffalo bones are loaded in a CPR boxcar in Moose Jaw, ca. late 1880s. Glenbow, NA 4967-10

country. He suggested that the foothills and adjacent grasslands constituted a 'special region' where pastoral activity could flourish but where arable agriculture was doomed to failure.

> The very dryness of the climate, which causes grasses to be produced which retain their nutritious properties during the winter and prevents exuberant growth of wood, renders agriculture precarious or impossible except where irrigation can be resorted to with facility. In addition to this, the elevation of much of the stock-raising lands is such that summer frosts not infrequently occur, rendering the growth of crops uncertain.[35]

Macoun pointed out the enormous potential of the coulees and river valleys for providing natural hay, and argued that the colder winters actually meant that stock would gain more fat. He contended that cattle from Texas demonstrated better weight gains during a northern winter than they did during a southern summer. McEachran discerned, in general terms, the differences between the mixed grasses of the drier plains and the fescue grasses of the Aspen Parkland, "The nearer you approach the mountains the more luxuriant it [pasture] becomes." He went on:

> The grasses are most luxuriant, especially what is known as "bunch grass," and wild vetch or peavine, and on the lower levels, in damper soil, the blue-joint grass, which resembles English ryegrass, but grows stronger and higher. On some of the upland meadows wild Timothy is also found. These grasses grow in many places from one to two feet high, and cover the ground like a thick mat. Nowhere else has the writer seen such an abundance of food for cattle.[36]

Later, exploring westward from the government farm at Pincher Creek, McEachran again drew attention to the grass which had not been grazed by wintering buffalo for several years. "There is an inexhaustible growth of nutritious grasses. In some places it is so thick and so long as to impede the progress of horses."[37] Alexander Begg, an incorrigible booster, contrasted this potential with the crowded and overgrazed conditions in the United States.

> In a few years it will be difficult to find vacant range in Wyoming, Nebraska, or Montana, suitable or capable of sustaining 5,000 head of cattle. The Dominion of Canada, on the other hand, has "limitless" ranges waiting to be taken up and occupied.[38]

Perhaps surprisingly considering their long and arduous journey, Cochrane and McEachran spent only a week at the newly chosen site for the headquarters of the Cochrane Ranch under the shadow of the "Big Hill." They approved the layout of the ranch buildings and met the men who would be looking after their affairs. They were well satisfied with what had been achieved. They had seen with their own eyes the potential of the region for stock rearing and had established personal contact with important westerners. While Baines was left behind as manager, it was time for the principals to return to Montreal to report to potential investors and to badger the Department of the Interior to publish the grazing lease regulations and award the first leases.[39]

Fred Stimson lingered rather longer in the west. He spent some time at the Winder ranch and rode with his brother-in-law up to High River and westward around the Porcupine Hills. It was early fall before he headed back east, and he stopped in Chicago to purchase some good-looking purebred Shorthorn bulls which had been imported from England. He also hired nineteen-year-old Herbert Millar, from nearby Chicago Heights, to tend the stock through the winter, and arranged for him to take them by rail and steamboat to Fort Benton and thence northward to the North West Cattle Company's lease near High River during the spring of 1882.[40]

The sudden interest of the eastern Canadian investment community in the grasslands of western Canada, exemplified by our group of "explorers," was ignited by the perception that there would be spectacular returns on the risk capital they were prepared to provide. Nor was this dream ill-founded. In the short term, government contracts to supply the North West Mounted Police and the Indians would be worth hundreds of thousands of dollars. In the longer term, the building of the Canadian Pacific Railway offered potential for ongoing prosperity, for it would provide a link to the transatlantic cattle trade and the lucrative markets of the old country. Clearly by this time Fred Stimson was committed to the western ranching enterprise and had even selected a region in which he planned to obtain grazing leases. He appreciated that the first ranches on the Canadian range would grab the lion's share of potential returns. He hurried back to Montreal to finalize his lease applications, to incorporate the North West Cattle Company, and to convince investors to back their promises with hard cash.

2

Fred Stimson and the North West Cattle Company, 1882–1902

"Certain aspects of reality can be captured only in narrative. Paradox and parable must in this sense be enacted or witnessed to, not analysed away."[1] – Paul Baumann

Fred Stimson played a key role in the establishment of the North West Cattle Company and managed the outfit from its incorporation in 1882 until it was wound up in 1902.[2] We know enough about his origins and his family connections to demonstrate that they were important to the history of the Bar U, but the man himself remains something of an enigma. Alan McCullough, who has probed deeply into the history of the Stimson family, reaches the conclusion that this larger than life extrovert and peerless raconteur was never wholeheartedly embraced by westerners – he was certainly not a 'folk hero' as was George Lane.[3] His photographs show a big, powerful man with a pensive look. He never seems to have relished the hard, gritty, dangerous work of "cowboying," but he was a careful, conservative, and dedicated manager. Duncan McEachran once remarked, "In no other business does management play so important a part, and a ranch will be profitable or otherwise just in proportion to the goodness or badness of the management."[4] If this was indeed the case, then Stimson must have been an exceptional manager, for the Bar U is acknowledged to have been one of the most successful big outfits in the grazing country.

2.1 Fred Stimson, 1882. Fred was forty years old. The previous year he had made his first journey to the west. This photograph may have been taken before he left to take up residence there as manager of the North West Cattle Company. Notman Photographic Archives, 63720 - B 11, McCord Museum of Canadian History, Montreal

Fred Stimson came from a well-established family in the Eastern Townships of Quebec. His father was a farmer and merchant who owned a store, a considerable amount of land, a mill, and interests in mortgages. When Arba Stimson died in 1863, Fred inherited both wealth and responsibilities. His brother Charles was four years his senior and was making a career for himself in Montreal as a merchant and manufacturer of leather goods. It fell to the eighteen-year-old Fred to manage the family farms in Compton County. For the next fifteen years he gained invaluable experience in every facet of practical stock rearing.

The Stimson family had important connections to several other well-to-do families in the region, each of which was to play a prominent role in the development of the cattle industry in western Canada. For example, Fred's uncle, Samuel Greeley Smith, was Matthew Cochrane's partner in a shoe factory from 1854 until he died in 1868. By the mid-1880s, Cochrane had taken Charles Cassils into a partnership, and their firm employed some three hundred people and had a turnover of more than half a million dollars. Charles Stimson, Fred's brother, was a partner to

2.2 Mrs. F.S. Stimson, 1874, aged twenty-nine years. Mary Stimson had been married to her husband Fred for fifteen years before he went west for the first time. Although her younger sister Julia was already living in Fort Macleod as the wife of William Winder, Mary was reluctant to leave her family and familiar roots in the Eastern Townships of Quebec. Notman Photographic Archives, 98,730 - B 1, McCord Museum of Canadian History, Montreal

Archibald M. Cassils in a competing leather and shoe business at the same time. Thus the Cochranes, the Smiths, and the Stimsons were linked both by their neighbouring landholdings in Compton County and by their common business interests in Montreal. This being the case, it is not surprising to find that Matthew Cochrane was an executor of Arba Stimson's will.

Fred Stimson married his first cousin, Mary Greeley Smith, in 1866, when both bride and groom were twenty-one years old. One of the witnesses was Alexander Rea Allan, the son of Sir Hugh Allan, the owner of the Allan Steamship Company and one of the richest men in Canada.[5] Five years later, Sir Hugh loaned the Stimson brothers $6,000 to modernize their farm properties. When this loan was successfully repaid, the stage was set for the Stimsons to approach the Allans ten years later to buy shares in a ranching venture in the North-West Territories.[6]

Fred Stimson's wife, Mary, had a younger sister named Julia who married William Winder in 1869. Four years later Winder joined the North West Mounted Police. He was active in recruiting for the force in Quebec and then took part in the march westward. His wife joined him at Fort Macleod in 1876, and they both witnessed the signing of Treaty Seven. In 1881 Winder retired from the force and formalized his existing interests in cattle by founding the Winder Ranch, with Charles Stimson being one of the shareholders. Charles had married Melinda Lemoyne in 1871, and her brother, John McPherson Lemoyne, another prominent stock farmer in Compton County, was also a shareholder in the Winder Ranch.

The alluring prospects of the cattle industry created a bullish turmoil in the investment community of eastern Canada during the early 1880s. As has already been mentioned, the Conservative government under Sir John A. Macdonald had been returned to power and was committed to building the transcontinental railway. The transatlantic trade in live cattle between Quebec and Ontario and Great Britain had grown into a multi-million-dollar business over the space of a few years. Cattle now ranked third among items exported from the Dominion.[7] Rumours of returns of 20 to 30 per cent from some newly formed cattle companies found their way into the St James Club of Montreal, and were repeated on the farms and estates of the Eastern Townships. English and Scots newspapers and American commercial journals spelled out the profits to be made by linking the unlimited free grass available in the west to the growing demand for beef in the new industrial towns of Europe.[8] The exciting potential of an investment in ranching was highlighted by the sombre state of other sectors of the Canadian economy and by the stubborn refusal of immigrants to respond to the blandishments of agents and come in large numbers to the Great North West.[9] It seemed clear that herds of cattle presided over by a few cowboys could earn magnificent returns while at the same time providing a profitable freight for the planned railway. No wonder that there was a frenzied rush to get in on a "sure thing."[10]

Groups of potential investors came together to discuss prospects in urban centres from Montreal to Halifax. Partnerships were formed and companies incorporated, only to fall apart as even more enticing prospects emerged. In this volatile investment climate it is not surprising that it took three attempts to establish a corporate framework for the Bar U Ranch. Indeed, this was typical for many other enterprises. It is worth examining how potential investors came and went around a nucleus of promoters who were committed to the endeavour.

2.3 Fred Stimson was never much of a cowboy, but that did not stop him from trying to look the part! Here he poses at Montgomery Ward's store in Chicago, 1882. Parks Canada, Andrews Collection

In June 1881, the *Canada Gazette* reported that a group of businessmen had applied for letters patent to incorporate the Rocky Mountain Stock Company. The potential investors were: Andrew Allan, Frank Stephen, Thomas D. Milburne, Robert A. Smith, John Cassils, and Walter Wilson, all of Montreal, and Frederick Smith Stimson of Compton. Allan, Stephen, and Milburne were to be directors.[11] A few days later, as we have seen, Fred Stimson left Montreal for the Canadian west. He left Thomas Milburne his power of attorney, particularly with regard to his request for a grazing lease, which had already been forwarded to the Department of the Interior. It must have seemed imperative, given the excitement of the times, that nothing be done to delay approval of the all-important lease.

On his return in the fall of 1881, Stimson reported to his partners what he had seen and learned on his trip to Fort Macleod and the foothills of Alberta. He explained that he was particularly impressed with the High River range west of the Whoop-Up Trail, about two-thirds of the way between Fort Macleod and Fort Calgary. Quite possibly his assessment was coloured by what he heard from his travelling companions. It seemed that Senator Matthew Cochrane was going to press his prior claim to obtain a huge lease along the Bow River west of Fort Calgary, and that Duncan McEachran was interested in the range drained by the upper Oldman River. There was still an enormous amount of land within the favoured 'chinook zone,' so it made good sense not to challenge powerful friends who continued to be involved in negotiations with the government.

Since his departure in the early summer, some potential investors had withdrawn, while others clamoured to be admitted to the enterprise. To accommodate these changes a new company was organized called the High River Stock Company. It had a capitalization of $200,000, and the principal shareholders were Thomas D. Milburne with 300 shares and Frederick Smith Stimson with 200, together these two men held half the shares subscribed. Other shareholders included Andrew Allan with 100 shares, Robert A. Smith with 150, John Cassils with 50, Frank Stephen with 100, and Walter Wilson with 100. At this stage it is clear that Stimson and his associate Milburne had a controlling interest in the proposed company and the Allans were merely one of several smaller investors.

Over the next three or four months there were some more far-reaching changes. Several of the smaller shareholders dropped out, while the Allans consolidated their hold over the company. In March 1882, the North West Cattle Company was established. It was to last twenty years and enjoyed considerable success, both on the range in Alberta and from the point of view of the eastern investors. The company had an authorized capital of $150,000 divided into 1,500 shares. Both Sir Hugh Allan and his brother Andrew held 250 shares, Thomas D. Milburne 150, Fred Stimson 150, and Charles Stimson 50.[12]

Fred Stimson's position had changed radically before he left for the North West for the second time in the spring of 1882. Instead of being the manager of a company in which he had the controlling interest, he had become an employee in a company controlled by others, although he was still a major shareholder. How did Stimson feel about this change in his prospects? The record only allows us to speculate on this intriguing question. On the one hand, his freedom of action was drastically curtailed; on the other, the scale on which he could plan had increased beyond his wildest dreams. Frederick and Charles Stimson had inherited a substantial estate and had added to it with considerable acumen, Charles in the city of Montreal and Frederick by managing their varied interests in the countryside of the Eastern Townships. However, their total assets would have been numbered in the tens of thousands of dollars. In contrast, the Allans were probably the richest family in Canada at the time. On his death in 1882, Sir Hugh Allan's estate was valued at between $6 million and $10 million.[13] Thus it was a coup on the Stimsons' part that they were able to parley their long-term family connections with the Allans into a substantial financial commitment. At the stoke of a pen the vision of a ranch of perhaps fifty thousand leased acres and a herd of a thousand head of cattle – similar to that of his brother in law William Winder – was enlarged to encompass a huge enterprise which could stand alongside that of Senator Matthew Cochrane's. The arithmetic of the 'Beef Bonanza' emphasized the economies of scale. The bigger the ranch the bigger were the percentage returns on investment. In exchange for the responsibilities and financial risks of being in control, Stimson emerged in the enviable position of being able to plan on a more lavish scale, backed by huge capital reserves, and with his practical independence as 'on the spot' manager guaranteed by the long distance to the head office in Montreal.

Why did Fred Stimson, at the age of thirty-nine, choose to uproot his family from their prosperous and secure "country seat" in Quebec, to risk all on the isolated frontier where most of the elements of comfortable middle-class life were noticeably absent? This is an interesting question, and it is easy to put forward suggestions based on our knowledge of the man gleaned from the remainder of his life. Nevertheless, it is tempting to hypothesize that Fred felt increasingly trapped in his role as the manager of the home farm. He had assumed these responsibilities at the age of eighteen, had married at twenty-one, and had become a father at twenty-three. He was close enough to the world of politics and commerce to know that he was living during a heady time of frontier expansion and opportunity. But the buzz and excitement of the city and the corridors of power only reached him at second hand through his brother Charles and from his neighbours.[14] If he took the safe course and 'stayed put,' he would always be on the sidelines, tolerated because of his family connections, but taking no direct role in developments. Could he perhaps have envisioned in 1881 that in the west he could become a key figure in the small emerging elite? That he would be sought after for advice by the Deputy Minister of the Interior and by the Minister for Indian Affairs? That he would hobnob with judges and police commissioners and would be pressed by his peers to make their wishes known to the government back east?

Probably, as is the case in most of our lives, the life-changing decision was in fact a series of logical steps, each with only limited consequences. Montreal and Compton County were alive with talk of investments in ranching. William Winder, Fred's brother-in-law, was looking for financial backing for his venture. Charles and Fred Stimson determined to get in on the ground floor of this investment opportunity, but somebody had to investigate the prospects carefully to make sure that the speculative enthusiasm had some substance and justification. Fred could be spared more easily because neighbouring family members could assume his management duties in Compton. In any case, he was the brother with knowledge of cattle. Once he was in the west, he was drawn by the obvious opportunity to build something from nothing and by the freedom from habit and tradition.

In many ways Stimson was far removed from the stereotypical westerner. Unlike so many of the young men who moved west to settle in the foothills, he was not an avid hunter or fisherman. He was prepared to go on long journeys where necessary, but he did not embrace discomfort willingly. One looks in vain for a photograph of Stimson mounted, perhaps in a group of riders during a roundup, for such mementos were popular. He did attend these pivotal events when his presence was absolutely necessary, but in the main he left the practical management to his foreman. And yet, in a broader sense Stimson does seem to have been something of a risk taker. Not only did he leave

T. C. Power & Bro., Fort Benton, M. T. — John M. Sweeney, Helena, M. T.

OFFICE OF

T. C. POWER & CO.,

Dealers . Agricultural Implements,

WHOLESALE AND RETAIL.

Helena, M. T. July 28 1882

2.4 In July, 1882, Fred Stimson wrote a triumphant note to Thomas Power (on the latter's headed paper) telling him that he had "...got all my stock, 3000 head for less than $20.00..." Montana Historical Society Archives, MC 55, Box 166, f. 6, Stimson to Power, July 28, 1882

Compton when he was thirty-nine; twenty years later, when the Bar U Ranch was sold out from under him, he boldly took a job in Cuba, managing a ranch for William Van Horne. From there he moved to Mexico, until ill health forced his re-turn to Montreal in 1912. These were not the moves of a dull traditionalist.

Perhaps Fred Stimson was eased or pushed out of his family home by cultural and economic forces of which he was only dimly aware.[15] The English-speaking, largely Protestant population of the Eastern Townships was beginning to give way before a wave of French-speaking Catholic infiltration from the seigneuries along the St. Lawrence. At the same time, farmers from Ontario and even the midwestern states were providing stiff competition not only for grain but also for stock.[16] At home then, the future seemed uncertain and the mood was bleak, while the prospects in the North West were alluring. As the story of the Bar U Ranch unfolds and Stimson's role in its development is chronicled, we will, I hope, have a chance to deepen our perception of this complex and interesting man and reach our own conclusions as to his motivation.

A Herd for the Bar U

Shrewdly, Stimson grasped the advantages that would accrue to the first Canadian ranches to stock their ranges. He had done all he could to ensure that the North West Cattle Company would be granted leases in due course. His immediate priority was to obtain a herd of cattle. Fred Stimson hurried back to the west early in the spring of 1882. He travelled by train and stagecoach and arrived in Montana several weeks before the Missouri river opened.[17] To help him in his search for a foundation herd, he enlisted the aid of a typical frontiersman, Tom Lynch. Born in the Pend Oreille Lake country of Idaho, Lynch had roamed widely on both sides of the line. He had played a role in trailing stock through the mountains from Washington and Oregon into Montana and Wyoming.[18] For the previous few years he had teamed up with George Emerson, and together they had brought bunches of cattle into southern Alberta from Montana.[19] He knew the country well and was an experienced drover

By the end of April 1882, Stimson and Lynch were in the Deer Lodge Valley of western Montana. This was a centre for progressive cow-calf ranches, and they were offered cattle at $23 per head. This was considerably more than the sum paid for the Cochrane herd the year before.[20] They decided to push on into Idaho to look for a better deal. In the Lost River region (north of Pocatello, Idaho) they were able to buy three thousand head of cattle from Colonel Champs for "less than $20.00 a head."[21] They were fortunate to find a suitable herd, for buyers were scouring Idaho for marketable cattle. During May 1882, some ten thousand head of cattle and fifteen hundred head of horses were driven from the counties of the Idaho panhandle.[22]

When the business was concluded, Fred Stimson headed back toward High River, leaving Tom Lynch to put together a trail outfit to bring the herd some seven hundred miles to their new range. Lynch would have needed between eight to ten hands to handle the herd.[23] He engaged Abe Cotterell to be his second-in-command, or the "trail boss," as Lynch knew that he himself would have to spend a lot of time riding in advance of the herd to make arrangements with regard to bedding grounds, feed, and supplies. Cal Morton filled the vital role of cook, and he must have been a wizard, as bad weather and bad food were the two factors which could destroy morale in a crew, and there was no hint of that during this drive.[24] Bill Moodie, John Ware, Preston, and Deaver were four of the riders, but the names of the others are not known.[25] Undoubtedly there would have been considerable turnover during the four-month drive. A man might help with the herd for one leg of the trip lasting a couple of weeks and then return to his home town. The important thing was that there was always a nucleus of experienced and reliable hands who acted like non-commissioned officers and supervised the efforts of the newcomers.[26]

Although no detailed description of the herd has survived, a few facts about it can be inferred. First, this was a mixed herd. It included a good proportion of breeding cows, and, as it was spring, many of them were accompanied by calves. There were also considerable numbers of two- and three-year-old steers. This part of Idaho had been 'breeding up' from basic western cattle for some time, and the herd was in all probability a Durham-Shorthorn cross which was showing some Hereford characteristics. The herd was trail-branded with a double circle on the left hip

2.5 These cattle, being herded through the streets of Barkerville, B.C. in 1868, were "westerns," probably a Durham-Shorthorn cross. The roaring mining towns of the gold rush days provided one of the first markets for stockmen. Provincial Archives of British Columbia, Victoria B.C., A-03787

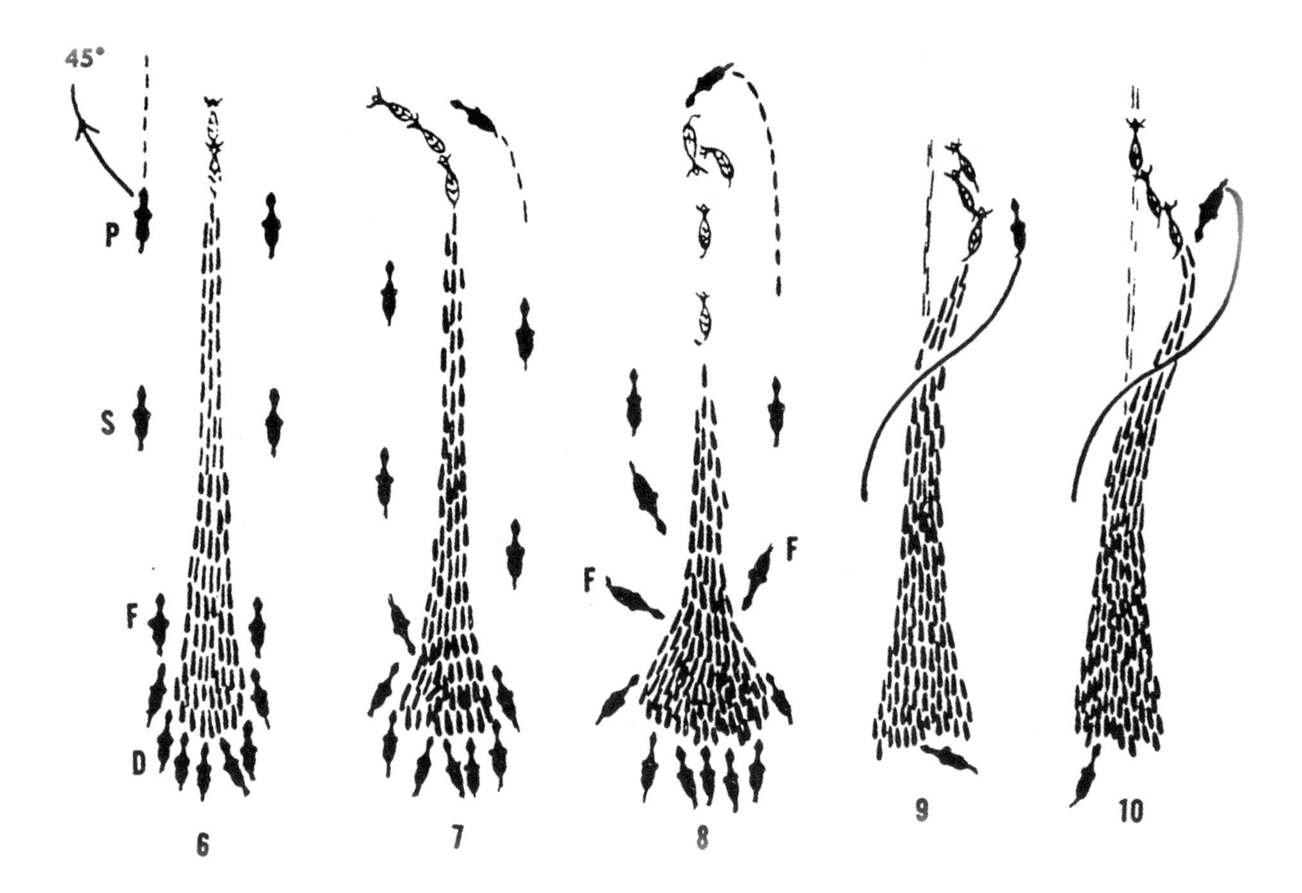

2.6 Fay Ward sketches the positions of riders round a trail herd: P for point; S for swing; F for flank; and D for drag. The series of sketches show how a herd could be slowed or checked so that the "drag" could catch up. Faye E. Ward, *The Cowboy at Work* (Norman: University of Oklahoma Press, 1987) With permission of Hastings House/ Daytrips Publishers

and all was ready to start the drive by the middle of May.[27]

The first major obstacle to be overcome was the gradual but bleak Moneida Pass which led northward out of Idaho and into the Beaverhead Valley of Montana. Here the spring grass was breaking through, and the cattle had forgotten their accustomed range and were getting used to being on the move. The leisurely rhythm of the trail asserted itself. After a breakfast in the dark of a spring dawn, the cowboys would choose a horse from their string and saddle up. The cattle would be moved off the bedding ground and allowed to graze for a couple of hours. As the chuckwagon and the remuda set off down the trail so the herd would string out behind them. Often a "pecking order" would develop among the lead cattle and a particular beast would become an acknowledged leader. Lynch had a particularly ornery old steer which had lost a horn in some confrontation. "Old Yellow" was so useful in this regard, especially for river crossings, that Lynch took him on several drives, and later pensioned him off on his own ranch.[28] Around the middle of the day, the point riders would come across the chuckwagon, and would turn the herd aside to stop and enjoy another period of grazing. By mid-afternoon the herd was on the move again until it came to the bedding ground selected by Lynch for the night. Time was not pressing; sometimes the herd would drift ten to twelve miles a day, but when they encountered good grass at some distance from a ranch they might stop over for a day or two.

The route led down the Jefferson River to the Three Forks of the Missouri and then northward through the Gates of the Mountains. The outfit skirted the mining town of Helena, but very probably "the boys" were allowed into town to enjoy the bright lights. Tom Lynch was an inveterate gambler and loved

2.7 This photograph of a herd of Bar U cattle, en route from the foothills to the Big Bow Ranch in 1917, shows a typical drive formation with four riders in the "drag" on the left of the photograph and others acting as flankers along the sides of the herd. Parks Canada, Hugh Paulin Collection, 96.11.01.29. See also p. 242.

to play poker. On a subsequent drive over much the same route, he and his cowboys shot up a bar, and compensation had to be paid before the drive could continue.[29] The herd's slow passage was marked by the rivers which had to be crossed. First the Sun, west of Fort Shaw, and then northward to the Teton and the Marias, before moving rather more swiftly across the alkaline flats of northern Montana to reach the Milk River in Canada. The herd was inspected by Canadian officials on 31 August. It consisted of 3,014 head of cattle and a remuda of 58 horses.[30] The editor of the *Macleod Gazette* welcomed the herd, and provided a useful description of it:

> The herd of this company [the North West cattle Company] passed through here ... en route to their ranges on High River ... it consisted of 3,000 head of the best grade cattle in Idaho, among which are 70 pedigree cows and 10 thoroughbred bulls. Several of our leading officials and stockmen visited the herd and were surprised at the good condition of the cattle after so long a drive. The number of calves was conspicuous to everyone. This is evidently one of the most successful drives ever made in this country and Captain Stimson is to be congratulated on having secured the services of such an experienced stockman as Tom Lynch. This enterprise has our best wishes.[31]

The editor was right; it had been a drive handled with consummate professionalism and blessed with good luck. However, scarcely had the herd arrived in the foothills than a severe fall blizzard struck. It lasted several days and the newly arrived herd was broken into small bunches which were pushed southward with their backs to the wind. When the storm finally abated, cowboys rode south and found that Bar U

2.8 River crossings were always hazardous to both cattle and riders on a long drive or a roundup. This photograph shows a herd of Shorthorn cattle crossing the Bow River. Glenbow, NA 1047-2

cattle had drifted as far as Fort Macleod and the Oldman River. They collected the herd and moved it northward onto the flanks of the Porcupine Hills. Here chinook winds had cleared the westward-facing slopes, and there was good shelter and water. Once again the herd had survived a crisis, and it spent the remainder of the winter rustling well to the south of the ranch headquarters.[32]

To trail thousands of cattle hundreds of miles with few losses was a remarkable achievement, even if it was not regarded as anything out of the ordinary by contemporary observers. However, the epic nature of cattle drives is memorialized by the many stories they contribute to the mythology of the range. One such gem concerns Fred Stimson. He had left Lynch in Idaho and hurried back to High River. Here, he put some men to work establishing the ranch headquarters, and then moved on to Calgary to attend to ranch business. When word reached him that the herd was in Canada and heading for Fort Macleod, he went down to meet it. After a jubilant welcome, the snowstorm hit the herd and Fred is reported to have retreated to the hotel in High River, where he "took to his bed" as the blizzard intensified and he became convinced that all was lost.[33] He only recovered when he was told that his newly arrived herd had escaped with minimal losses. This is another of those stories which depict Fred in a less than favourable light.[34] The "cowboy code" would suggest that both success and disaster should be accepted with unblinking taciturn stoicism. Hiding under the bed covers was really not on!

The Bar U cattle drive was also significant in that it brought the legendary John Ware to the foothills country. His request of Tom Lynch for "a little better saddle and a little worse horse," led to the first of many displays of horsemanship from this amiable giant.[35] The drive was also the setting for the first of Ware's confrontations with unsavoury elements on the range. He is credited with tracking and recovering some missing cattle. MacEwan recounts the story in the following words:

> ... he came upon the two men heating branding irons in a coulee. Alongside was a crude corral holding cattle with a familiar appearance. Nervously, the men stood erect and drew their guns. Their faces betrayed sin but if they were like other rustlers of that time, they'd shoot rather than accept inquisition. John was in danger, but using his head. He asked where a thirsty man could get a drink of water, and, before there was time for an answer, his horse bounded forward at the touch of spurs, upsetting

> the two men and knocking the revolvers from their hands. Leaping like a cat John was on the ground gathering the guns and covering the desperadoes. It took some maneuvering, but what Lynch and the men saw later was John returning, driving the stolen cattle and leading the two rustlers at the end of his lariat.[36]

This and many other stories about larger-than-life characters were part of the oral tradition of the range and were improved upon and embellished in the telling. Often there are several versions of the same tale recorded by later chroniclers like Kelly and MacEwan.

It was not until the spring of 1883 that the main Bar U herd from Idaho was gathered and returned to the lease lands along Pekisko Creek to be branded. At the ranch they encountered for the first time the purebred bulls which had been purchased in Chicago in the fall of 1881 and brought to the ranch by Andrew Bell and Herb Millar during the early summer of 1882.[37] As soon as the Missouri River was open for navigation they moved the valuable herd to Bismarck and thence on a sternwheeler to Fort Benton.[38] Here they were able to link up with some other purebred cattle heading for Canada. The safe arrival of these bulls in the High River area well before any hint of a cow herd gave rise to much merriment among the old-timers who were interested to see how this eastern-based ranch would fare with an all-bull herd![39]

The purchase of the Bar U herd in Idaho and its slow drive northward to Pekisko Creek was part of a much wider movement of cattle onto the Canadian range. There were only some 9,000 head of cattle in the entire North-West Territories in 1881. Five years later there were between 90,000 and 104,000 cattle in the "grazing portion of the northwest."[40] This estimate had swelled to 194,000 in 1891 and grown to nearly half a million in 1901.[41] This rapid increase was spurred by flows of stock into the region from a variety of sources. The earliest flow was from south to north, but the cattle acquired by Canadians in Montana and Idaho were themselves "migrants" from Oregon and the Pacific North West, or from the southern Great Plains. By the mid-1880s, Montanan cattlemen were moving herds onto newly acquired leases on the Canadian range or drifting cattle across the border from their ranches in northern Montana in a clandestine manner[42] By this time too, the first bands of "dogies" or "pilgrim" cattle had begun to arrive from the farms of Manitoba and Ontario, using the newly finished railway for transportation. So, the twenty-year history of the North West Cattle Company unfolded during a period of very rapid expansion and mixing – at various speeds and to varying degrees – of several genetic pools. We have to be suspicious of generalizations which purport to summarize the 'quality of the herds' on the Canadian range, unless their scope is limited with respect to place and time. As long as the total number of cattle was relatively small and controlled by a few corporate ranches it is possible to speak with confidence about general trends, but as the number of cattle increased

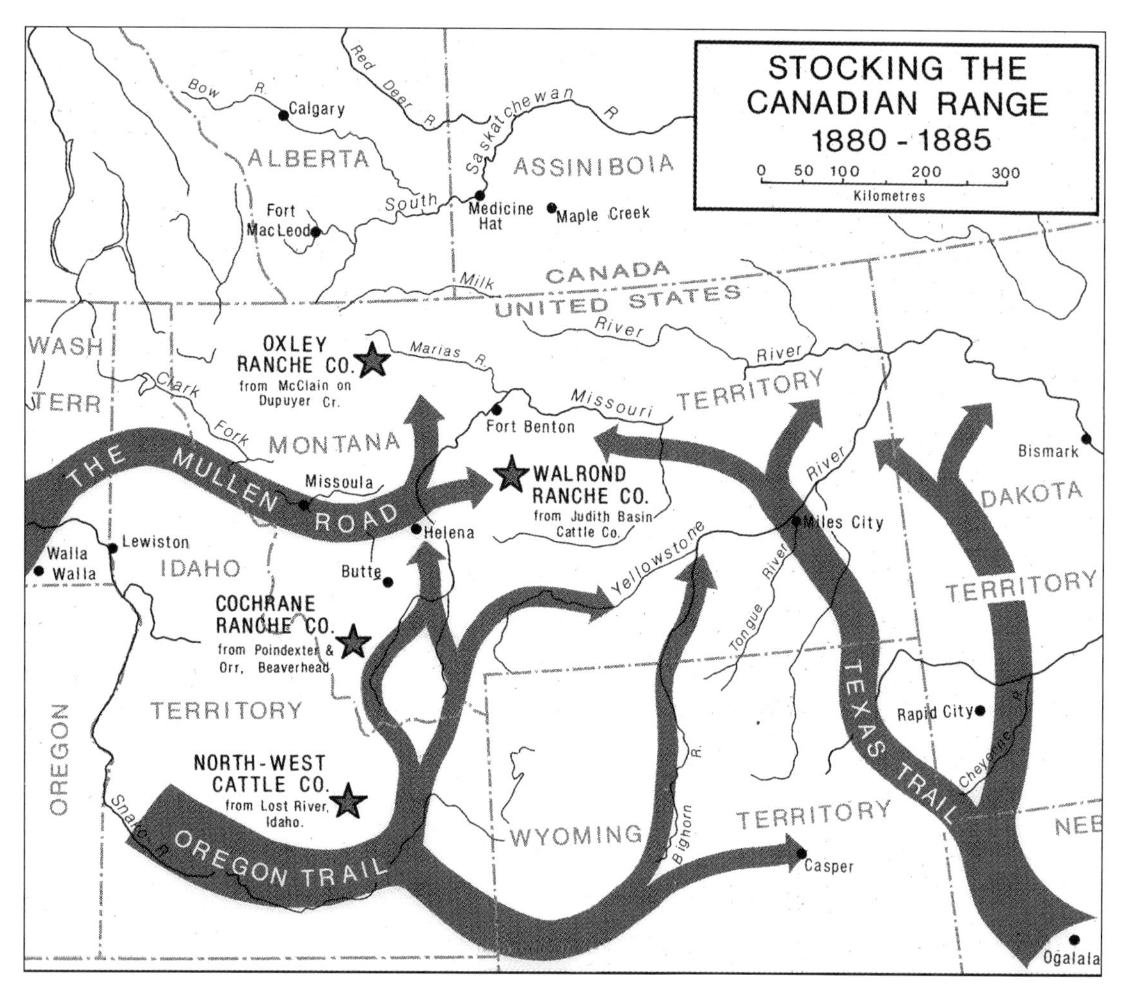

Map 2.1 The principal cattle trails into the northwestern interior states of the United States, and the locations where four major Canadian ranches obtained their foundation herds.

during the 1880s so too did the number of possible origins and the spectrum of quality. Moreover, the goals and strategies of ranchers were varied; while some sought to improve bloodlines, others looked only for size and rustling qualities in their cattle. Several of the larger cattlemen and companies ran calf-cow operations in one location, while running steer herds as speculative ventures in another. During this period of rapid growth the incoming herds "milled" and mixed on a vast scale.[43]

The early 1880s witnessed a flow of cattle onto Canadian grasslands that dwarfed the small-scale movements which had proceeded it. Imports rose from 1,352 head in 1880 to 6,284 in 1881 and 16,282 in 1882.[44] Most of these cattle were purchased in Montana. Indeed, sales to Canadian cattlemen helped to inflate cattle prices and fuel the "beef bonanza" in the northwestern United States. During this period the border was to all intents and purposes an open one. The regulations which had established the grazing leases also empowered leaseholders to import herds free of duty. This privilege was extended first until 1885 and then until September 1886.[45]

Ranching had taken root in valleys and basins in the Rocky Mountains of western Montana as early as the 1850s. The Deer Lodge, Bitterroot, Jacko, Beaverhead, and other valleys were being used as winter pastures by 'road ranches' which supplied beef to stopping houses along the Oregon Trail. As the mining boom provided a new and much larger market during the 1860s, so these valleys filled up with ranches and cattle. Terry Jordan makes it clear that a "thriving midwestern derived ranching industry" had occupied the region by the 1870s.[46] Hay was put up in huge quantities; irrigation was used where practical; and there was an emphasis on improved breeding. Most of the cowboys were from the midwest rather than from

Texas or California.[47] It was eastward diffusion from these valleys that established ranches in the Montana foothills along the Marias, Teton, and Sun Rivers and in the Judith Basin.

The cattle on these ranches had originated in the Pacific northwest. Early herds with marked Iberian characteristics had been transformed into 'westerns' by several generations of selective breeding [Map 2.1].[48] The versatility of the Durham-Shorthorn cross had made it popular and ubiquitous. The breeding area of the Willamette Valley had benefited from the location there of the Reidville stock-breeding farm, and the Columbia Plateau was another important nuclear area. Thoroughbred bulls had been imported into eastern Oregon, eastern Washington, and southern Idaho. Hereford bulls were used to upgrade some herds, but the Durham-Shorthorn cross remained the favourite.[49] Englishman Baillie-Groham reported that western cattle were far superior to the Texas cattle strains and outnumbered them in Wyoming by three or four times.[50] Governor John W. Hoyt of Wyoming echoed these sentiments; he pointed out that 'westerns' were better suited to the northern ranges. They were acclimatized to the severe winters, they knew how to rustle, and they were excellent mothers.[51]

The contribution of this eastward flow of stock to the cattle industry in Montana is illustrated by the photographs of L.A. Huffman. His biographers, who collected and published so many of his photos, concluded that "Huffman's pictures indicate that stock brought into these ranges, in the main, were not the long horned rangy animals which had made up the first trail herds out of Texas fifteen years before. Much of the early stock, of good Shorthorn and Durham breeding, was trailed from Oregon, Washington and Idaho."[52]

2.9 Range cattle, photographed by L.A. Huffman in Montana about 1880, are typical of those which made up the foundation herds imported into western Canada. White faces bear witness to the use of Hereford bulls to improve the herd. Glenbow, NA 207–74

This is an interesting observation, since Huffman's base of operations was at Miles City on the Yellowstone River in eastern Montana, the acknowledged goal for herds moving up the Northern or Texas Trails. Thousands of head of Texas cattle found their way north to the ranges along the Little Missouri, the Powder, the Tongue, and the Musselshell rivers. The volume of the influx and the speed of the transformation from virtually uninhabited grassland to overstocked range made a deep impression on contemporary cattlemen. Granville Stuart commented:

> It would be impossible to make people not present on the Montana cattle ranges realize the

removed from the near feral cattle which had moved northward in the 1870s.[56] These outfits from the southern plains maintained to a greater or lesser degree the characteristics and technology of their region of origin. As the country north of the Bear Paw Mountains and up to the international border filled up during the 1880s and 1890s, "Texas Cow Culture" began to play a part on the Canadian range.

In contrast, a number of Montana "cattle kings" were actively engaged in upgrading their herds during the 1870s. Conrad Kohrs repeatedly brought in good bulls from Iowa, and even went as far afield as Toronto in search of prime stock. Poindexter and Orr also obtained the nucleus of their fine breeding herd in eastern Canada.[57] T.C. Power's Judith Basin Cattle Company was putting up hay and planting timothy to sustain thoroughbred cattle by the early 1880s.[58] It was to these men, the leaders of improved breeding in the northern plains, that the incoming Canadian entrepreneurs from eastern Canada and Great Britain looked for their foundation herds.

The stock brought into Canada by large ranch companies between 1881 and 1884 was carefully chosen from among the most select herds in Montana Territory. One of the factors which drew risk capital to the Canadian west to establish ranches was the flourishing trade in live cattle which had grown up between Ontario and Britain during the 1870s. The prospects for expansion seemed limitless if the

2.10 These cattle pictured on a ranch near Calgary in 1883, are large animals with straight backs and rectangular well-fleshed bodies displaying both Shorthorn and Hereford characteristics. They were a complete contrast to the stereotypical "Texas Longhorns." Provincial Archives of Alberta, E. Brown Collection, B - 114

> rapid changes that took place on those ranges in two years. In 1880, the country [central and eastern Montana] was practically uninhabited ... but by the fall of 1883, there were 600,000 head of cattle on the range.[53]

Some of these herds were made up of yearlings or two-year-olds to be "double wintered" and finished for market on northern grass.[54] But others, like Andy Adams' "Circle Dot" herd, contained breeding stock as well as fat steers.[55] There can be no doubt that cattle from Texas and the southern Great Plains contributed to the gene pool of range cattle in eastern Montana. But Huffman's pictures and contemporary accounts suggest that Texan cattle, still identified by the general name of 'Longhorns,' had been improved by culling and breeding, and were several generations

right product could be produced. Both their aesthetic appreciation of fine stock and their business acumen encouraged Canadian cattlemen to purchase the best available cattle for their herds. Nor was the exchange only one way. The Cochrane Ranch advertised in the *Benton Weekly Record*, and sold purebred bulls from the Hillhurst estate in Quebec to Montana ranchers. The Department of Agriculture heralded the movement of purebred stock to the newly established ranches of the North West with great satisfaction. The minister remarked: "Considerable numbers of store cattle have been taken to the ranches in the North-West of the Dominion near the Rocky Mountains, together with a number of very choice animals which have been imported with a view to improving breeds."[59] MacEachran, the Dominion Veterinarian, who was in a position to inspect all imported cattle and who was also a director of the Cochrane Ranch Company, went into more detail in his report:

> The importation of Hereford and Polled Angus bulls, 136 in number, by the Cochrane Ranche Company, for use on their ranche in the Bow River District of the North-West Territories, form an interesting feature in beef producing developments in Canada. They are the pioneers of pure blood in that great grazing belt, which, for its richness of soil, mildness of climate and abundance of food and water all the year round, will, at no distant day become the source of meat supply not only for the Dominion, but largely for export as well. The value of this important enterprise and of these purebred bulls in laying a good foundation for the future stock of that new country cannot be estimated. The numbers of Herefords and Polled Angus cattle imported this year verifies the statement which I made in my last report that "from all appearances these two breeds are going to be given a preference to all others in the stocking of the vast prairies of the west."[60]

These importations of purebred bulls continued during the next few years.[61]

In the spring of 1884, a list of stock owners and their herds was published in the *Calgary Herald*.[62] There were some forty-four thousand head of stock on the range, and the four big companies, the Cochrane, Oxley, Walrond, and NWCC, owned about half of this total. The heterogeneous herds of the early traders and small ranchers were outnumbered for a time by much better quality stock. These cattle owed more to Durham-Shorthorn ancestors than they did to Longhorns, and many of them had been improved by the introduction of pedigree Hereford or Polled Angus bulls. Indeed it seems likely that the quality of stock on the open range reached its zenith in the period between 1884 and 1886, while the growing cattle industry was dominated by a few large outfits. This fine genetic base established by Canadian and British cattlemen could not be sustained, however, given the methods of the open range and the rapid expansion of ranching.

"Natural Increase" at the Bar U

The Bar U herd grew through natural increase. In the fall of 1884 the ranch was visited by members of the British Association who were on a "field trip" to the North West after their annual meeting in Montreal.[63] The Brandon immigration agent who accompanied the visitors reported:

> The stock on this ranche consisted of 4,500 head of cattle, besides this year's increase of 1,200 calves. They sold in July 800 steers at $65.00 each for beef. They have also 300 horses, besides the broncho ponies for cowboys' use.[64]

Among the British visitors was John Prince Sheldon, a professor at the Royal Agricultural College, Cirencester. He cast his experienced eye over the stock, and made the following comment:

> The immense herds have been greatly improved by the Shorthorn and Hereford, until nearly every trace of the native blood has been obliterated, and among them are a number of pedigree cows and bulls of Shorthorn blood. The pre potency of the imported blood is seen so prominently as to hide entirely the old foundation, and the herds are essentially Shorthorn and Hereford in character, the former greatly predominating. The stock are all healthy, and most of them fat – grand steers of 1,200 to 1,800 pounds, rolling in obesity.[65]

The calf crop suggests that the breeding herd consisted of about two thousand cows. In addition, at least a thousand of the remainder would have been yearlings born in the spring of 1883. The sale of eight hundred steers is about double what one would expect of a herd of this size if only four-year-olds were sold, but would have been less exceptional if both three- and four-year-olds were sold. It is also possible that the company purchased steers from other ranchers to fill its contracts with the government. The sale of eight hundred head at $65 each would almost have recouped the initial cost of the herd.[66]

In 1886, the NWCC reported 5,105 head of cattle, and later the same year it acquired the Mount Head herd which numbered 1,595, for a total of 6,700 head. When Sheldon visited the ranch again the following year he mentioned an estimated herd size of 9,000. This increase was about what one would expect given an average calving season. But it must be remembered that the winter of 1886–87 was one of great severity, and one would expect this to have had a devastating effect. Sheldon, visiting the ranch during the summer after the killing winter, does mention in general terms that weaker cattle and calves might die in considerable numbers during a severe winter and points out that he advocated putting up hay during his first visit to the ranch in 1884. He does not mention that the Bar U had experienced abnormal losses. Moreover, the Walrond Ranch cattle book shows that calf branding there was rather delayed, but it was not abnormally low in numbers.[67] A tentative conclusion might be that acclimatized range cattle, at sheltered locations within the foothills, produced a normal calf crop, while ranches stocked with 'pilgrim' cattle on more exposed mixed-grass prairie to the east experienced severe losses. In general, this slowed down but did not reverse the buildup of stock on the Canadian range, indeed, it was during the summer of 1887 that production exceeded the demands of local markets for the first time, and an experimental shipment of cattle was made to Great Britain.

In 1890, the Department of the Interior asked William Pearce to furnish a census of stock on the range.[68] This task should have been a simple one since leaseholders were supposed to submit a yearly report on the number of stock they held. Not all of them complied, however, and Pearce was forced to estimate the returns for some ranches while relying on information from the North West Mounted Police and other government agents for other regions.[69] The total number of cattle reported was 137,098. The eighty-six leaseholders accounted for 80 per cent of the total, the remainder being held by 'stockholders' who had either abandoned, or had never held a lease.

There were apparently 10,140 cattle and 832 horses at the Bar U. This figure would be consistent if the company had continued to build up its herds since 1887. On the other hand it was considerably in excess of the number required to stock the lease according to the regulations.[70] Pearce, in a confidential letter to the Department of the Interior, opined that the major companies had inflated their numbers:

> In addition to what I have reported to you officially, I might add that the popular supposition is that the Walrond Ranche has not 7,000 head of cattle on it. The North West Cattle Company is also supposed to be very much overestimated, and the same remark might be applied to some extent to the Cochrane Ranche. I imagine that the Oxley Ranche is also very much reduced in stock.[71]

However, dangerous as it may be to dispute the judgment of the "Czar of the Prairies," it seems quite possible that the Bar U was running ten thousand head. As we have already remarked, there was no constraint on the available range at this time. The only limit on the size of a herd would have been the capacity of an outfit to manage it efficiently. A year or two later, the Superintendent of the Calgary Division of the North West Mounted Police attached a list of ranches and their stock holdings to his report, and again the North West Cattle Company was credited with ten thousand head of cattle and seven hundred horses.[72]

One important factor which the patchy stock returns do bring out is that the North West Cattle Company was running a substantial herd of breeding horses for most of its twenty-year history. While the fame of George Lane's Percherons has always been an integral part of the Bar U story, the fact that Stimson was breeding horse for sale on a considerable scale is far less well known. By 1884, there were three hundred head of horses on the ranch, apart from the 'remuda' of work horses used by the cowboys and the teams of heavy horses used for pulling supply wagons and for haying.[73] Sheldon described a band of four hundred mares on his second visit to the ranch in 1887:

> The great majority of mares are of superior quality, and while the few inferior ones are being weeded out, imported sires of good blood are being mated with the better ones. In this way superior carriage and army horses, and even

> hunting horses for the English shires are being produced. These horse have been bred together for several years, and it is amusing to see a mare with three of her own offspring, colts and fillies alike, bearing each other company.[74]

Fred Stimson brought one of his favourite stallions from Quebec to the Bar U in 1883. "Bruce" travelled on the newly laid CPR track as far as Medicine Hat, and proceeded across the prairie to the ranch. Although he was not big, the "Canuck" stallion had great power and endurance. Other sires of note were Terror, the Mexican-bred Spooner, and Ben Lomond.[75] J.H. (7U) Brown brought a bunch of good B.C. horses to the ranch through the mountains and started a long association with the Bar U.[76] Stimson maintained good relations with the Quorn Ranch, his northern neighbour, and benefited from C. W. Martin's knowledge of the British market and occasional imports of livestock from the United Kingdom arranged through the Quorn.[77] While the grand vision of shipping first-rate cavalry mounts and hunters back to Britain was never realized, the demand for horses with a variety of attributes remained strong as the population of the prairies rose and the margins of settlement expanded.[78] Moreover, the cowboys' working horses were bred up too. The working horses had to handle large, heavy cattle and tended to be bigger than American cow ponies. The breeding and sale of horses remained an unsung but economically significant part of operations at the Bar U, and a tradition of handling superior horses had been developed during the twenty years before George Lane purchased the ranch.

A number of factors conspired together to reduce the quality of cattle on the Canadian range during the last decade of the nineteenth century. Chief among these was the importation of eastern farm-raised cattle, or 'dogies.' The idea of shipping young cattle westward to fatten on "free grass" had become widespread in Montana during the early 1880s. Indeed, between 1882 and 1884, there were as many cattle moving westward to be fattened as there were eastward to market.[79] A similar strategy was adopted by some cattlemen in the Canadian foothills. Why not, they argued, avoid being involved in the risky and time-consuming business of raising calves, when one- and two-year-old cattle could be obtained cheaply from the east? Near neighbours to the Bar U, like the Quorn Ranch and A.E. Cross at the a7, tried this and obtained young cattle during the summer of 1886. Unfortunately, the newcomers had no time to get to know their new range, or to acclimatize to the different ecology, before the killing winter of 1886–87 was upon them. Grant McEwan described what happened in dramatic terms: "… the soft and infantile specimens from Manitoba and Ontario seemed ready to drift like tumbleweeds in a wind. Having neither fear nor cow sense they refused to be turned or guided by men on horses."[80] They died in their hundreds, and this poor showing put a temporary damper on the westward movement of stockers in Canada.

In 1892, Canadian cattle imports to Britain were "scheduled," because of the possibility that they might carry the dreaded pleuro-pneumonia disease. This meant that cattle had to be slaughtered within ten days of landing; they could not be sold to British farmers as 'store cattle' to be fattened for a few months and sold at an opportune time. Ontario farmers, who had been profiting hugely from this sideline, were without a market for their young cattle. They turned to the cattlemen of the western ranges. In 1895, the Deputy Minister of Agriculture mentioned this new development in his report:

> Thousands of yearlings and two year old steers have been brought into Alberta, and should this experiment prove successful, thousands more will be shipped there next summer.[81]

The following year sixteen thousand head of young cattle moved westward into the grazing region, while an ever-increasing number of yearlings moved from farm regions within the Territories to be fattened on the ranches.[82] This figure rose to some thirty-five thousand head in 1899. Most of these stockers were of dairy strains and "could never, with any system of feeding or management attain the high quality which is essential to make the business of maturing cattle a success."[83]

There is no evidence that Fred Stimson adopted this strategy of buying young cattle to fatten. He had a flourishing "cow-calf" operation at the Bar U, and his long experience as a breeder, both in Quebec and in the west, may well have made him hesitate to take an action which would have threatened the quality of the fat, grass-fed steers he was producing for the British market.[84] Ranchers, he said, did not want Holstein or Jersey blood in their stock. He was intent on maintaining the quality of his breeding herd and had recently brought in several carloads of purebred Shorthorns. He was quoted as saying that Shorthorns were a sure thing if good quality bulls were used. He also liked a Highland cross because of their virility and the thriftiness of their offspring. He had no objection to Herefords and found that they rustled well and on grass alone were hard to beat.[85]

Another source of cattle which tended to degrade the gene pool in Canadian cattle country were the small bunches of cattle brought across the border by incoming settlers. These beasts would constitute an important source of young stock in the years to come, but they were of abject quality. D.H. McFadden's detailed reports of the cattle quarantine at Emerson include a case where a farmer had to carry some of his stock to the quarantine ground.[86] William Pearce could hardly believe that "such wretched stock would be kept and bred from."[87] Commissioner Herchmer of the North West Mounted Police suggested that settlers should be encouraged to dispose of their stock and purchase fresh animals when they arrived in Canada. He pointed out that this would "provide a good market for our people and prevent the whole country becoming flooded with a lot of inbred cattle, which years of careful breeding would not improve sufficiently to equal our own stock."[88]

The presence of herds of American "tramp cattle" from Montana also had a considerable impact on the quality of Canadian herds, especially in the border country. Thousands of head of cattle drifted northward over the Milk River Ridge and into the Pot Hole country, while to the east, the herds of the Turkey Track and the T Bar Down were pushed up the Whitemud River into the Cypress Hills.[89] Some of these American herds were composed of southern cattle of indifferent quality. When they mixed with

Canadian herds, the tough range bulls from Montana tended to drive off the smaller purebred animals and hurt the heifers, adversely affecting both the quality and the size of the calf crop.

The last decade of the North West Cattle Company's history was played out against an overall decline in herd quality on the Canadian range. This was one factor eroding Canada's competitive advantage in the European meat market. Canadian cattle were failing to hold their own, particularly when compared with stock from Argentina.[90] If a rancher was determined to maintain the quality of his herd, his only recourse was to forego the extensive, free-and-easy techniques of the open range, in favour of more intensive management within a fenced range. But for several decades such a strategy ran counter to the hopes and dreams of Canadian cattlemen.

"Give me land, lots of land ..."

The compelling arithmetic of the "Beef Bonanza," which had so captivated Stimson and his peers, was based on the assumption of "free grass." In the west lay "the boundless, gateless, fenceless prairies of the public domain, covered with grasses, which hundreds of observers had declared to be the most nutritious that livestock ever fed on. With no operating expenses save that of a few cowboys, some corrals, and a branding iron, one might transform these leagues of free grass into steers at top prices."[91] Young stock purchased for $5 a head could be sold for $45 to $60 after four or five years of grazing on the public domain.[92] In addition, the amazing fecundity of nature was reported to ensure that cow herds doubled in size after only two years. "Cattle is one of those investments men cannot pay too much for since if left alone they will multiply, replenish, and grow out of [even] a bad bargain."[93] Moreton Frewen stressed to the English shareholders of the Powder River Cattle Company the importance of free land. He wrote:

> I am convinced that results will show that companies will pay dividends throughout, in direct relationship to the amount they have sunk in land, and so called improvements; I can imagine no more mistaken policy than to pay money for land which you can possess for nothing.[94]

He went on to argue that the purchase of a defined range might lead to overgrazing and degradation of the grass, while the immensity of the North West, "up to the British possessions on the Bow River, is so vast as to insure land for a generation of stockmen yet unborn."[95] In his view, the ability to move ever onward to new grass would ensure the recovery of once-occupied range. Brisbin also stressed the inexhaustible extent of the plains, while an English observer of the Great Plains wrote:

> The immensity of the continent produces a kind of intoxication: there is moral dram-drinking in contemplation of the map. No Fourth of July orator can come up to the plain facts contained in the land Commissioner's Report.[96]

In Canada, too, the theme of 'free grass' was implicit in the carefully prepared estimates of expenses and returns which were promulgated in lectures and publications to inform prospective ranchers and to encourage investors. Professor W. Brown, of the Ontario Agricultural College in Guelph, proposed a hypothetical situation where three young men might each contribute $5,000 to establish a ranch, on which they would provide most of the labour required. Most of the outlay projected was on livestock (64 per cent); wages and expenses accounted for 19 per cent of investment; and ranch infrastructure absorbed a further 16 per cent. The land cost amounted to less than 1 per cent of the total, to purchase a hundred acres for the 'cattle station.' The hypothetical ranch was a modest one of two thousand acres, but Brown went on:

> It will be evident that we are calculating circumspectly pro tem, whatever the future may bring about. Until grazing location becomes as regular as Ontario farms are to each other, our 2,000 acres may be 20,000 so long as neighbours don't push or out feed us in the number of stock.[97]

In his view the expansion was only limited by the capacity of herds for natural increase, and the initiative and drive of potential cattlemen to seize the opportunities with which they were presented. The larger the enterprise, the more spectacular the returns.[98]

"Accustomed Range" versus "Grazing Leases"

While the immense grasslands of the West seemed to offer equal promise to cattlemen on both sides of the international border, no more stark contrast exists between the cattleman's frontier in Canada and in the United States than in the differing roles played by the central governments of the two countries in the management of their western lands. In Canada, a comprehensive legislative package was introduced in 1881 which enabled individuals or companies to lease large acreages of grazing land for a period of twenty-one years.[99] At much the same time, the men who occupied the grasslands of Montana, Wyoming, and Colorado rejected proposals which might have resulted in the modification of existing land laws and in the formation of large landholdings, on the grounds that such a "land grab" was contrary to their theory of government.[100]

In the United States the cattlemen were regarded "as merely an advance screen ahead of the real conquerors of the land, the pioneer farmers."[101] The cattle boom, that explosive surge of men, stock, and capital onto the western rangelands, occurred outside the protection afforded by the law. As more and more herds moved north and west to appropriate for themselves, by prescriptive right, an "accustomed range," crowding developed. Illegal fencing, fraudulent acquisition of land, and range wars were the results. Even where the health of their stock was the issue, regional bickering between stockmen delayed the establishment of effective quarantine and inspection arrangements.

Those charged with administering the public domain in the United States were quick to appreciate the need for larger units of landholding in the semi-arid grasslands. As early as the 1860s, requests for leases to provide some security of tenure to ranchers were received, and in 1875 the Commissioner of the General Land Office urged Congress to modify the existing land laws.[102] The Secretary of the Interior pointed out that the existing homestead and pre-emption laws were entirely inappropriate as a basis for settling much of the remaining public domain.[103] President Hayes was persuaded, and in his annual message of December 1877 he declared:

> These lands [west of the hundredth meridian] are practically unsaleable under existing laws, and the suggestion is worthy of consideration that a system of leasehold tenure would make them a source of profit to the United States, while at the same time legalizing the business of cattle raising which is at present carried out upon them.[104]

The following year, Major James Wesley Powell submitted his seminal summary of conditions in the west and his proposals for how best to cope with the challenges of semi-arid environments. It contained, "both in its analysis of western conditions and its evaluation of consequences, the classic statement of the terms on which the west could be peopled."[105] In 1879, Congress established a commission on the public lands. The commissioners spent the summer holding public hearings throughout the western plains. The unifying theme which runs though their testimony and transcends regional contrasts is that the people already in the west regarded any proposals for radical changes with the utmost suspicion. The Commissioners summarized their findings:

> There is a deep-seated conviction in the minds of the majority of the people in this country that a system which tends toward monopoly or even permits the aggregation of the very large tracts of land into the ownership of a single person is unjust.[106]

If the settlers of the west and their territorial and state representatives were reluctant to see sweeping changes in the land laws, Congress was certainly not going to initiate them. The stock grower was to remain a tenant by sufferance. An opportunity had been lost. The cattlemen who solved so many problems were unable to find a way to control the use of the grassland commons during the cattle boom.

The eastern public, who knew nothing of the realities of the west, were bombarded first by stories of the fantastic profits made by the "cattle kings" and the great foreign-owned companies, and then by reports of illegal fencing and fraudulent acquisition of land on a vast scale. Agrarian discontent, which was expressing itself in attacks on railroads, the grasping corporation, and the power of Wall Street, was turned upon the cattle country "land grabbers."[107] The central government could make no concessions to those who were apparently showing an arrogant disregard for the rule of law. In retrospect it seems true that:

> To expect the federal Government to pass legislation that would assure each grazier his

> share of the grass was to call for too wide a departure from the ideas upon which our public lands policy was founded.[108]

In Canada, a very different course was adopted. The federal government played a major role in promoting and sustaining the range cattle industry. In a real sense, ranchers, like the police, Indian agents, and railway men, were agents of the National Policy. The Conservative government of Sir John A. Macdonald was confronted with a number of pressing geopolitical problems. Above all, the prime minister was deeply suspicious of the United States' intentions along the Upper Missouri frontier. Already the region was in the economic hinterland of Fort Benton, and in the thrall of Montana-based trading companies. It had been the depredations of Montana-based whisky traders and the excesses of American wolfers in the Cypress Hills which had led to the formation and deployment of the North West Mounted Police.[109] Since that time a number of legitimate settlers had moved across the line. Unless something was done, the region might become American by default. In Macdonald's view it was imperative to have some visible and effective settlement on the ground to hold the western grasslands for Canada. What could be more statesmanlike than to encourage men who had already proved themselves as stockmen, and who were well aware of the requirements of the British market for beef, to establish large ranches in the west? At one blow, this would provide a cheap source of meat for feeding the police and the Indians and loosen the stranglehold of Montana-based companies on the economy of the North-West Territories. In the longer term, mature stock would provide freight for the railway and make a contribution to the growing export trade in live cattle.[110] In the short term, the work force of the ranches – hard-riding cowboys – would be a useful supplement to the police should trouble erupt on the frontier. For a time, and for the purposes of the Dominion, the Canadian government set aside the cherished image of the homestead settler and the family farm and created a "big man's frontier."[111] In Canada, the frenetic growth of the five years between 1881 and 1885 took place within a legal framework and was regulated on the spot by agents of the federal government. Those who held grazing leases could rely on the support of the North West Mounted Police, while the Department of Agriculture was actively promoting the sale of their product in Great Britain.[112]

Rules for the Canadian Range

The grazing regulations, promulgated by Order-In-Council in December 1881, were shaped by almost a year of dialogue between Senator Matthew Cochrane and the government of Sir John A. Macdonald.[113] An amendment of the Dominion Lands Act of 1872 had recognized the possibility that there might be regions within Canada which were suited to grazing rather than to arable farming.[114] Under its provision, lands might be leased to individuals or companies without any commitment on their part that agricultural settlement would follow. Cochrane requested that the general provisions of the amendment be translated into precise operational regulations. He pointed out the numerous advantages which would follow from the establishment of a robust ranching industry in the west. His case was presented to the cabinet by John H. Pope, Minister of Agriculture and Cochrane's friend and neighbour in Compton County, Quebec.[115] The regulations were discussed by the Privy Council, and finalized in an Order-In-Council published in December 1881.[116]

The terms offered to prospective graziers were generous in the extreme. Leases of up to 100,000 acres were made available for a period of twenty-one years at a rental of 1one cent per acre per year. The wording of the regulations created the impression that the security of tenure thus conferred was absolute. Only in the overarching statement, "under the provisions of the Act 44 Vic., Chap 16, and the subsequent regulations approved by Order In Council," was the ultimate power of the government veiled. For the act referred to contained the following provision:

> ... every such lease shall among other things, contain a condition by which, if it should thereafter be thought, by the Governor in Council, to be in the public interest to open the land, covered by such a lease, for settlement, or to terminate the said lease for any reason, the Minister of the Interior may, on giving the lessee two years notice, cancel the said lease at any time during the term.[117]

In spite of this important *caveat*, Cochrane appears to have been satisfied. He had argued that large scale investment in the west would only be forthcoming if it were clear that government support would be ongoing. Although the letter of the law made it clear that leases could be cancelled after only two years, nevertheless, the lessee had considerable *de facto* control over his holding. A prospective homesteader, seeking to locate on a river bottom within a grazing lease, would have to wait two years, and then prove to the Department of the Interior that it was in the public interest to open the land in question. The big corporate ranchers, with their lawyers and powerful connections in Ottawa, were able to convey the impression that their tenure was legal and absolute. Most incoming farm families bypassed the "grazing country" and settled further north. For example, Mormon homesteaders, seeking land in Alberta close to the international boundary line, almost despaired of finding suitable land for settlement. It was "all being taken up under grazing leases."[118] The situation of those who had established small ranches prior to the passage of the lease regulations was a different problem and continued to plague ranchers and administrators for a prolonged period.[119] As long as the government tacitly supported the lessees, the "closed lease provisions" were secure; however, the Department of the Interior retained the right to react to changing circumstances and to direct the manner in which the North West should be settled.[120]

The lease regulations envisioned large-scale enterprises, for they referred to townships, or portions of townships, rather than to sections.[121] The upper limit of a lease was established at 100,000 acres. In some cases this provision was circumvented by the simple procedure of taking out leases in the names of several officers of a company, or by establishing a subsidiary company. Whether this was a planned strategy or merely the result of the scramble to "get in on a good thing" is not clear. Section 18 of the regulations tied the size of the area leased to "the quantity of the livestock kept thereon, at the rate of ten acres of land to one head of stock."[122] This provision was an attempt to prevent speculators from holding leases in the hope that they would be able to trade them to later comers. The original terms gave the lessees three years to meet their obligations, but this was amended in 1885 to read:

> That the lessee shall in each of the three years from the date of the Order in Council authorizing the issue of the lease, place upon the tract of land hereby determined not less than one third of the whole amount of stock which he is requested to place upon the said tract. . . ."[123]

This change was used to cancel a large number of speculative leases during 1887 and 1888. Of course, the stocking ratio of one steer to every ten acres was too high to be realistic under open range conditions. The limits were changed from ten to twenty acres in 1888, and even this ratio was not always met by *bona fide* ranchers during the first decade that the regulations were in force.[124]

The terms of the lease were set at a rate of $10 per thousand acres, or "one cent per acre per annum." This has a fine ring to it, and has encouraged the perception that the rents collected were of nominal value. Certainly, this rent seems to have met with the approval of those interested in establishing ranches in the North West. It was not the focus of long debates or complaint. However, for a ranch comprising

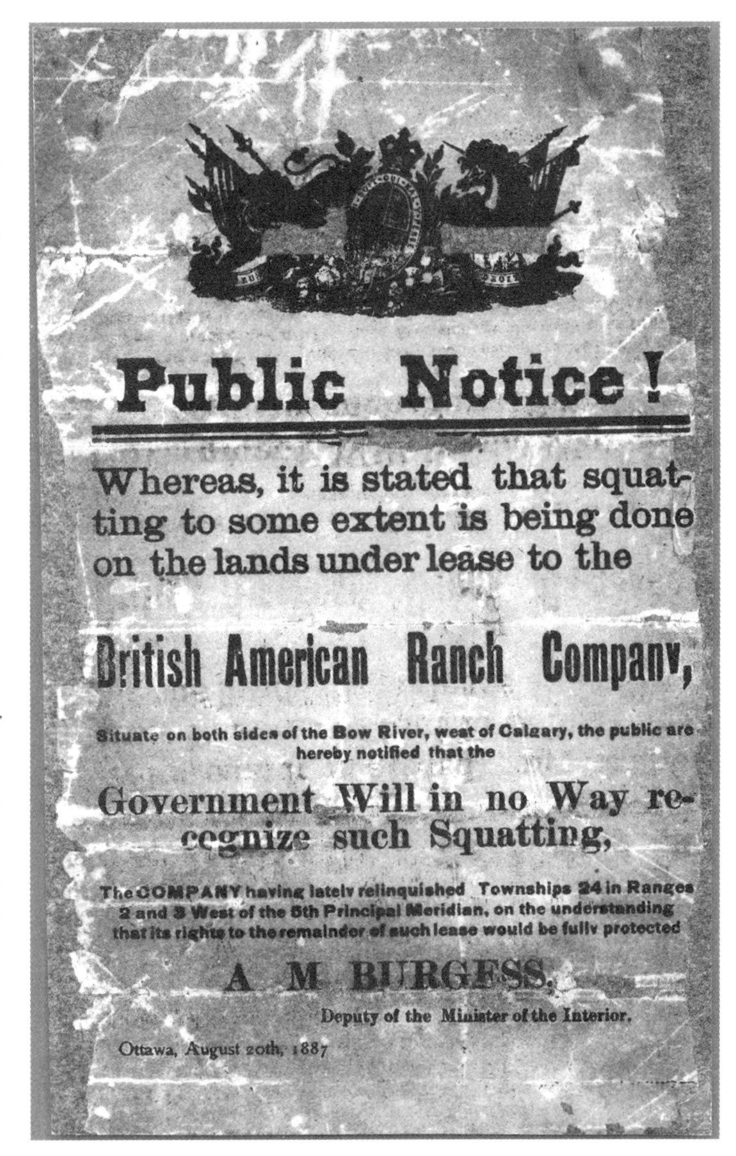

2.11 The lease system gave large Canadian investors some security of tenure to their ranges. This was in marked contrast to the situation in the United States where a variety of abuses and serious overgrazing were evident by the 1880s. National Archives of Canada, RG 15, B2a, vol. 5 f.137261

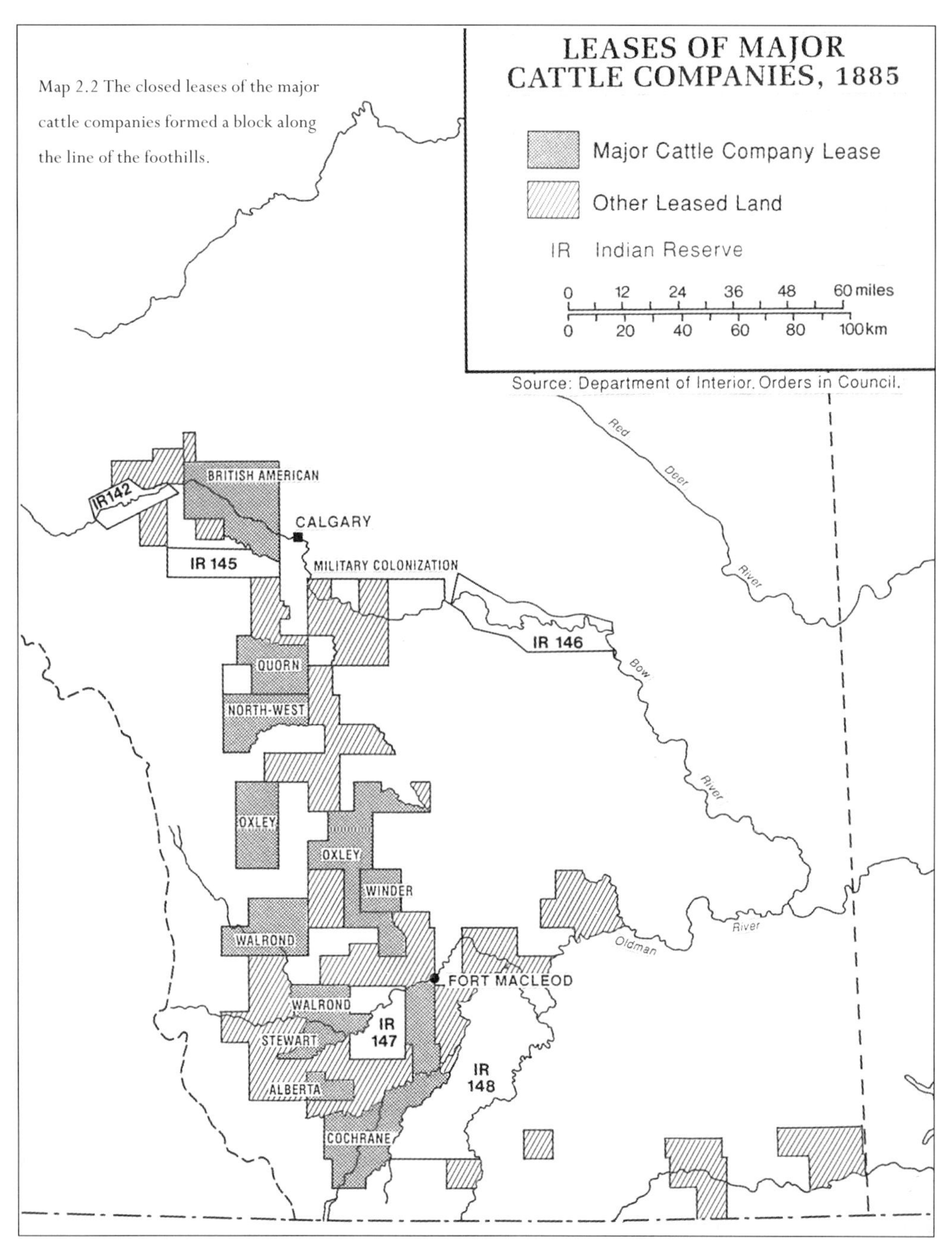

Map 2.2 The closed leases of the major cattle companies formed a block along the line of the foothills.

100,000 acres, the rental was $1,000 a year, not a trifling sum. On a fully stocked lease the lease fee amounted to twenty cents per head per year. From the government's point of view the receipts from the lease of grazing lands were by no means negligible. They far outweighed all the cash revenue derived from the sale of Dominion lands in the 1870s, and this income was earned by land which could be reclaimed by the Crown after only two years' notice.[125] This carefully crafted package of lease regulations ensured that the frantic growth which followed its introduction took place within a legal framework and was regulated on the spot by agents of the federal government.

The Homestead Act framed in the United States was adapted for use in Canada, but no such prototype existed for grazing lands.[126] Faced with the problem of framing regulations to provide for the orderly establishment of ranching in the west, the Canadian government turned to Imperial experience. It is to legislation evolved in Natal and the Australian colonies that the origins of the lease legislation must be traced.[127] In Cape Colony, for instance, the fact that a "viable unit" for agriculture varied greatly in size for ecological reasons was incorporated into the Crown Lands Act.[128] In Australia, graziers were licensed and "tickets of occupation" were purchased from the state government at rates dependent on the size of the flock pastured.[129] The Imperial Waste Lands Occupation Act of 1846 envisioned a threefold division of New South Wales into settled, intermediate, and unsettled areas. Lands sales proceeded in the normal manner within the settled area, while in the other zones, lands were withdrawn from settlement and leased for eight or fourteen years, respectively. No limits were imposed as to the size of holdings, and rental was assessed according to the carrying capacity of the land. Western Australia adopted a similar system.[130] The Australian land reform movement of the 1860s demonstrated the flexibility and utility of the lease system, and attracted widespread attention throughout the British Empire.[131] It is unlikely that these innovative ideas could have escaped the notice of Canadian authorities as they wrestled with similar questions.

Taking the Bait

The Canadian government must have been gratified by the immediate response to its policy. The promulgation of the regulations coincided with a surge of interest among eastern entrepreneurs in the development of the West. They vied with each other to obtain leases. Many investors became involved in both land colonization companies and ranch enterprises. The Department of the Interior started to receive requests for leases in 1880, well before the regulations had been finalized. In 1882, when this backlog was tackled and new requests were received, a total of 154 applications were processed and seventy-five leases were issued covering somewhat more than four million acres.[132] In 1883 and 1884, nineteen new leases were approved as small stockmen who were already running herds hastened to acquire some legal title to what they regarded as their habitual range. The Northwest Rebellion of 1885 did little to check eastern interest in ranching as 113 new applications were received, although the leased acreage dropped slightly as speculative leases were cancelled. Another marked quickening of interest occurred in 1886, which more than doubled the number of leases issued and brought the total acreage involved to 8.5 million acres. This was heady progress indeed [Map 2.2]. By 1884, Deputy Minister Burgess was able to report to the Minister of the Interior that "the success of the cattle industry in the Fort Macleod region, and generally along the base of the Rocky Mountains to Calgary, may now be considered settled."[133] The Minister of Agriculture boasted:

> It is an established fact, that Canada is destined to become one of the most important cattle raising countries of the world. The business of cattle raising here is capable of unlimited extension.[134]

The flocks and herds with which the leases were being stocked represented an influx of capital to a region "which would otherwise have remained unnoticed for years."[135] This remark stands the test of hindsight. The grazing regulations, so astutely framed by the Canadian government, acted as a catalyst to lure eastern risk capital to the ranges of the North West. The very details of the regulations suggested that the faraway grasslands had been surveyed; that bureaucratic machinery existed to collect rents; that the North West Mounted Police could be relied upon to provide security; and that the lessee was assured of absolute control of his lease for twenty-one years. The promulgation of the grazing regulations altered investors' perceptions of the North West and reduced the risk to an acceptable level. As is so often the case,

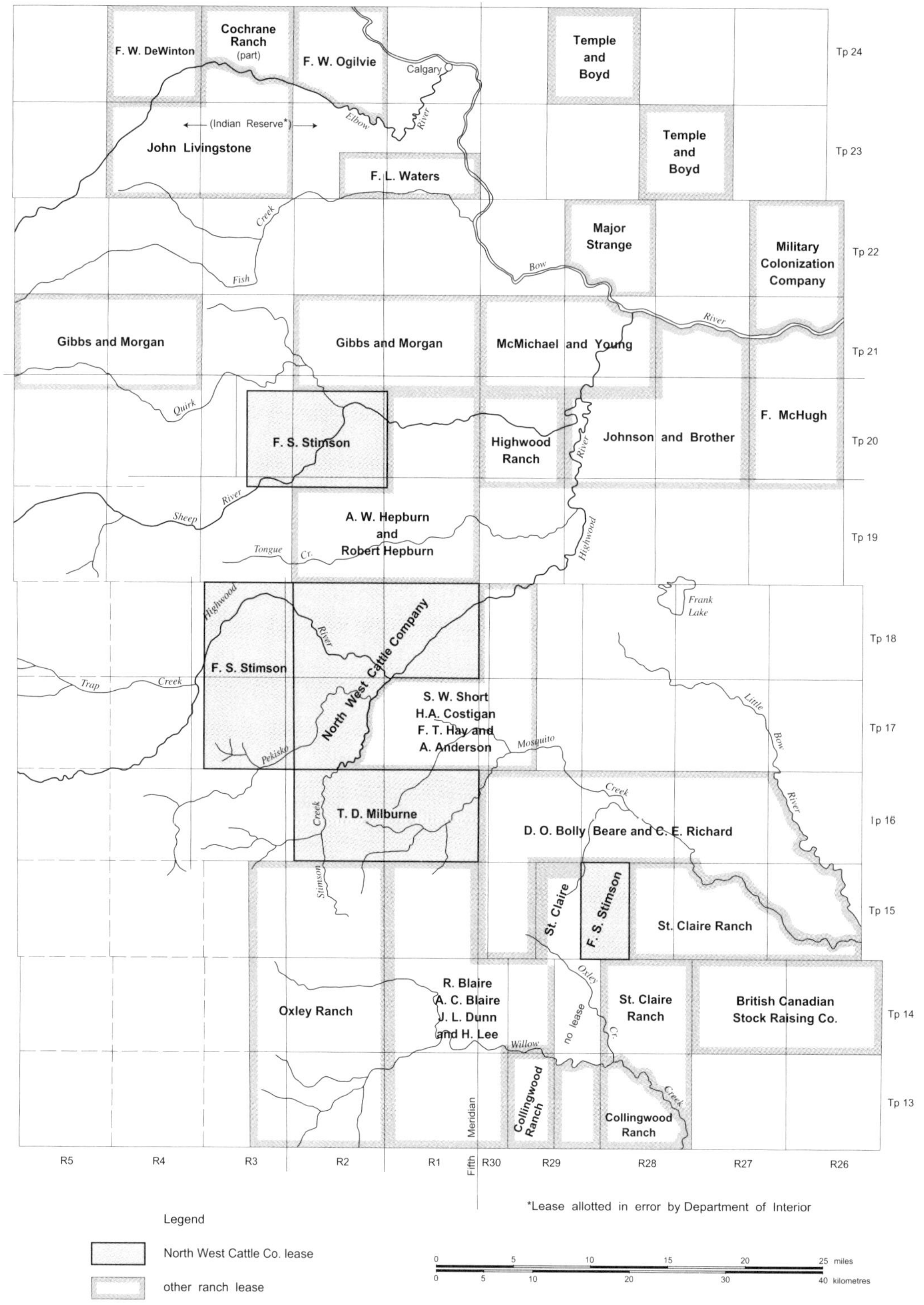

F. W. DeWinton
Cochrane Ranch (part)
F. W. Ogilvie
Calgary
Temple and Boyd
Tp 24
(Indian Reserve*)
John Livingstone
Elbow
River
F. L. Waters
Temple and Boyd
Tp 23
Creek
Fish
Major Strange
Bow
Military Colonization Company
Tp 22
Gibbs and Morgan
Gibbs and Morgan
McMichael and Young
River
Tp 21
Quirk
F. S. Stimson
Highwood Ranch
River
Johnson and Brother
F. McHugh
Tp 20
Sheep
River
A. W. Hepburn and Robert Hepburn
Tongue
Cr.
Highwood
Tp 19
Highwood
River
North West Cattle Company
Frank Lake
Tp 18
Trap
Creek
F. S. Stimson
S. W. Short
H.A. Costigan
F. T. Hay and
A. Anderson
Little
Bow
Tp 17
Pekisko
Mosquito
Creek
River
Creek
T. D. Milburne
D. O. Bolly
Beare and C. E. Richard
Tp 16
Stimson
St. Claire
F. S. Stimson
St. Claire Ranch
Tp 15
Oxley Ranch
R. Blaire
A. C. Blaire
J. L. Dunn
and H. Lee
Oxley
no lease
Cr.
St. Claire Ranch
British Canadian Stock Raising Co.
Tp 14
Willow
Creek
Collingwood Ranch
Collingwood Ranch
Meridian
Tp 13
R5
R4
R3
R2
R1
Fifth
R30
R29
R28
R27
R26
*Lease allotted in error by Department of Interior
Legend
North West Cattle Co. lease
other ranch lease
0 5 10 15 20 25 miles
0 5 10 20 30 40 kilometres

Map 2.3 Leases applied for by the North West Cattle Company and its officers, 1882 and 1883. Drawn from Orders In Council.

the fact that this new vision was based on partially erroneous information did nothing to dampen the enthusiasm of potential investors. Ranching would inevitably have taken root and spread from nuclei already established in the foothills during the 1870s. The lease regulations, by provoking a sudden influx of capital, changed the course of development completely.

Acquiring the Bar U Range

It was in this competitive atmosphere, while the Department of the Interior was struggling with a backlog of applications for grazing leases, that the North West Cattle Company acquired its first leases.[136] The company was granted fifty-nine thousand acres, covering land to the north and west of the southern branch of the Highwood River [Map 2.3]. In addition, the managing director of the company, Fred Stimson, obtained a lease for eighty-eight thousand acres. Half of this range was located along the western margin of the company lease, while the remainder was in a block a few miles to the north. Only one year later, the northern block of Stimson's lease was assigned to James Moore and Charles Martin, who were developing the famous Quorn Ranch along Sheep Creek.

To place these lease grants in context, and to help us grasp their size and significance, it will be helpful to review briefly the way in which western Canada was surveyed. Even before Rupert's Land was finally acquired from the Hudson's Bay Company, a system for surveying the new lands had been approved.[137] It involved a rectangular grid based on the east to west baseline provided by the international boundary, and on a series of prime meridians running north to south. Because the whole of prairie Canada was acquired at one time, and, in the main, the survey preceded settlement, it was possible to implement "one vast system of survey, uniform over the whole of it."[138] The basic unit of the survey was the township, a square area measuring six miles by six miles. This unit was subdivided into thirty-six sections, each one mile by one mile. Some sections were reserved for the Hudson's Bay Company and to support local schools. The remaining sections were each divided into quarters to provide four "quarter sections," each of 160 acres. These were the units of farm settlement offered as free homesteads under the terms of the Dominion Lands Act.[139] Every township could be identified in terms of a township number from south to north; and a range number from east to west. These numbers are shown on the map of leases. It is clear, for example, that the North West Cattle Company leased township 18 (Twp. 18) in ranges 1 and 2 (R1 & 2), west of the fifth principal meridian.[140]

In 1887, an important company merger took place which increased the leased acreage of the Bar U Ranch. Earlier, in 1883, the Mount Head Ranch Company had been incorporated in Britain. Among the shareholders – mostly drawn from the Anglo Irish aristocracy – was T.D. Milburne, who had played an important role in the formation of the North West Cattle Company and had held a power of attorney for the affairs of Fred Stimson while he was away in the west. It was Milburne who obtained the lease land for the Mount Head; forty-four thousand acres on the southern margin of the North West Cattle Company's lease. The Mount Head established its headquarters on Sheppard Creek, engaged Godfrey Levine to manage the outfit, and started ranching with a thousand head of cattle bought from Emerson and Lynch.[141] By 1885, the company had 1,595 head of cattle and was apparently flourishing. However, the shareholders decided to sell out to the North West Cattle Company the following year. Milburne negotiated the deal by which shareholders in the Mount Head acquired shares in the North West Cattle Company and a cash dividend. It must have been a satisfactory arrangement because all the shareholders took the 'stock option' and held their shares in the North West Cattle Company until it was wound up in 1907. Perhaps this important transaction should be regarded as a 'friendly merger.' It may well have been motivated by the desire of the British investors to pass decision-making responsibilities over to a company which had a respected manager on the spot and a good financial record. They were looking for a promising investment, not the headaches associated with absentee management.[142] Absorbing the Mount Head added two townships to the leased range of the North West Cattle Company, bringing the total acreage controlled by the company to 158,000.[143] The broad valley bottom land included in the transfer from the Mount Head became one of the best haying areas on the ranch and was used as a winter camp for calves throughout the company's history.

The leases of the North West Cattle Company, along with those of the seventy or so other lease holders, formed a compact block on contemporary maps of the grazing country and provided an illusion of precise organization. Nothing could have been further from the real situation. There were virtually no fences on the range during the 1880s, and very few, of limited extent, during the 1890s. Stock roamed at will, and it was the responsibility of the small holder to fence his garden or his small herd of dairy cows. Over a period of years cattle learned the dimensions of their home range, which tended to correspond closely to watersheds because of their need for water. In early summer, herds were drifted slowly eastward out of the foothills, towards the Little Bow and the Bow rivers. After the fall roundup, ranch herds returned to the sheltered valleys within the foothills which had enjoyed several months of rest from grazing.

Many of the leases granted, and shown on the map, were purely speculative. The Minister of the Interior was instructed to "satisfy himself of the good faith and ability of the applicant to carry out the undertaking involved in such applications."[144] In practice it was impossible to separate speculators from those with a genuine interest in raising cattle. This irritated the Prime Minister, who fumed, "Some 8 or 9 companies got Ranches on giving the assurance that they were both able and willing to stock them. It turns out that they all lied and merely got their leases for the purpose of selling them – other parties were prevented from getting and stocking these lands. A year or so has been lost and not a hoof or home put on any one Ranche."[145] Nearly half the leases authorized were never occupied or stocked, and these leases were cancelled between 1885 and 1886.[146] By the time the two million acres which had been held by speculators was once more available, the regulations had been changed and all new leases were opened to homestead entry. In fact, this huge area acted as a cushion which eased many problems as the country began to fill up. Without this large reserve of uncommitted land the disputes between small cattlemen and large, and between stock interests and those of the farm homesteader, would have been even more difficult to handle [Map 2.4].[147]

The extensive leases held by the North West Cattle Company were like an island of legally held grazing land in a sea of free grassland. To the eastward, lease 4, issued to Short, Costigan, Hay, and Anderson, was never stocked and was cancelled in 1886, as was lease 7, held by Beare and Richard, so no competing outfit blocked the eastward movement of Bar U herds in summer. To the north, J.S. Moore and C.W. Martin had taken over the Hepburn lease as well as the range assigned from Stimson, and were establishing the Quorn horse ranch. Tom Lynch trailed in the first five hundred horses in 1886. But the central axis of the Quorn was Sheep Creek, while the Bar U's northern boundary was the height of land just north of the Highwood River.[148]

Map 2.5 shows the extent and location of the Bar U lands in 1888 superimposed on a modern road map. The contemporary landmarks, like the course of the major highways and the position of communities, help us grasp the extent of the leases. If you are travelling toward the Bar U Ranch from the north down Highway 22 through Black Diamond, you would first encounter the boundary of the original lease as you climb over the shoulder of Longview Hill and look down on the small community. To your right, the course of the Highwood River, deeply incised into level plain, can be traced westward toward the mountains. Beyond the river, forested north-facing slopes give way to gently rolling hills. All this was Bar U land. Indeed, even if you are driving briskly southward toward the Crow's Nest Pass, it will be some twenty-eight kilometres before you leave the lease. This 'exit point' is marked by the entrance to the Mount Sentinel Ranch on your right, about three kilometres north of the Chain Lakes Reservoir.

An alternative route to the ranch is to use Route 540 westward from Highway 2, between High River and Claresholm. This beautiful country road parallels for part of its course the Bar U Trail, the old wagon road from High River to the ranch. About fifteen kilometres west of Cayley you will notice a gravel road crossing the highway with a signpost indicating that it follows the line of the Fifth Principal Meridian. All the rolling farmland leading down to the river on your right, and the more undulating land on the far side of the river rising up to Longview Hill, was Bar U summer grazing. As the road turns southward, you can see the sinuous meanders of Stimson Creek below you. This creek was the eastward margin of the lease in this area. Then the road winds westward again, and you pass the old Pekisko Store and post office, identified by a sign, and descend from a bench to cross Stimson Creek. From this point, it is about sixteen kilometres to the westward margin of the lease beyond the rounded, partially forested foothills which you can see outlined in front of the shining wall of the Rocky Mountains. These leases constituted a staggeringly extensive "estate."

Drainage basins are the best units to use to evaluate the Bar U lands. Grass more than two miles from water is seldom grazed during the summer. The Highwood River flows from west to east across the northern tier of townships within the leases, and it is the major drainage feature of the area. However, throughout

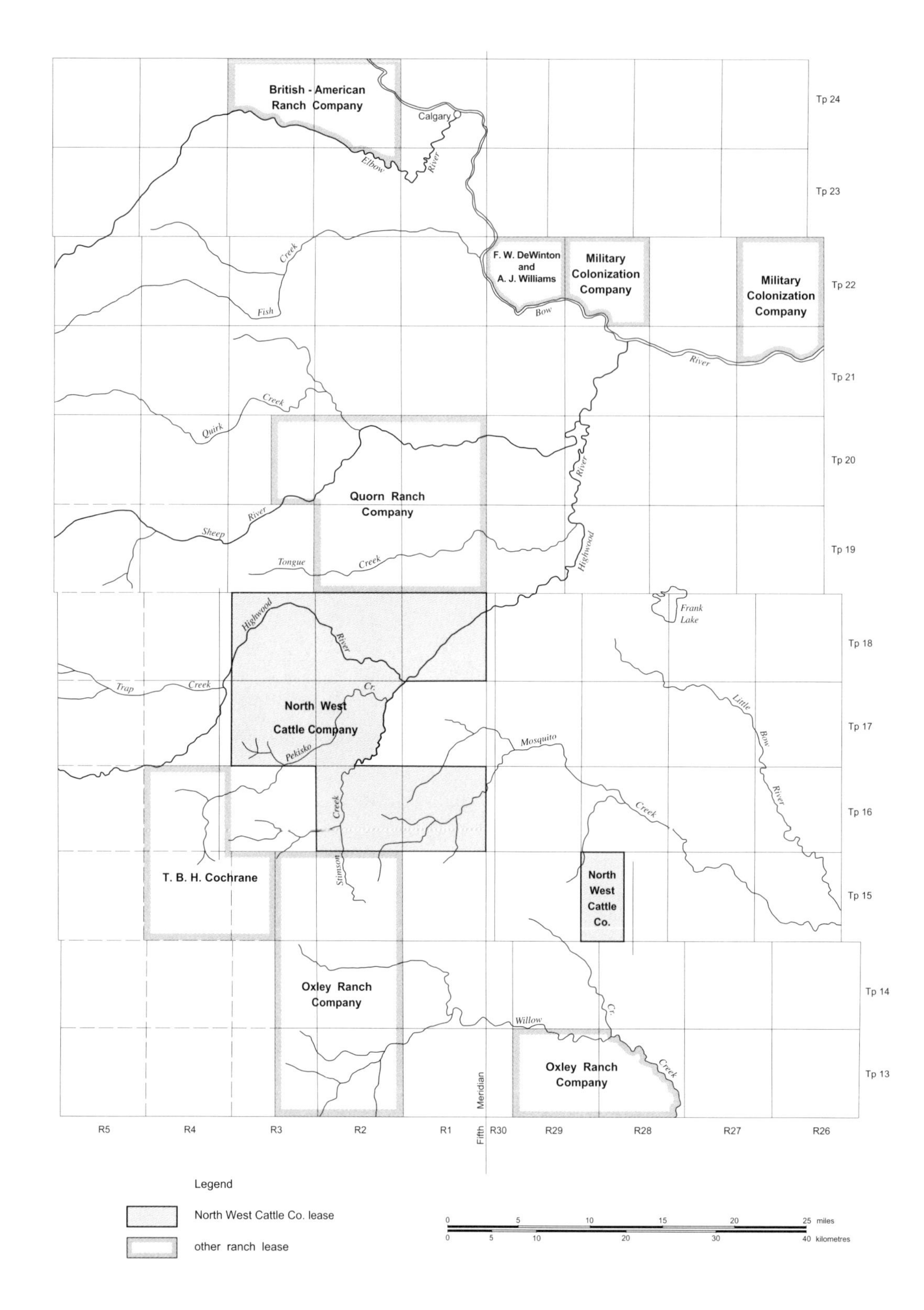

Map 2.4 Remaining closed leases after speculative leases had been cancelled, 1888.

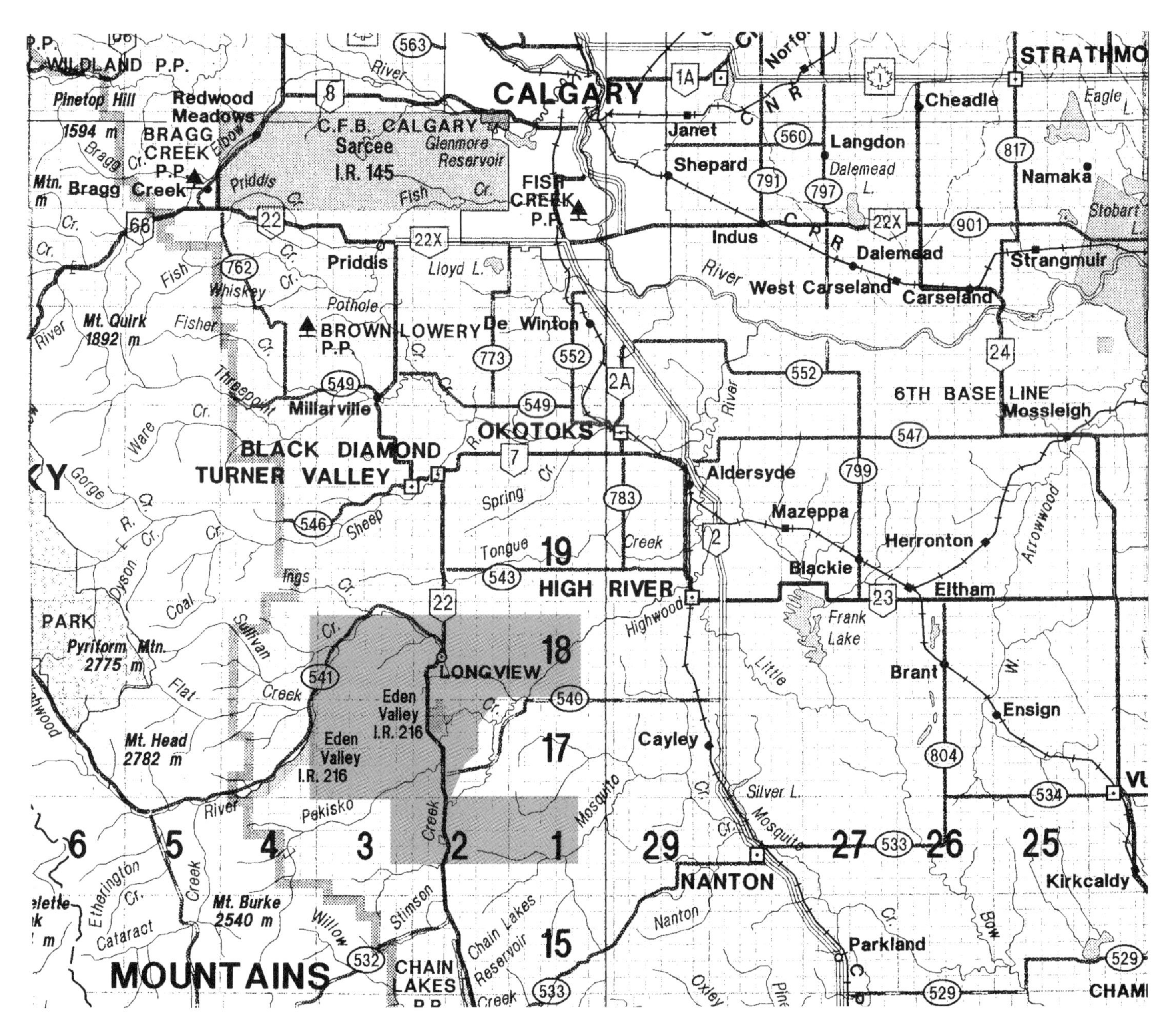

Map 2.5 Bar U leased land shown on a modern road map to show its extent. The leases covered almost seven townships, or 158,000 acres. Base map provided by Alberta Sustainable Resource Development.

2.12 The Highwood River looking south and east to the rolling hills at the northern end of the Bar U lease. In spite of the rugged topography shown there are a number of fords where horses and cattle can cross the river with relative ease. Parks Canada, Alwood Rousseau Collection, Rosettis, 6398-18, 94.12.01.25

this part of its course the Highwood is incised quite deeply below the level of the land through which it runs. Thus, the river was a marked obstacle to the north-south movement of wheeled traffic. The Bar U Trail stayed on the higher ground to the south of the Highwood River all the way from High River. There was no road along the line now adopted by Highway 22 until the 1960s.[149] Of course, there were a number of places where riders or stock could ford the river. An early visitor to the Bar U had this to say:

> We had to cross the North Fork [Highwood River] and then ride across some of the foothills to the Middle Fork [Pekisko Creek] where his [Emerson's] ranch is. It was very steep going down the river bank and up the other side but the horses seemed to think nothing of it and Charlie went at break neck pace it seemed to me; they call it a "cut bank" which means almost perpendicular.[150]

Once cattle had been pushed across the river in late spring, they tended to stay there. As settlers moved in during the 1890s this became a bone of contention as a petition forwarded to M.P. Frank Oliver makes clear:

> In order to enlighten you as to what has been the practice of the cattle companies and large individual cattle owners living south of the river: that it has been their custom annually for the last ten years or more (back to 1886) to gather their cattle, some 12,000 (this is a small estimate) and drive them from the south to the north side of High River and to keep them there all summer by placing riders along High River to keep them from going south.[151]

The settlers complained that the big herds ate out the grass and lured away the domestic stock with which they mingled.

The broad, terraced valley of Pekisko Creek was probably the reason why Stimson chose this site to locate the head-quarters of the company. It provided water, shelter, and wood. Naturally, the flat land close to the ranch buildings was used for labour-intensive activities like calving, grazing the valuable imported bulls, and for the saddle-horse herd. Dense stands of willow provided effective protection from winter winds. To the south of the ranch headquarters, the flat terraces along Stimson Creek were used for haying during the summer. Stacks were put up wherever the wild grass grew most thickly, and calves and weak cows were moved to the hay during the winter. The headquarters of the Mount Head Ranch became a calf camp and was occupied for many years by Henry Meinsinger and his family. Finally, the springs and numerous small streams leading into Bull Creek played an important role in enhancing the utility of the undulating hill country to the north and west of the ranch headquarters. This was excellent winter grazing for mature range steers, for the chinook winds tended to clear the westward facing slopes and expose the cured fescue grass. Willow and alder thickets in the bottoms provided shelter and many of the springs remained open all winter. It was a huge and varied land base which generations of foremen learned to use to its best advantage for different purposes as the seasons came and went.

The lease system was of fundamental importance to the North West Cattle Company during its two decades of existence. The lease boundaries, so carefully drawn on maps, had no finite expression on the ground. Nevertheless, the lease regulations played a crucial role in luring risk capital to the Canadian west during the 1880s. *Bona fide* cattle companies not only had a legally defensible land base on which to fall back as settlement began to press in, they also bargained aggressively with the government for the right to purchase a part of their leases to protect

2.13 The extensive flat land in the foreground could be used for wild hay, while the rolling foothills and sheltered wooded coulees provided ideal winter range. Glenbow, NA 67-6

their "improvements."[152] Eventually, this enabled them to acquire large acreages of deeded land at bargain basement prices when the closed leases were terminated in 1896. The major cattle companies made a smooth transition from a purely leased land base to a mixture of leases and freehold tenure. Thereafter, the capital assets of these companies were based on their land holdings as well as on their stock, and, whatever happened to their ranching operations, they had the option of selling their land after it had appreciated in value, and thus compensating their shareholders generously. It is to trace how the North West Cattle Company fared in this transition from leased to deeded land that we must now turn.

Buying the "Home Place"

For almost a decade after its foundation, the Bar U Ranch flourished without having freehold tenure to a single acre. Barns, bunkhouses, the ranch house, and extensive corrals had been built, but all on leased Crown land. It was not until 18 March 1891 that Fred Stimson turned his attention to the acquisition of a deeded land base.[153] He applied for homestead entry for the land on which the original ranch buildings had been constructed. As soon as patent was granted on this parcel of land, he applied for a homestead preemption to cover the remainder of the headquarters site to the east. Here he ran into a problem, since the land in question was in section 8 and was therefore owned by the Hudson's Bay Company.[154] The North West Cattle Company claimed that they had occupied the site long before the land had been surveyed and they asked the Hudson's Bay Company to surrender the land in exchange for the same amount elsewhere. This was successfully arranged.[155] By the end of 1899, the company owned 320 acres and had freehold tenure to their 'home place. [Plate 1, colour section]

Stimson must have remained confident that his huge closed leases would remain inviolate well into the next century. At least he made no obvious efforts to buy land for the company. However, he must have been all too aware that the lease system was in considerable trouble. It had been established partly for geopolitical reasons to meet an emergency situation. However, Canada's hold on her western domain had been tested in 1885, and she had successfully responded to the challenge. Now the railway spanned the country and the pace of incoming farm settlement was beginning to quicken. The political costs of supporting the large ranches was growing, and the Conservative government was faced with increasingly vocal and intense criticism from 1885 onward. On the one hand, a flourishing industry had been established, which was represented in Ottawa by a powerful lobby; on the other hand, the lease regulations were presented by the opposition, the Territorial Assembly and by local interests, led by C.E.D. Wood, the editor of the *Macleod Gazette*, as an unwarranted obstacle to settlement.[156]

The government's first response to criticism was to attack the most glaring abuses of the lease system. Speculative leases were cancelled and 1.6 million acres were reopened for settlement. In 1886,the regulations were amended to ensure that future leases were "open to settlement or preemption entry", while a new category of small leases was offered to homesteaders who wished to graze some cattle close to their holdings.[157] The Deputy Minister of the Interior could argue that "the department now offers the strongest possible inducement to settlers owning small herds of stock to become themselves leasees of areas of grazing lands proportionate to their means and the number of their stock."[158] These were far-reaching changes, and they reduced the government's vulnerability to criticism somewhat. But other problems loomed. The 'open' leases now available offered no security of tenure, and many ranchers took to running their cattle on the range without a lease. William Pearce, the federal government's man on the spot, reported that this trend was widespread as early as 1886.

> I think that experience in the past points to the probability that within two years, provided that there is no distinction in the charge upon stock between leaseholders and non leaseholders, that at least 90% of the leases will lapse through non-payment of rent. At present non leaseholders run stock in any number, and in many cases do not keep bulls but trust to the services of these animals being furnished by neighbouring leaseholders.[159]

Stock returns collected in 1889 record 145 "stockholders" who had no leases. More than half of these outfits were located with some precision by Pearce, which suggests that they had taken out homesteads, or were in the process of doing so. Although the majority of these stockholders ran modest herds, there were several who reported more than five hundred head. It was this class of settlement that was identified by contemporary sources as the main threat to the large corporate leaseholder, and the cattle compact were not slow to complain to the Department of the Interior about these 'pirate herds.'[160]

The government were responsive to these protests from legitimate leaseholders and added a section to the grazing regulations which stated:

> No person shall be allowed to graze stock of any kind on the public domain, without the consent of the Minister of the Interior being first obtained. Grazing of the same will render them liable to seizure and for feiture by the owner.[161]

It seems unlikely that this measure had any marked effect, for while there is ample evidence that the men concerned maintained their stock interests, there is none of any prosecutions. During the 1890s Pat Burns was grazing thousands of head on unused grass without any leases at all.[162]

2.14 William Pearce, the Federal Government's agent on the spot during the 1880s. Nicknamed the Czar of the Prairies, he was convinced that ranching was the optimum land use for much of the foothills and southern Alberta. Glenbow NA 339-1

The changes in the lease regulations which affected those seeking new leases had done nothing to discomfit the privileged few who still held the original closed leases. The dominance of the "big four" ranches, the Cochrane, Oxley, Walrond, and the North West Cattle Company, was hardly impinged upon; they still owned almost a third of the stock on the range. This fact was not lost on local opponents of the large companies, or the opposition party in the House of Commons. During the spring and summer of 1891, tension mounted in the range country as Duncan McEachran adopted an inflexible attitude toward squatters on his company leases. In particular the eviction of the Dunbar family incensed public opinion and thoroughly embarrassed the government.[163] Something had to be done to resolve the confrontation.

In December 1891, the Minister of the Interior, Edgar Dewdney, invited representatives of the western cattle industry to meet with him. Although he assured the delegates that nothing would be done to disrupt such a large and successful industry, he made it clear that the government were not prepared to tolerate any further turmoil on the ranges and that the closed leases would have to go. He pointed out that the Calgary and Edmonton Railway was being extended south of Calgary toward Fort Macleod. It ran close to several of the leases and was bound to attract farm settlement. Moreover, the government was obligated to subsidize the railway company with extensive land grants, which would include parts of the closed leases. To compensate the leaseholders for the termination of their closed leases only ten years into a twenty-one year agreement, the leaseholders were to be permitted to purchase 10 per cent of their leases at $2.00 an acre – a sum which was later reduced to $1.25 per acre. Moreover the big ranchers were to be protected from any sudden dislocation in their business because the final termination of the leases was put off until 31 December 1896, four years away. Historian David Breen concluded that there was another important element in the compromise, "the government's unwritten promise of a gradual but vast extension of the region's stock-watering reserves."[164] Critical sections of land throughout the grazing country were to be withdrawn from homestead settlement so that stock from the open range would be able to gain access to water. Breen demonstrated that the number and extent of these reservations did increase rapidly through the 1890s, and suggested that they were "in large part responsible for the survival of the cattlemen's empire throughout a decade and a half of Liberal rule after 1896."[165] It was true that the uncertainty concerning which sections of bottomland and which springs had been reserved meant that many farm families simply bypassed the foothills grazing country and moved further north. John Craig claimed that stock-watering reserves practically excluded settlement in some districts.[166]

In the short term, too, the government's agents in the west did everything in their power to ensure that the transition from closed leases to a deeded land base was accomplished with a minimum of disruption. William Pearce spent much of the summer of 1894 visiting ranches and inspecting the lands which had been chosen for purchase,[167] and the Deputy Minister of the Interior reported that more than half of those who held closed leases had applied to purchase a portion of their leases.[168] By 1895, the minister was able to claim that the process was well nigh completed.

> The majority of the persons and companies holding the old leases have accepted the offer made to them, namely, that they purchase 10% of their leaseholds. This will afford them sufficient holdings in fee simple on which to continue their business.[169]

The government and the ranchers were not the only parties concerned in this gradual transfer from leased to deeded land. One of the factors prompting the negotiated settlement of 1892 had been the government's desire to settle the land claims of the Calgary and Edmonton Railway Company. William Pearce explained how these varied interests were unscrambled:

> There being nothing to prevent government from giving the C and E the even as well as the odd sections, an arrangement was arrived at agreeable to all three interests, the leaseholder, the railway, and the government, to grant them the even as well as the odd sections in the area which they had undertaken to sell to leaseholders, and the railway company sold at prices varying from $1.00 to $1.25, possibly in some cases as high as $1.50 an acre. By this means one of the largest ranching companies in the country viz., the Cochrane, acquired nearly its whole leasehold, and it was to the company a most profitable venture. The Bar U and some other leaseholders also acquired lands to as great an area as they desired, and all could have under similar conditions. Most of them within five years greatly regretted that they had not availed themselves of the privilege to the fullest extent.[170]

The change over to deeded land had much to offer cattlemen. They could start to manage their ranges in a rational and progressive manner. Fences could be constructed to protect breeding herds from scrub bulls; the onset of calving could be more carefully ordered; hay and fodder crops could be put up in increasing quantities; and improvements in water supplies could be undertaken. But, most important of all, the land was obtained at a bargain price. At the very time when interest in western lands was beginning to quicken and land and development companies were mushrooming, the cattle companies were offered an opportunity to purchase large quantities of prime land at prices which were only about half the 'going rate.'[171] Thus they were able to acquire land assets which offered them an even more reliable prospects of capital gains than had cattle in 1881.[172]

The North West Cattle Company did not move aggressively to take advantage of the opportunities implicit in the new agreement. During the fall of 1893, Fred Stimson initiated the process by writing to the Deputy Minister of the Interior indicating that the company he represented had decided to purchase all the land to which it was entitled; 10 per cent of their 158,000-acre lease.[173] However, this action was repudiated by the head office of the company in Montreal. The secretary-treasurer wrote to the Department of the Interior to say that no decision

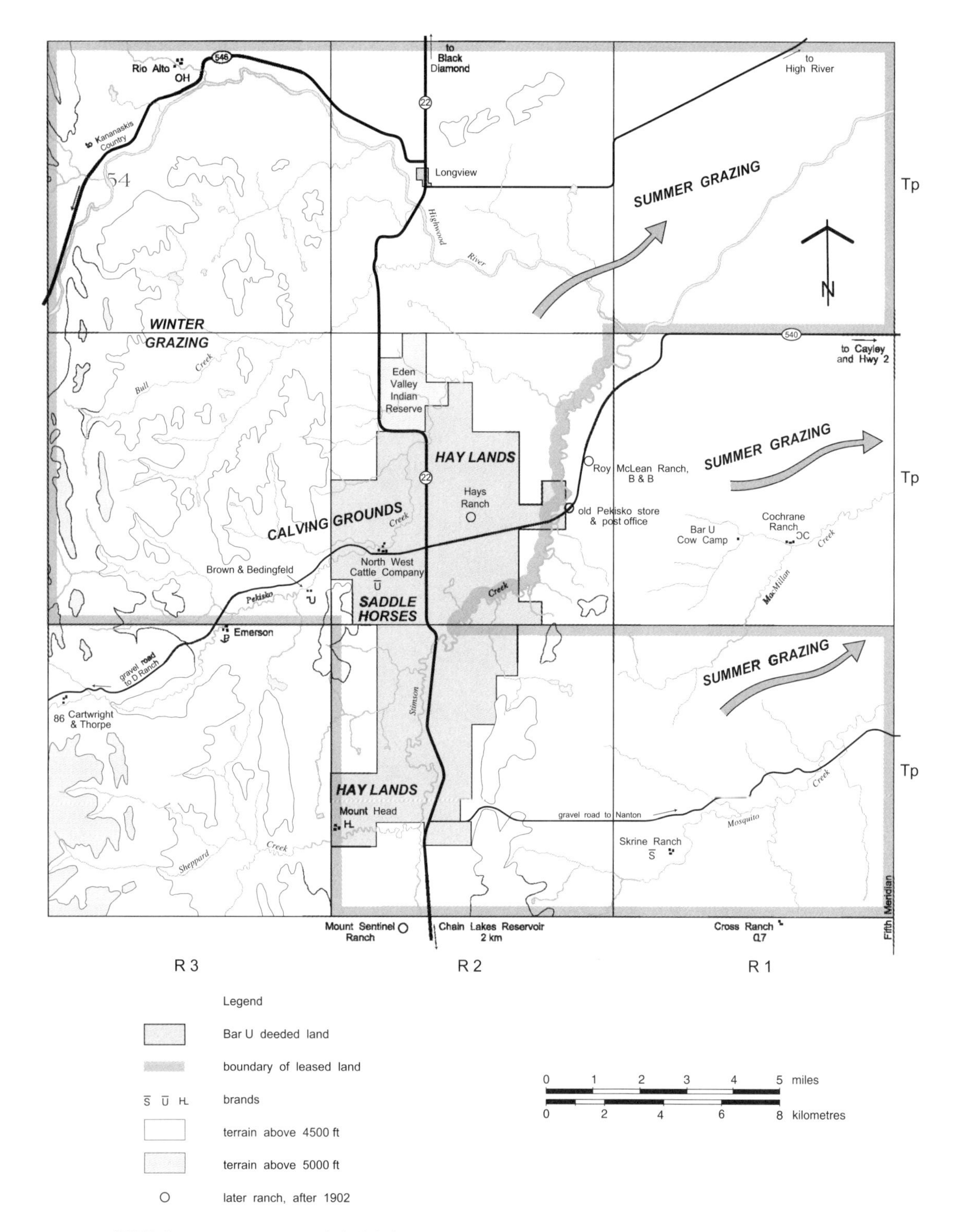

concerning the purchase of land had been finalized and that nobody had been authorized to make an application to purchase land.[174] This was a bitter slap in the face for Stimson and presaged the ructions of 1902. The inference must be that relationships were already strained between the local manager and the investors in Montreal. It seems likely that the future of the company was being reviewed, and the option of winding it up was being considered. The business interests of the Allan family had matured during the decade since the ranch company had been founded. Banking, insurance, and the provision of urban utilities now constituted the heart of their financial empire. Perhaps their relatively small investment in a far-off ranch seemed an anachronistic leftover from another era. William Pearce, who was monitoring the situation for the government in the range country, pointed out that if the company did liquidate, there would be an excellent opportunity going begging. He wrote to a local land company, "if the company decides not to purchase, Mr. Stimson or anyone who understands the situation would have no difficulty in organizing a company to acquire the priveleges [*sic*] of the North West Cattle Company and take over the business, and it

Map 2.6 Leased and deeded land in relation to topography, drainage and land use, of North West Cattle Company. See also colour section.

would be a calamity to allow the business to be wound up."[175]

Having weighed the possibilities carefully, the board of directors of the North West Cattle Company reached the same conclusion as Pearce and decided that further investment was justified. In May 1895, the company secretary appealed to the Department of the Interior on their behalf. The land under consideration was fit for nothing but grazing, he argued, and therefore the price of $1.25 an acre was too high. He suggested that $1.00 an acre would be a fair price. The government, which had already concluded several land deals with other companies, rejected this claim, and pointed out that the deadline for lessees to purchase land (31 December 1896) was fast approaching.[176]

Finally, in September 1895, the company applied to purchase almost sixteen thousand acres of land.[177] This area included some school lands and some land owned by the Hudson's Bay Company, but some fourteen thousand acres were available for immediate purchase. The North West Cattle Company acquired a roughly rectangular block of deeded land covering twenty-three sections. This land was to remain the core of the Bar U land holdings throughout its history [Map 2.6]. It was prime ranching country, almost all flat or gently rolling prairie, and it included the gently terraced flood plains of Pekisko and Stimson creeks. The surveyor of Township 16, B.J. Sanders, had written "excellent land" across the centre of these newly acquired lands.[178]

There were some important loose ends to tie up. There was a disparity between the number of acres that the Bar U had actually bought and the number the company was entitled to under the terms of the 1892 agreement.[179] Stimson was able to buy another section of school land, and some energetic follow-up work by lawyer James Loughead resulted in the acquisition of some more lands to round out the Bar U's deeded holdings.[180] The company entered the new century with a solid deeded land base which had been acquired very cheaply. However, it was cattle, not land, that constituted the real wealth of the North West Cattle Company during its early years.

Dollars and Cents

Markets were of crucial importance to the establishment and growth of the cattle industry in western Canada. Not only did the returns from sales pay cowboys, sustain ranching families, and satisfy eastern shareholders, they also gave credence to the claims of the "boosters." The investment fever which gripped investors in eastern Canada and Great Britain was based upon the difference between the price of a steer on the western grasslands and the price for which the same beast could be sold in Liverpool or London. It was the elusive mirage of this potential profit which spurred the "beef bonanza."[181] The perceived market, the market of the mind, was the hook which lured risk capital to the undeveloped western grasslands.

The newly formed corporate ranches of the North-West Territories could rely, during their tentative first few years of existence, on local markets for their beef. This undoubtedly played a major role in their survival and eventual success. As the plight of Native peoples deepened and near starvation became endemic on the

Mounted Police Contract
Calgary
64,000 lbs Beef @ 8 3/4 per lb.

Beef to be delivered on foot, animal by animal as required to be slaughtered by the police. The head feet & hide to be returned to Messrs I. G. Baker & Co. The Dept. of the Interior paying for the four quarters of meat only.
The necks of the cattle slaughtered for beef shall be cut off at the fourth vertebral joint, and the heart trimmed down. The shanks of fore-quarters shall be cut off from three to four inches above the knee joint and of hind

2.15 Frank White, newly appointed bookkeeper of the Cochrane Ranch, made notes of contracts, beef prices, and stock returns in a small spiral notebook. These pages deal with Indian and Mounted Police beef contracts. Glenbow Archives, M1303, "Cochrane Ranch Notebook, 1881-1882"

recently established reserves, so Ottawa responded reluctantly to meet their treaty obligations.[182] During the late 1870s the Montana based trading firm, I.G. Baker and Company, earned some $250,000 annually from their beef contracts with the Canadian government.[183] During the years between 1882 and 1886, almost $300,000 worth of beef was provided annually for distribution in the Treaty 7 area.[184] Thereafter, purchases declined, but they still averaged about $140,000 annually until 1900.

Fledgling Canadian and British corporate ranches were able to share in this lucrative market as soon as they had mature steers to offer. In 1882, the Minister of Agriculture reported that progress had been made in stocking the ranches of the North West and that "in room therefore of importing beef cattle from the United States for feeding the Mounted Police and the Indians of the North-West, after another season these supplies may be obtained at home and at a much cheaper rate."[185] The Cochrane Ranch subcontracted with I.G. Baker to supply beef to the police and the Indians during their first season, even while they were struggling to stock their Bow River leases.[186] In 1884, the North West Cattle Company sold eight hundred head of fat steers for $65 each, while the Oxley and Stewart ranches were also reported to be taking an active part in fulfilling beef contracts.[187] The ongoing significance of government beef contracts for one large corporate ranch is revealed by the account book of the Walrond Ranch.[188] The books show that total sales of cattle amounted to $676,000 between 1883 and 1897, and almost half of these sales were made directly to the Indian Department, while a further 15 per cent were made to individuals and firms known to be government contractors.

The crucial importance of contracts to supply Native peoples with beef, and the competitive and politicized atmosphere of the times, is illustrated in a letter Fred Stimson wrote to Thomas Power on 1 January 1882. This letter predates the final establishment of the North West Cattle Company. It demonstrates that the lucrative returns from "Indian

Contracts" played a role in encouraging investment in ranching. Because of its significance, it is worth quoting in full:

> Dear Power, I have just returned from Ottawa and find your chance to get the contracts are A1 if only you come and attend to the thing they are all down on the I.G.B. [I.G. Baker and Company] My friends are going straight for you but you must not go in with the Conrads and divide it up – if you will go into the thing for yourselves in a bonafide way there will be no difficulty in your getting them [the contracts] in spite of all they [the Conrads] can do. Write me at once what you think of doing – I would like to see you before I return to Montana. If you get the contracts which you can if you only stir yourselves you must give me the Blackfoot and Sarcee to supply with beef. The Stoneys I will also take if we can make arrangements. If you intend doing anything there will be no time to lose as the Government are ripe for a change. With the compliments of the season I remain very truly yours, Fred Stimson[189]

2.16 Blackfoot Indians un-loading beef at the ration house, c. 1883–84. An engraving from an unpublished book about the work of the Anglican Missionary Reverend John W. Tims. Glenbow, NA1033-4

In spite of the plots of Stimson and his friends, the Conrad brothers of the I.G. Baker company adapted efficiently to the new political realities in the Canadian west. In 1882 they established the Benton and St. Louis Cattle Company, to formally separate their ranching business from general trading. The new company had a mandate to pursue the cattle business on both sides of the International boundary. Between 1881 and 1883 they imported more than six thousand head of cattle to Canada, and paid duty on them because they did not hold a grazing lease at that time. The company retained the lion's share of Canadian government beef contracts

2.17 Blackfoot Indians receiving rations, c. 1883–84. Glenbow, NA 1033-3

most of the area over which their herds had been accustomed to graze since about 1878. The "Circle outfit," as it was called, remained an important presence on the Canadian range until 1911.

The Canadian government's purchases of beef underwrote the development of large corporate ranches in the North West in a very real way. A crude but telling comparison shows that the $2.5 million that was invested in these enterprises was balanced by sales to the Indian Department of $2.4 million between 1881 and 1887.[190] This was extremely important. The ranches represented an outpouring of risk capital into a new area for investment, beyond – for the moment – the reach of well developed transport, communications, legal, and banking systems. It was important that investors be reassured by some immediate returns. Ultimately, the hopes and expectations of shareholders were based on the prospects of trade in live cattle with Britain. But it was clear that it would be some time before a system

through the 1880s, although they subcontracted some business to emerging Canadian and British ranches, as we have seen. It was not until the 1890s that direct returns to the firm from government contracts fell below $100,000 annually, and by that time a variety of alternative markets were beginning to emerge. In 1886, they took out substantial leases between the Bow and the Belly (Oldman) rivers which covered

for moving cattle from the western ranches across a continent and an ocean would be perfected. In the interim, the local non-native population was small, and construction crews on railway projects provided but temporary markets. Some cattle and beef were shipped to Winnipeg in 1884, and some were moved westward to the coast the following year, but at this time it was really only the government beef contracts which could absorb the growing output of western ranch companies.

There were some obvious advantages for the ranchers in this system. First, as fat cattle could be driven to the points where they were to be slaughtered, there were no transport costs to be absorbed. Secondly, the contracts stipulated that bunches of cattle should be delivered throughout the year. This brought in regular returns. Thirdly, the demand remained more or less consistent over a twenty-year period. What had started as a response to an emergency in 1877 and 1878 became a permanent feature of government policy with respect to the aboriginal peoples of the plains. Typically, the outfit which held the meat supply contract for a reserve would establish a corral and a crude slaughterhouse at some convenient point. The routine on slaughtering days, usually twice a week, was well established. The cattle for butchering would be driven up and corralled, and at the agreed hour they would be shot. The beasts were then skinned and dressed by the Indian women and divided up between all the families who were eligible to receive rations.

The prices obtained by ranchers were high during this period. The sudden demand for cattle in Montana to stock the ranches of the North West, bid up the price of young stock from \$20–\$25 in 1881–82 to \$40–\$45 in 1884–85. After three or four years these steers were sold for \$50–\$65. The editor of the Macleod Gazette explained:

> The cause for this high price … is the very simple fact that there has been and is no reason for lowering it. High prices were paid to Montana ranchmen in the first instance, and it was not very likely that our ranchmen would of their own free will cut them down.… Thus far there has been ample demand for all marketable steers at the above prices and will be until the supply begins to get largely in excess of local demand.[191]

C.E.D. Wood went on to point out that there was already a small surplus over and above the requirements of the government contracts. As this surplus increased he argued that the prices would have to drop to more realistic levels. He suggested that the prices in Winnipeg would in all probability be well below those offered at Chicago (perhaps \$50–\$55), and the Alberta ranchers would have to absorb the transport costs to new markets. Thus the "protected" or closed system of supplying a local and somewhat inflated market would soon give way to a much more volatile situation in which Albertan cattlemen would become small players in a highly competitive and politicized world meat market.[192]

Beef for Victorian Britain

What an audacious undertaking it was to move live range cattle from the foothills of Alberta, almost four thousand kilometres to a seaport, and then another four thousand kilometres across the stormy North Atlantic! And both ranchers and shippers expected to make a profit as well![193] The first shipment took place in 1887, only two years after the Canadian Pacific railway was completed. At that time it was a remarkable feat to have ordinary trains running more or less on schedule from coast to coast. Cattle trains were far from "ordinary," for the freight they carried was highly perishable, and unforeseen delays could prove disastrous. Stock in transit had to be unloaded and fed and watered every two or three days. This required a high degree of organization, and the establishment of considerable infrastructure in the way of loading chutes, corrals and watering troughs. In retrospect, it is not surprising that complaints by ranchers about the CPR, the middlemen and the shipping companies became endemic. What is truly amazing is that this trade flourished for a quarter of a century, proved to be a reliable outlet for top grade steers, and made fortunes for those who did so much to make the trade run smoothly.

It was in the interests of the Canadian government to portray the cattle trade in the most favourable light possible.[194] The growth of the cattle trade was one bright spot during a period of agricultural depression.[195] The successful establishment of corporate ranches would open the way for farm settlement. Thus the sanguine accounts of the Ministers of Agriculture and the Interior have to be sifted carefully before some appreciation emerges for the difficulties, uncertainties, and setbacks which had to be overcome before the shipment of range cattle to Great Britain became a routine matter.

The transatlantic trade in grass-fed cattle from the ranches of the North West was grafted onto the highly successful and rapidly growing trade in cattle from the farms of eastern Canada. A unique opportunity existed in the meat markets of Great Britain, for, as population and buying power increased, so the beef herds of the United Kingdom and Europe were subjected to a series of devastating epidemics.[196] Demand was rising rapidly and traditional sources of supply were seriously diminished. For several decades, part of this deficit was met by an influx of live cattle from North America.[197]

The Canadian Minister of Agriculture summarized the remarkable growth of the cattle trade in his annual report for 1883:

> The cattle trade of Canada has now ceased to be an experiment. It has grown to be one of the greatest lines of trade of the country, and hundreds of thousands of dollars have been invested in it by shrewd and practical men.... Within the two last years the value of our exports of live stock has been not less than $3,500,000 annually, while the total value of the cattle shipped from Canada six years ago was little more than $36,000.[198]

It was this spectacular launch of a new branch of trade with the mother country that assured the continuing attention of the Dominion government in every facet of the cattle trade. However, the success of the trade from eastern farms posed some initial difficulties for ranchers shipping from western Canada. Until close relationships had been forged with agents, insurance

2.18 Loading big grass-fattened steers into box cars for the traumatic ten-day journey to Port Levis, Quebec on the St. Lawrence River. This was followed by another ten days on shipboard crossing the Atlantic. Glenbow, NA 1368-13

companies, and the shipping companies, stock from the west might languish in poorly equipped holding pens at the ports, or be shipped in tramp steamers which had been quickly converted to handle cattle, and offered only the most rudimentary facilities. The rapid expansion of the trade does much to explain the contradictory reports concerning conditions on the ocean crossing and the state of cattle when they arrived at Liverpool or Glasgow. Small lots of valuable pedigree stock, used to being stall-fed and handled frequently, were assured of excellent conditions on regular shipping lines. For example, when Senator Cochrane sent some of his prize-winning shorthorns back to Britain in 1877, the Canadian agent at Liverpool arranged a reception committee of several prominent agricultural corres-pondents to watch the unloading. He goes on to describe their reactions:

> They, one and all, expressed their surprise at the extraordinary good condition the cattle were in, and their appreciation of the accommodation on board ship for their transport. Some of these gentlemen frankly admitted that these cattle had suffered no more from the long ocean voyage than the Irish cattle crossing the channel.[199]

On the other hand, as the volume of the trade increased so conditions on some vessels deteriorated. Cattle from the range, unused to any contact with humans and already traumatized by a rail journey which had taken ten days, could be subject to grievous overcrowding, abuse, and neglect.[200]

The first shipment of seven hundred finished steers from western Canada to Great Britain was made in the fall of 1887 by the North West Cattle Company.[201] The cattle were held in the new stockyards which had just been completed at Calgary. An observer described the shipment:

> The cattle which have just come off the range are in magnificent condition and range from 3 to 5 years of age, and will weigh at the present time about 1,500 pounds each. There was some little difficulty getting the first car loaded, but under the superintendence of George Lane and John Ware, the boys soon got into the hang of it and rushed them in. The rest of the shipments – about 700 – will follow as soon as the CPR can provide cars to take them.[202]

Duncan McEachran, responsible for veterinary inspection of cattle leaving from ports along the St. Lawrence and a major investor in the Walrond Ranch, heralded this experiment with enthusiasm. In his view it was an unqualified success, and he suggested that "the quality of the stock, their size, and the delicious flavour of the beef, will command a market at paying prices, and experience will teach how to transport them from the prairie to the butcher's stalls with very little trouble or risk." Indeed, later in the report he added, "... no trouble was experienced in conveying them in cars, or shipping them on the ocean steamers. They learned very soon to eat and drink out of troughs, and they landed, except on steamers which encountered very severe weather, in good condition."[203] From the grazing lands along the foothills, William Pearce was equally euphoric. He reported that "the class of beef furnished ... has been an agreeable surprise to those who had not been acquainted with the excellent quality thereof," and he added that cattlemen had received a return of between $40 and $50 a head. He noted too that "the Canadian Pacific Railway Company have, it is stated by cattlemen, done everything possible to promote this traffic."[204] The Minister of Agriculture concluded that the question of a market for western beef had been settled.[205]

In the long run these enthusiastic and optimistic assessments were to be proven true. The trade in live cattle grew until it was valued at $10 million by the end of the century and ranked third among items exported to the United Kingdom.[206] In the short term, however, there remained a number of seemingly overwhelming problems. The uniformly high quality of the first shipment from the west was not maintained in subsequent years. The bland confidence of the Canadian authorities was strangely at odds with the eyewitness report of Mr. Graham, the Canadian agent at Glasgow. He commented:

> There were a couple of consignments of animals from the ranch districts of the North-West. These animals were, as a rule, of good size and many of them of fair quality, but they were exceedingly wild and difficult to handle, shooting having to be resorted to in most instances in killing them. From the long distance which they had traveled on land before the sea voyage they were very much pulled down in condition and the internal fat did not meet expectation in most instances.[207]

He went on to express doubts as to the practicality of shipping cattle which were not used to being handled, and stated his conviction that none of the shippers had made any money on the deal.

Part of the problem was that cattlemen were trying to handle the arrangements for the complex journey themselves. They tended to put off any thought of marketing until after the fall roundup, and their cattle moved eastward through the railway system late in the season. So it was difficult to find space on transatlantic vessels. Premium freight rates had to be paid, and then the ships faced equinoctial gales. Battered and half-starved cattle came ashore in Britain only to find the market glutted and prices low.[208] Western producers desperately needed intervention from experts who knew every facet of the trade and wielded enough power, through the volume of their business, to ensure the best possible conditions and rates.

The Winnipeg firm of Gordon, Ironside and Fares grew to fulfill this role.[209] Robert Ironside began shipping cattle from Manitoba to eastern Canada as soon as the CPR was completed. The experiment of shipping them on to Glasgow was a natural follow-up. He realized that well finished cattle shipped to Britain early in the season yielded the best returns. By 1891, the firm was shipping 1,500 head drawn from as far afield as Prince Albert, Battleford, Edmonton, and the ranching country of the foothills. The business grew rapidly. In 1893, about 5,000 head of cattle were purchased in the foothills for $200,000. In 1894, they shipped 19,335 head to Britain; in 1895, 33,907; and in 1896, 27,057 head. This total accounted for more than 25 per cent of all the cattle shipped from ports along the St. Lawrence, and the firm continued to dominate the live cattle export business until the First World War.

The North West Cattle Company on Pekisko Creek and other outfits like it, both great and small, might expect a visit from a buyer representing Gordon, Ironside and Fares at the spring roundup.[210] Cattle would be bought for delivery in the fall at various railheads. William Henry Fares joined the company in 1897 and assumed responsibility for purchasing cattle. He also managed a number of ranches for the firm where cattle could be held and fattened until the most opportune time for shipment. Gordon handled the railway transport and the packing plant at Winnipeg, while Ironside managed the export trade and spent much of his time in Montreal. Together the three men made a formidable team with considerable political and economic power.[211] They contracted not for single box cars, but for space on whole trains and ships, realizing significant economies of scale. Under their astute and experienced management, holdups and shipping problems became the exception rather than the rule.

Nevertheless, good organization could only mitigate the effects of the long journey on cattle shipped directly from the range. A typical cattle train would take five days of actual travel time to cross the continent, and each of these "runs" would be from twenty to thirty hours in duration. In addition there would be four layovers of about twenty-four hours each for feeding and watering. The steers selected for export were mature animals at least four years old. During their lives they had had very little contact with humans, and almost no experience with unmounted men. They were extremely wild, powerful, quick, and dangerous. They were packed seventeen to nineteen in each boxcar so that they had little room to move and damage each other. Everything was unfamiliar to them, the noise, close confinement, the water troughs, and the hay they were fed. A layover in a safe, well-maintained yard provided with good hay and adequate water might revitalize and sustain the cattle for the next stage of their ordeal. Too often, however, the yards were inadequate in size, the hay rank, and the water troughs inadequate. Moreover, local people came in large numbers to the yards to watch the cattle being loaded or unloaded, and they did everything they could to add to the excitement by shouting and baiting the nervous stock.

Conditions on the cattle boats on the passage across the Atlantic varied enormously. On large, fast ships owned by the major lines, and operating on regular schedules, facilities for cattle were good and the care provided was professional. Tramp steamers, speedily converted to carry cattle on wooden decks above the general cargo, offered much less. Cattle were crowded, and feeding and watering became a nightmare in rough conditions. Actual losses at sea were usually low, but range cattle continued to depreciate during the week or ten days which they spent on shipboard.[212] This is the key to understanding the varied and contradictory descriptions provided by eyewitnesses. For example J.T. Gordon, who had shipped so many thousand head of cattle, observed:

> ... I visited the lairages in Liverpool ten years ago [1890] and also last year [1900], and I was amazed at the great improvement there was in the transportation facilities during that time; I saw a whole ship-load of Ontario domesticated cattle landed there, and they looked as if they had come from their own stables.[213]

In contrast, an inspector on the docks at Glasgow commented:

> The impression one gets of the Canadian ranch bullock is that it has shrunk heavily on the trip; in conversation with the European representative, Mr. Ironside, jr., of Gordon, Ironside and Fares, he remarked on the great amount of hay it took to get ranch cattle filled up after landing.... From the condition of the ranchers on arrival a person is warranted in assuming that the methods of handling and shipping can yet be greatly improved upon. However good the condition the ranch bullock is in on leaving the west, it never seems to reach the old country markets in any better than "store condition."[214]

Cattle from the western states of the United States seldom moved directly to eastern ports. Instead, the steers were sold at the Chicago market to midwestern farmers, and spent at least ninety days being fattened with grain. During this period they lost some of their wildness, and many were dehorned. Finally, after this finishing, only the best beasts were selected for export. Canadian cattle were wild: "... when they are unloaded from the cars, they rush out in spite of all efforts to restrain them and are very frequently bruised against the sides of the doors, and in the yards they rush about on the slightest disturbance, hitting against posts and gateways."[215]

The Chief Veterinary Inspector of Canada, J.G. Rutherford, having received opinions about the trade from a number of expert witnesses, concluded that even the best-bred steers in prime condition shipped direct off grass to Europe were going to shrink en route, even if transport facilities were almost perfect.[216] This was a significant conclusion which was largely ignored in the years of acrimonious debate between cattlemen and the CPR, and one which has also been overlooked by students of the cattle trade. The surprising thing is that in spite of shrinkage both ranchers and shippers made money on the trade, while Gordon was boasting of paying $40–$45 per head for fat stock in Alberta, steers were fetching $90–$100 in Britain.

The expansion of the export market for beef during the 1890s was paralleled by a steady growth in demand in western Canada. Pat Burns followed the rails westward, obtaining contract after contract to supply construction crews with fresh beef.[217] In 1890, he was buying quite large lots of cattle to supply camps along the line between Edmonton and Calgary, and, no sooner was that link completed than work started on the line between Calgary and Fort Macleod. Moreover, the population of Calgary was already 3,800, and Burns' makeshift slaughterhouse was absorbing several hundred head of cattle a month to supply his butchers shops. A year or two later, Burns began to ship cattle into the mining districts of the Kootenay, overcoming all manner of transportation difficulties to reach this hitherto inaccessible area. Such was the growth of demand that the *Macleod Gazette* commented, "It will take pretty nearly all the beef cattle southern Alberta has for export."[218]

The North West Cattle Company, during the twenty years of its existence, profited from sales to a variety of markets. During its first years, government contracts to supply beef to the Indians played a vital role. After 1887, the British market absorbed the best of the Bar U steers, and the wild range cattle enjoyed a privileged position in the holds of vessels of the Allan Line. The growing local market, dominated by Pat Burns, provided a complementary outlet for fat cows culled from the breeding herd and for the less well-bred steers.

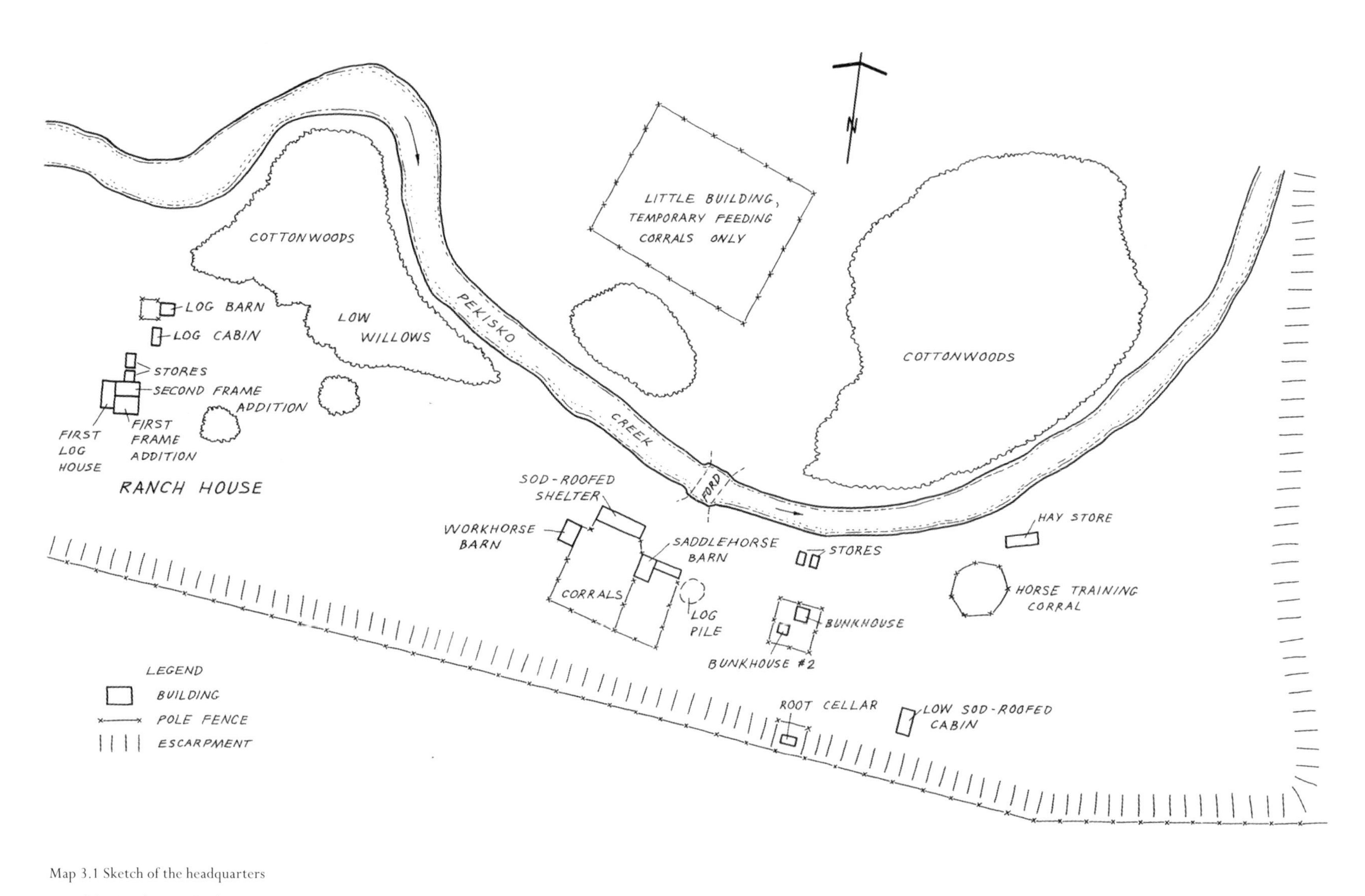

Map 3.1 Sketch of the headquarters site of the North West Cattle Company, c. 1900.

3.1 The "Village Street" of the North West Cattle Company's headquarters described by John Innes. The cookhouse on the left faces four storage cabins, while the saddle horse barn and the work horse barn can be seen on the left in the middle distance. The dark building at the end of the street, framed by the other buildings, is Pekisko House, the home of the Stimsons. Glenbow, NA 466-12

3
Building the Home Place: The Community and the Work They Did

"History is not the past, but a map of the past, drawn from a particular point of view to be useful to the modern traveller."[1] – Henry Glassie

John Innes' description of the Bar U headquarters in 1900 depicts a well-established community; the culmination of seventeen years of steady growth [Map 3.1]. He wrote:

> The river leaves the foot of the high bluffs and takes a sweep northwestward, then turns east once more. This forms a cozy corner of level land, about a quarter of a mile long east to west, and nearly one hundred yards wide, on average north to south.... At the western end, near the ford, stands the manager's house, with its long straggling line of outbuildings. About one hundred yards east from these were the breaking stables and the corrals, the team stables, wagon sheds and the men's quarters, all being built in a straight line along the north side of the trail that passes through the center of the ranche. Opposite this main group were several log shanties used as store houses; giving the whole the appearance of a little village street.... It is a truly well thought out and workmanlike arrangement.[2]

The decision to establish the ranch headquarters on the middle fork of the Highwood River, close to the centre of the company's leases, was made by Fred Stimson during the summer of 1882.[3] He had arrived back in the west early in the spring and had successfully acquired a foundation herd of cattle for the North West Cattle Company. He hurried back to the High River area to establish a base and to

3.2 The cookhouse and the bunkhouse. The Lanes occupied the house on the left. Note the rail fences to keep out stock; on the extreme left of the picture are a number of hay-boxes to fit on wagons; a huge pile of logs provided fuel for cooking and heating. Summer c. 1898. Glenbow, NA 466-13

arrange for hay to be cut to feed his purebred bulls during the winter.[4] The site chosen combined several advantages: it was on Pekisko Creek, a reliable source of water; the coulee provided some shelter from the prevailing westerly winds; and the trees of the flood plain provided fuel and shelter for the stock. Stimson dispatched a few men to build a shack and put up some hay. Phil Weinard arrived at the site during the third week in December 1882. He had been engaged to drive the ox team which was used to haul hay through the snow to the bulls. His description suggests that it was not a Christmas to remember fondly!

> All there was of the Bar U at that time was a cabin fourteen by sixteen feet in size and a hay corral ... the cabin newly built out of green cottonwood logs felt a bit damp, there were some double deck bunks on one side, a fireplace in the center of the north wall, a window on the east wall, a door in the south wall, dirt floor and roof. Jim Meinsinger and Cal Morton were getting out logs for a stable, Emerson and me did the hay hauling, and when we got the corral full, we used to snake out the logs with the oxen. Herb Millar was cook. Our living was pretty rough as he was not a cook and there wasn't much to cook.
>
> We had to go six miles for hay once when it snowed ten inches and it was no fun to walk twelve miles on "coffee" and a chunk of baking soda bread (rare). One night there was just bread and dried apple sauce and Jim Meinsinger came in. He picked up his gun and went out where the cattle were feeding and shot a yearling bull. Our supper consisted of liver and brains.[5]

Herb Millar must have prepared the feast with mixed feelings; the bull was one of the purebreds which he had brought up from Chicago in the spring.

Construction of the saddle-horse barn proceeded whenever the weather permitted, and a good deal had been accomplished by the spring of 1883, when the main herd and the remainder of the crew arrived at Pekisko Creek. The township boundaries were surveyed by R.C. Phillips during that summer. He wrote in his notes: "The North West Cattle

Company have ranch buildings on the north-west quarter of section 8, and a large number of cattle feed on the plains."[6] The township map which he drew shows an L-shaped structure, which was probably the saddle-horse barn with an open feeding shed constructed at right angles to it and the rectangular cabin in which the men were living.

3.3 "Pekisko House," the home of the North West Cattle Company's resident manager, Fred Stimson, and his family. The newly added frame wing of the house masks the original log house behind it, but the square log construction is visible on the extreme left of the photograph. The small verandah along the south side of the house was to undergo many modifications over the years, as was the number and location of the sheds to the north of the house. Glenbow, NA 1459-22

The process of building and improving existing structures went on more or less continuously during the next few years. When men were not totally involved in stock-rearing activities they were kept busy with construction. A two-storey log ranch house was built, chinked with mud and white washed, to accommodate the ranch manager and his family. About 150 metres to the east a log cookhouse, with sleeping quarters on the upper floor, replaced the original sod-roofed cabin as quarters for the men. When George Lane, the ranch foreman, married Elizabeth Sexsmith on Christmas Eve, 1885, his bride recalled, "We went to live in a log house which was built for us. We lived there about four years."[7] The scant evidence of the few existing photographs of the site at this time suggests that the Lanes were the first occupants of the log house built adjacent to the main bunkhouse.

Early in October 1886, the ranch was visited by a routine police patrol. A young visitor from Britain, J.L. Douglas, had been invited along, and he recorded his impressions of their arrival at the Bar U in his diary.

> We breakfasted again on bacon (sourbelly as they politely term it) and started soon afterwards for Stinson's [*sic*] ranch about 25 miles away up the forks of High River and close to the foothills of the Rockies. It was a very pretty drive as the country round about is more varied and hilly, and we got there about 12 o'clock and had dinner with Mr. Stinson. A girl called Bowen lives with Mr. and Mrs. Stinson, and the ranch really looks like a "home" which is more than can be said of any of the others I have seen at present; there are nice comfortable chairs, pictures, curtains, etc. so it looks very well.[8]

At this time the family were living in the original two-storey log building, onto which two wings would be added.

This patrol would have been a routine visit to the police outpost which was located close to the Bar U. The Commissioner described the post: "at Stinson's [*sic*] Ranch, on the Middle Fork of High River, about 55 miles south-west of Calgary, it commands a trail leading behind the Porcupine and the Foothills, by which criminals can escape Fort Macleod and its outposts getting into the United States near Chief Mountain...."[9] During the late 1880s there were five constables and their horses at the post. They had occupied a cabin vacated by William Iken, which was located close to the present bridge where Highway 22 crosses Pekisko Creek.[10] This party of police was big enough to look after itself. There was always one man "at home" to maintain the security of the post. Nevertheless, one can imagine that there would have been frequent visiting with the men at the ranch. During the fall of 1889 the detachment was reduced to two constables who bunked at the ranch when they were not on patrol. Rent was paid to Stimson for "Board and horse feed." This arrangement worked well until the demands of growing towns prompted a redeployment of the police to meet new priorities soon after the turn of the century.[11] This police presence both reflected the importance of the Bar U as a major node of population and added to its significance as a centre of the district.[12]

A photograph taken a year or two after Douglas's visit shows a group of log buildings huddled on the flood plain of Pekisko Creek. Along the river the cottonwoods are for the most part quite young, although a few mature trees can be seen toward the western end of the open site. Two store cabins had been constructed across the trail from the bunkhouse and backing onto the creek. A long, sod-roofed animal shelter can be seen to the east of the saddle-horse barn, and a framed addition had been constructed at right angles to the original ranch house. Details of this structure are shown in a particularly fine photograph taken in the early 1890s. At this time the ranch house had a modest south-facing verandah, and a number of storehouses and sheds were roughly aligned with the new wing of the house.

The original buildings constructed at the ranch headquarters have a story to tell in their own right. Terry Jordan and his colleagues have demonstrated how much information about origins, diffusion and connections can be wrung from examination of "the archaic greater artifacts found in the ordinary human-built landscapes."[13] But the correct analysis of old buildings, barns, weathered granaries, and relict fences depends on a wealth of comparative experience. The authors of *The Mountain West* inspected some two thousand log structures from New Mexico to Alaska. Although the Alberta foothills lay outside the primary focus of their study, Professor Jordan has visited the area on several occasions and he explored the Bar U Ranch in 1992. Later he summarized some of his findings:

> Inspecting the log structures at the Bar U I discovered that their distinctive carpentry is unquestionably derived from Ontario rather than the American Midwest as revealed in details such as end only hewing and the dominant use of dovetail notches.[14]

We can therefore argue that not only was the entrepreneurship, capital, and labour force which brought the ranch into being Canadian, but so too were the origins, layout, and building techniques used in some of the original log buildings at the site.[15] However, Professor Jordan would be the first to advocate caution before jumping to hasty conclusions. The eight log buildings that have survived at the ranch were built over a period of more than thirty years. Edward Mills suggested that vernacular building traditions associated with Métis and French-Canadian log craftsmen were important in some of them.[16] Moreover, the layout of the work horse barn, with its huge hayloft, seems very similar to the "Mountain Horse Barn" found throughout the ranching country of the interior plateau of British Columbia. This would point to rather different origins, as Jordan himself comments: "invented most likely in western Montana, the barn provides a visible index of Midwestern American influence in western Canadian cattle ranching."[17] Each of the extant buildings at the ranch has been subjected to rigorous examination by the Architecture and Engineering Services of Parks Canada.[18] However, the focus of these reports was the conservation of precious old buildings, and no definitive attempt has been made to weigh the origins of the log structures and to plumb what they can tell us about connections and diffusions. This would be a worthy project for somebody with the requisite skills.

The Bar U Community in 1891

The nominal rolls of the 1891 Census of Canada provide the first detailed description of the Bar U Ranch community.[19] During April 1891, William E. Holmes, the enumerator for the High River District, set out to visit the scattered farms and ranches of the neighbourhood. He collected information on each inhabitant: name, age, country of birth, countries of father's and mother's birth, religious affiliation, and occupation. On 6 April, Holmes visited the Bar U Ranch. He recorded that seventeen persons were living at the ranch, the largest concentration of population between Fort Macleod and Calgary! The Stimson household, at the main ranch house, consisted of Fred and Mary Stimson, two unmarried cousins, Joseph Stimson and Nellie Bowen, and a domestic, Mary E. Bigland. Twelve male employees lived in the two bunkhouses. Everett Johnson, the foreman, and Samuel Leighton, the ranch bookkeeper shared the second bunkhouse which had been built for the Lane family, but they took their meals in the main bunkhouse. James Barbour, the cook and the oldest member of the community at forty-eight, had his own room adjacent to the kitchen, while the remainder spread their bedrolls on straw-filled sacks in the "dormitory" upstairs.

At this time Fred Stimson was forty-seven years old and at the height of his power and influence. "He was a big, fine looking, powerfully built man of tremendous energy and driving power with a smiling countenance and mischievous look in his eyes indicating that there was another side of his character which it might be dangerous to arouse."[20] Mary Stimson is a rather shadowy figure about whom we know little. She moved to the ranch during the summer of 1885. Oral tradition suggests that she was never truly reconciled with being uprooted from her home and family in Quebec. Apparently she did not like the isolation of the ranch, and she spent much of her time visiting friends in High River and Calgary. It is possible that

Mary's health was not robust. Mary Bigland, although described as a "general servant" in the census, was in fact a registered nurse who had been engaged to help Mrs. Stimson recover from a bad bout of scarlet fever. Nevertheless, Mary Stimson was postmistress at Pekisko from 1 August 1886 to 1 February 1902.[21] The Stimsons had one child, Bryce, who would have been twenty-two years old in 1891. He was in Montreal completing his schooling when the census was taken, but it is clear that he spent some time at the ranch from the warmly affectionate letter which he sent to Herb Millar years later.[22] The other permanent member of the Stimson household was Nellie (baptized Ella) Bowen, the daughter of Fred's aunt Sarah (Stimson) Bowen. She seems to have been a family retainer in the tradition of Victorian novels for she lived with the Stimsons for thirty years in Alberta and Mexico and was with Fred when he died in Montreal in 1912.[23] Nellie's younger brother Percy Bowen also worked at the Bar U for some time.

Everett C. (Ebb) Johnson was Bar U Ranch foreman in 1891, having replaced George Lane in 1889.[24] He was a worthy successor to his illustrious predecessor for he had packed a lot of experience into his thirty-five years. He was born in Virginia and moved to Minnesota with his parents after the Civil War. He matured early, for at fifteen he was driving a stagecoach in the Black Hills of Dakota. At nineteen he was one of three foremen of Moreton Frewen's Powder River Cattle Company in Johnson County, Wyoming.[25] Here, he acted as guide and hunting companion for Owen Wister, and the family believe with some justification that Ebb's character and some of his exploits were blended into "The Virginian," the protagonist of Wister's immensely successful western.[26] When Frewen sought to escape the dry, overgrazed range in Wyoming for untouched grass in Canada, he sent Johnson ahead to scout out a lease in Alberta. Ebb was able to find a block of fine grazing along Mosquito Creek, and the "76" outfit moved there during the fall of 1886.[27] A year or two later, D.H. Andrews, who was running several ranches for the Canadian Land and Ranch Company, recommended Johnson to the North West Cattle Company in the following glowing terms: "A first rate cowman, in fact I think about the best all round cowman in this country, and he is very good with young horses."[28] In November 1891, Johnson married Mary Bigland, and soon after left the big ranch to work for Gordon, Ironside and Fares as a cattle buyer.[29]

The other position of some importance at the ranch headquarters was that of bookkeeper, held at this time by thirty-three-year-old Samuel Leighton. He was an Englishman who had replaced Duncan Cameron, the bank clerk from Montreal,[30] who had been sent west by the Allans to look after the company books in 1882. The other men living in the bunkhouse included two who called themselves cowboys, two horse breakers, a cook, four teamsters, and a young "chore boy." One wonders why there were four "teamsters" who would presumably have spent most of their time with heavy horses and wagons rather than as riders on the range. Perhaps the enumerator had visited the ranch when a delivery wagon had arrived and the men were staying over, or perhaps these men had been employed to feed thoroughbred bulls, weaned calves, and weak cows. Lusk and Longabaugh, the two horse breakers, would have doubled as experienced riders when required, but there were eight hundred head of breeding horses to look after as well as the working cow ponies. This meant that Johnson could rely on only two full-time cowboys and himself to manage eight to ten thousand head of cattle. This would have been an impossible task, and leads to the conclusion that the four "teamsters" were part of the permanent establishment and worked with the cattle when they were needed.[31]

The name of Henry Longabaugh stands out from the others, for he was the notorious outlaw and member of the "Wild Bunch" gang, a.k.a. the Sundance Kid.[32] Longabaugh apparently used Canada as a safe haven when there was a warrant out for his arrest in the United States. After completing an eighteen-month sentence for horse stealing in 1889, he was soon in trouble again for threatening Sheriff James Swisher in Sundance, Wyoming. He fled north of the line, and worked for the McHugh brothers on the H2 Ranch before he joined his old friend Ebb Johnson at the Bar U. They had ridden together on the Powder River Ranch.

For the most part Sundance was well liked. Fred Ings of the Midway Ranch called him "a thoroughly likable fellow … a general favorite with everyone, a splendid rider and a top notch cow hand."[33] Herb Millar – who had been riding the ranch "rough string" for some years, and who was credited with being a superior rider and horse breaker in his own right – remained suspicious of the American. They apparently had a confrontation when Millar claimed that he had seen a hacksaw blade hidden between Longabaugh's saddle and his horse blanket.[34] In range country such tools were connected with cattle rustling, since they could be used to modify branding irons. Friction

3.4 Everett Cyril (Ebb) Johnson, Cheyenne Wyoming, 1882. Ebb was a friend and hunting guide for the author Owen Wister and is thought to have been the inspiration for his famous hero "The Virginian." Glenbow, NA 2924-12

$8,000.00 REWARD!

In ADDITION to the reward of $1,000 EACH offered by the Union Pacific Railroad Company for the capture of the men, dead or alive, who robbed Union Pacific train No 3, near Tipton, Wyo., on the evening of August 29th, 1900, the Pacific Express Company, on the same conditions, hereby also offers a reward of $1,000.00 for each robber.

One man is described as about five feet 10 inches in heighth, smooth face, sandy complexion, grey eyes, talks very fast.

Second man---About five feet seven inches, sandy complexion, talks very coarse, wore canvas coat, corduroy pants, shoes badly worn.

Third man---About five feet nine inches, dark complexion, wore dark flannel shirt, no coat.

No description of fourth man.

Any information concerning them should be sent W. L. PARK, Superintendent, Union Pacific Railroad Company, Cheyenne, Wyoming; E. DICKINSON, General Manager, Omaha, Nebraska; J. W. ROGERS, Acting Superintendent, Pacific Express Company, Salt Lake City, Utah; or F. C. GENTSCH, General Superintendent, Omaha, Nebraska.

THE PACIFIC EXPRESS COMPANY.

August 31st, 1900.

Green River, Wyo., September 1st, 1900.

Dear Sir:-

Our money loss is $50.40, damage to car safe and express freight will amount to $3000.00. Robbers used Kepauno Chemical Co., Giant Powder, dated September 15th, 1899. Ashburn Mo. Works, forty percent.

Yours truly,

(signed) F. C. Gentsch, Gen'l Supt.

3.5 One of the many press notices generated by The Sundance Kid and the Hole in the Wall Gang. A reward poster relating to a train robbery at Tipton, Wyoming, in August 1900. Courtesy of Union Pacific Railroad, Union Pacific Historical Collection

between the two men may have been exacerbated by differences of opinion on how to train horses, and Millar, a long-time employee, may have been jealous of the new arrival and his friendship with Millar's immediate boss. It may be significant that, in August 1891, Sundance was arrested for 'cruelty to animals.' *The Calgary Tribune* reported:

> At the Barracks Police Court on Friday before Captain McIlree and Inspector Cuthbert, Henry Longabaugh was charged on the information of Inspector Snyder with cruelty to animals. Longabaugh, while breaking a bronco, was observed by the Inspector, who had him summoned. Mr. J. Bang appeared for the accused, who was discharged.

However, this incident did not affect Harry's position at the Bar U. He was best man at Johnson's wedding in November. Sundance spent the winter in Calgary as a partner in the Grand Central Hotel, located on Atlantic Avenue just across the street from the railway depot. He returned to the United States early in 1892, and became involved with a gang which ran stolen horses from northern Montana into Saskatchewan. On 29 November 1892, Longabaugh participated in the robbery of a train at Malta, Montana. The "take" was $19.02 in cash and a cheque for $6.05! However, not all his efforts were so poorly rewarded. Over the next nine years the Hole in the Wall Gang robbed five trains, three banks, and one mine payroll for a total haul of $200,000 (about $2.5 million in today's dollars).[35]

3.6 The Hole in the Wall Gang included the Sundance Kid seated at the left and Butch Cassidy at the right. The others, left to right, were Will Carver, Ben Kilpatrick, and Kid Curry. This was the photograph the Pinkerton Detective Agency used to hunt down the gang members. Denver Public Library, Western History Collection, Noah H. Rose, Z-49

Quite apart from the human interest stories of individuals, the collective picture of the community provided by the census is of considerable historiographic importance. On the one hand, there are various historians who have portrayed the Canadian ranching frontier almost as a "transplant" from Victorian Britain, complete with hunting, polo, and grand balls; other writers have emphasized continuity and similarity across the international border. They point out that much of the technology, the know-how, the dress and the language, came from south of the line.[36] The nominal census allows the American presence on the ranches of the foothills to be measured for the first time.[37] At the Bar U, the manager and his family were from Quebec, but the foreman and three employees had been born in the United States.[38] Much the same was true of another big corporate ranch, the Cochrane Ranch along the Belly River. It was managed by Senator Cochrane's sons, William and Ernest, who were from Compton County, Quebec, but James Patterson the foreman and six of the employees were born in the United States. It is difficult to avoid the conclusion that American cowboys played a key role in passing range skills and lore onto Canadians and British. This is not surprising, since ranching companies had enormous investments in cattle to protect and naturally sought out the most experienced and expert help they could find. Duncan McEachran, in his series of letters to Thomas Power concerning the purchase of a foundation herd for the Walrond Ranche Company, becomes increasingly insistent that Power find him an experienced foreman both for the drive and to run the new ranch.[39] Early in the spring he wrote, "Next to getting the cattle, the getting of a good ranch manager is most important."[40] As the purchase of the herd became imminent, finding a good man became even more of a priority. McEachran again wrote, "will you please write without delay and try and secure a first class cattleman for the Walrond Ranche Company of Fort Macleod who is competent to take charge of the drive from your Warm Springs Ranch, Montana to the ranch on the Oldman River, Fort Macleod."[41] It is clear that he felt that it was imperative to find an experienced man from Montana rather than looking for somebody in Canada.

However, skilled labour from the United States was not an absolutely necessary condition for a successful ranch. While 'middle order' ranches like the Winder and the Military Colonization Company did employ American cowboys, others flourished without. Walter and Fred Ings came from Prince Edward Island to establish the Rio Alto (OH) Ranch. Frank Bedingfeld and his mother Agnes came from England to join Joseph H. Brown from Ireland, in a partnership that built up the outfit which was to become famous as the EP Ranch. John Thorp too, had just arrived from England and settled on what was to become the nucleus of the Cartwright's D Ranch. Walter Skrine and "Billy" Cochrane had both come from England and established themselves along Mosquito Creek, while Ernest Cross, at the a7, was from Quebec. Among these emergent cattlemen the census shows that only Cross employed a cowboy from the United States, although it is true that most of these ranchers had served some kind of apprenticeship on a major ranch before going into business on their own.

Those employed in the cattle industry were described in a variety of ways by the census enumerators, as cowboys, stock rearers, stockbreeders, ranchers, ranch hands, horse breakers and cattle breeders. Of the 657 men in the Macleod District so described, 42 per cent were born in Canada, while a further 37 per cent came from the British Isles. Seventeen per cent were born in the United States, and 5 per cent were from other origins. These figures suggest that we can take remarks made by contemporary observers on the origins of the labour force on the Canadian range at their face value. The American correspondent of the *Times* of London, for instance gave an outsider's assessment of the Macleod area when he visited in November 1886: "The management of these ranches is generally in the hands of Englishmen and Scotch men with Ontario men, but the foremen, herders and cowboys are mostly from the States."[42]

"Mail Call and Roll Call"

Postal communication on a regular basis was a prerequisite for carrying out business during the 1880s.[43] Stimson was managing a large enterprise several thousand miles from the head office in Montreal. He had to be appraised of company policy; to be in constant touch with respect to markets, prices, and transportation; and to have a reliable link to the government as the details of the lease regulations were worked out.

During the exploration and establishment phases of the development of the North West Cattle Company all mail came west in closed bags through the United States. The Missouri River route was used during the summer, while during the winter season the mail moved by train to Ogden, Utah, and thence northward to the head of the railway near Dillon, Montana. From there it was carried by stagecoach to Helena and Fort Benton. The last link in the chain, from Montana up to Fort Macleod and Fort Calgary, was also the slowest, since the mail was carried by bull trains up the Whoop-Up Trail. In an emergency the nearest telegraph office was in Fort Shaw, Montana.

As the Canadian Pacific Railway was pushed across the prairies in 1883 great changes were eagerly anticipated. By the end of that year Calgary boasted a telegraph office and the delivery time of letters to metropolitan centres in the east had been slashed from weeks to a matter of a few days. It took some time for these revolutionary improvements to reach into the countryside. A formal post office was established in High River in 1884, and in 1886 Stimson managed to persuade the Post Office Department to establish another branch, to be known as Pekisko, at the North West Cattle Company's ranch headquarters. Not only was the Bar U the largest concentration of population in the district, but it was also a business enterprise which could provide a stable environment for the postal service. From the beginning it had a bookkeeper and some kind of office. Stimson got his post office, but he was responsible for getting the mail to and from High River without any cost to the government. Late in 1888, Stimson wrote to D.W. Davis, Member of Parliament for Macleod, arguing the need for a reform in the existing arrangements. He pointed out that the Mounted Police were helping by carrying letters to and fro when they were on patrol, but he explained that there was so much heavy mail that his company was forced to send a wagon and team into High River far more often than ordinary ranch business demanded. His plea was heard and money was provided to cover his expenses.

Pekisko post office served between twenty-two and twenty-five families at this time, and was used as a drop-off point for incoming mail and as a dispatching point for outgoing mail. For an isolated population, mail was of enormous importance; "Going for the mail" provided a chance to meet people and exchange news, as well as to make contact with faraway family. The regular arrivals of newspapers and journals through the mail were events to be treasured. From 1886 to 1902 Mary Stimson was nominally the Postmistress, but it seems likely that the mail was actually handled by the ranch bookkeeper and made available in his office adjacent to the main house.

By this time the Bar U had quite a number of neighbours, although they were widely dispersed. Only a mile from the ranch up Pekisko Creek were the Bedingfelds and their partner Joseph (7U) Brown. Mrs. Bedingfeld was a remarkable woman. The widow of a colonel in the Indian army, she was not prepared to embrace a future of dull and impoverished gentility in England. She brought her son to the new world with the intention of taking up some land. The pair spent a few months on a farm in the United States before arriving at the Bar U in 1884. Mrs. Bedingfeld had agreed to act as housekeeper for the Stimsons while her son Frank learned the skills of a cowboy. Frank Bedingfeld, at sixteen years of age, must have joined with relish the "sort of high class corps of amateur cowboys."[44] He had excellent teachers, for George Lane was foreman, Herb Millar a senior hand, and range celebrities like George Emerson and John Ware were frequent visitors. Things must have been harder for Mrs. Bedingfeld. The ranch buildings were raw and new and were constantly being enlarged or modified. In addition to the Stimson household, she had to cater to a constant stream of visitors. It must have been a far cry from being a "memsahib" in India with several servants at her beck and call. Mrs. Bedingfeld was one of a large number of female immigrants from Britain, although she was more mature than most. Lewis Thomas remarks, "the wholesale importation of young and eligible female friends and relatives as governesses, housekeepers, and companions, which was almost an industry in the nineties and nineteen hundreds, considerably relieved the burden of work which fell to the rancher's wife."[45] In 1886, Frank turned eighteen, and the pair filed homestead claims on adjacent quarter sections.[46] Over the next thirty years the Bedingfelds put down deep roots in Alberta and built up a successful ranch, concentrating mainly on raising fine horses.[47]

To the south of the Bedingfeld place was the Paleface Ranch. This had been established by Octavius Greig and ran some of the first purebred Hereford cattle in the region. Unfortunately, the climate (and perhaps the isolation) did not suit his wife and they returned to England, leaving the ranch in the hands of his brother Ronald and his sister Rachel. A mile or two further up Pekisko Creek was George Emerson's headquarters. Emerson, although only thirty-three years old, had already pursued many of the occupations associated with a frontiersman. He had mined for gold at Virginia City, Montana, and then worked for the Hudson's Bay Company as a teamster. After the arrival of the North West Mounted Police, Emerson went into partnership with Tom Lynch and started trailing herds of cattle from Montana to be sold to policemen and aspiring small ranchers. From the early 1880s, Emerson had been running cattle on the open range from his base on Pekisko Creek. He was a tough and crusty bachelor, but welcomed all comers to his house. Younger cowboys were drawn to him both because of his mastery of range lore and because of his colourful storytelling. In 1891 there were three men "batching" with him, a carpenter, a teamster, and a man called Green Walters who described himself as a 'hay contractor,' but who had worked on many foothills ranches as cook and man-of-all-work. Douglas came across Green at the "CC" Ranch and was impressed with his talents! He wrote in his diary:

> We had a most amusing evening as they have a very original nigger named Green there as a sort of general servant who is about the most original specimen I have ever come across; he is a very little fellow and built in the most extraordinary way imaginable; he sang plantation songs and other minstrel songs all the time and kept us in roars of laughter and the more we laughed the funnier he became; he would be worth a small fortune to anyone who could transplant him home as a music hall artiste.[48]

John Thorp had arrived from England with a letter of introduction from the Allan family. Stimson met him in Calgary and he stayed at the Bar U for a while. Later Thorp bought a cabin nine miles west up the creek and this eventually became the headquarters of the "D" Ranch. In 1891, Thorp was living in the cabin with the Gervais brothers from Compton, Quebec. There were several families strung out along the upper reaches of the Highwood River. James Minesinger and his wife and child had a cabin on what is now the Eden Valley Indian Reserve. John Sullivan and his wife and four children were located near the mouth of the creek which still bears his name, while Walter and Fred Ings had built the Rio Alto Ranch, with its famous OH brand, on the bend of the river.

Along the southern margins of the North West Cattle Company leases there were also a number of settlers. Charlie Knox had come out from England to work on the Mount Head Ranche. When this outfit was absorbed by the Bar U, Charlie started his own place where the Gardner's Mount Sentinel Ranch is today. In 1891 he employed two cowboys, one of them being Herb Millar's brother Charlie, who was to work at the Bar U for many years. Henry Minesinger and his family lived in a cabin close to the old buildings of the Mount Head, and from there he could look after the Bar U calf camp. A few miles further west, Henry Sheppard had built a cabin and some corrals. He married the young woman who had come out to be governess to the Greig children. Guy Pallister, who became a well-known figure on the Canadian range, was living with the Sheppards and working for them at this time.

Across the valley, on the headwaters of Mosquito Creek, was Walter Skrine's ranch. Two cowboys from England, John Graham and Edward Maldy, were helping him with his stock. Quite close to him over the ridge was Ernest Cross and the a7 Ranch, which was the home of three cowboys and a Chinese cook.

3.7 This photograph shows the relationship between the original log house and the wings which were added to the main ranch house during the 1890s. It was taken in 1914. Parks Canada, Mildred Lane Day Estate Collection, 95.01.01.03

The Maturing "Village"

The years before the turn of the century saw further expansion and adaptation of the ranch headquarters. In 1894 another frame wing was added to Pekisko House. Taken together, the two newer structures almost completely masked the original log house, when viewed from the east. The new addition was not quite as long as the earlier one, and the western facade of the composite building was offset by eight or ten feet. The new wing meant that adjacent storage sheds had to be moved and realigned. In addition, a small, free-standing log dwelling was built between the house and the barn to the north. This may have been where the ranch house cook lived. A photograph from the Lane collection, although taken some years later, helps one grasp the rather complicated "footprint" of the main house. It shows that the original building made of squared logs nestled behind the additions. None of the buildings had a full upper storey; each had attic accommodation lighted by windows in the gable. It is obvious that the three buildings which together made up the house now provided plenty of space for the household and for receiving guests. Norman Rankin recalled the interior of the ranch house when the Stimson's lived there:

> For comfort, its picturesque residence house makes one remember the original shack residence of manager Stimson at the Bar U. Outside it was a rough log hut, but inside a haven of rest and comfort, an oriental hall of luxury and elegance. Stimson's friends were legion; everyone that knew the Allans and was going west, secured a letter of introduction to the genial manager of the Bar U, and so delightful must have been their stay, that when they moved on, back came a present – a token of friendship – at first opportunity. Pottery from Egypt, brasses from India, temple cloths from Japan and China, and the thousand and one little ornaments and

> trinkets that "world roamers" pick up carelessly as they pass by, decorated its rooms. Such a collection of Indian work has perhaps not been seen in the west.[49]

The throwaway last line of this description raises more questions than it answers; what was the collection of Indian work referred to? The context of Rankin's comments does not make it clear whether he meant work from the Indian subcontinent or from the First Nations of western Canada. Recent work by Arni Brownstone of the Royal Ontario Museum clarifies this, for he has shown that Fred Stimson, during his twenty years at the Bar U, put together one of the finest collections of Plains Indian beadwork in North America.[50] Bryce Stimson, Fred Stimson's son, bequeathed the collection of some three hundred items to the museum during the 1940s.

This wonderful collection throws additional light on the man who put it together. Stimson developed a deep interest in Native culture because it was another fascinating aspect of the "west" he embraced with such enthusiasm. He learned to speak the Blackfoot language quite well, and was an occasional advocate for Indian rights, when such a position would have been unpopular, if not unthinkable, among the elite of the range country.[51] On the other hand, Stimson had an acquisitive streak and was probably forceful, not to say unscrupulous, in the methods he adopted to obtain "gifts" of beadwork, headdresses, leggings, and moccasins.[52] Nor were Stimson's motives those of the culturally sensitive anthropologist. He was by all accounts a raconteur, a storyteller, and an entertainer.

> His repertoire of stories was unlimited and he had stories to suit all occasions. He could entertain a drawing room audience, the elite of Montreal's society, with grace, charm and wit, and on the other hand, he could keep a gathering of tough and seasoned cowboys and ranch hands out on the roundups in roars of laughter with his stories of another type.[53]

Stimson's interest in and interaction with Native people provided him with an inexhaustible source of stories and information which would have been entirely novel and absorbing to visitors from the east, whose interest would have been piqued by his "wonderful collection of skins, heads, and Indian gear … which served to make the sitting room of his house one of the most interesting possible."[54] Stimson possessed a quick mind and burning curiosity. However mixed his motives, the fact that he built a remarkable collection of Indian artifacts sets him apart from his peers, who all too often tried to ignore the presence of the former occupants of their ranges.

3.8 Some of the results of twenty years of energetic collecting! Fred Stimson, 1893, wearing regalia now in the collection of the Royal Ontario Museum. The leggings are probably Cree, as they are identical to those worn by Chief Pokan in an 1886 photograph. The other artifacts are thought to be Blackfoot. Glenbow, NA 2307-33 [Plate 2, colour section]

3.9 An Indian tipi at the Bar U Ranch, c. 1890s. Glenbow, NA1215-2

Brownstone's efforts to establish the origin of the artifacts led him to review the aboriginal groups which lived and moved through the Bar U neighbourhood during the 1880s and 1890s. While most overviews would place the ranch squarely in Stoney territory, references to the presence of members of this tribe during this early period are scarce. It was not until the first decade of the twentieth century that Stoney families became a vital and integral part of some foothills ranches that were desperately short of competent part-time labour. In contrast, there are a number of reports of Blackfoot in the area during the Stimson era. Historian Hugh Dempsey talks of the [North] Blackfoot coming from their reserve to hunt in the foothills and to visit their children in the boarding school at the mouth of the Highwood River. William Holmes, the census enumerator, recalled that around 1885 several Blackfoot women visited the Bar U, and in general terms he remembered the area around the crossing being "Blackfoot country" before the Stoneys came to work on the ranches.[55] Mrs. Leitch, the daughter of Thomas Robertson, remembered a visit to the Bar U made when she was a child of seven or eight. She recalled that a Blackfoot family camped close to the Bar U, and helped with domestic rather than ranch work. The couple's teenage son, Caypey, was well versed in Indian dances and would perform to entertain visitors.[56] The varied and beautiful artifacts of the Stimson collection cannot themselves tell the story of their origin, but their association with Fred Stimson and the Bar U provides "the strongest reason to attribute the beadwork to the Blackfoot."[57]

The last few years of the nineteenth century saw some important changes in the working area of the ranch. The sod-roofed shelter to the west of the saddle-horse barn was removed and replaced with a solid work horse barn which is still in use. Two additional storage sheds were built opposite the bunkhouses, enhancing the effect of the "village street." Cattle-rearing techniques matured during the Stimson era and open range methods were adapted to ensure fewer losses and intensified production. More and more hay was put up; the time of calving was controlled; and weak cows and calves were winter fed. The North West Cattle Company had three feeding camps, and was capable of looking after fifteen hundred to two thousand head if necessary. When the company was sold in 1902, there were still more than a thousand tons of hay left over after the calves had been fed all winter.[58] The extensive feeding corrals at the ranch headquarters were situated across Pekisko Creek, while the branding corrals were located on the flat above the ranch buildings, where the visitor centre now stands.[59]

The 1901 nominal census provides an opportunity to weigh developments on Canadian ranches over the decade since 1891.[60] If Americans had played a significant role in guiding the Canadian range cattle industry through its inception, what had happened during the ensuing decade? On the big corporate ranches there was considerable continuity in management but an almost complete turnover in the work force. At the Bar U, the ranch had grown – as we have seen – into a small village with seven dwellings, and twenty-two barns, stables and storehouses. Fifteen people lived there in five households, while three other employees lived at a cow camp some eight miles to the east. The Stimson family, including the faithful Ella Bowen, still lived at Pekisko House, although they would have been mortified to know that it was to be their last year in residence.[61] A married couple, George and Josephine Brandt, lived with them. Josephine was cook at the big house and her husband, a teamster, was also a general odd job man. They probably occupied the log cabin adjacent to the ranch house. The ranch foreman was Charlie McKinnon, who hailed from Ontario and had already served for nine years with the Bar U. He got the foreman's job when Herb Millar was persuaded by George Lane to move to his YT Ranch to manage his horses. McKinnon lived with the bookkeeper, Joseph Fearson, in the smaller bunkhouse. The "muscle" was provided by the Thorne brothers and young Hiram Jones. Milt Thorne was the horsebreaker and received an additional $200 a year for his skill. Percy Cowan and his servant were also staying at the ranch; perhaps the young man from Quebec was learning the cattle business. Frederick Lature, his wife Marjorie, and their young daughter lived in a separate cabin. It seems very likely that Marjorie was the bunkhouse cook at this time. Finally, out at the cow camp were Francis Pike, lately returned from the Yukon, who was to become Lane's bookkeeper and played an important leadership role over the next decade;[62] and Charlie Lehr, the cook who fed the Bar U and Mosquito Creek wagons for many years on the roundup. Simon Hunter, at sixteen, must have been somewhat in awe of the two experienced men who were twice his age.[63]

On the Belly River to the south, William Cochrane was still managing the Cochrane Ranch, his permanent workforce somewhat reduced in number. Only one of his three riders was from the United States. Two young employees, one from Quebec and the

other from Russia, described themselves as "farmers" and were engaged full-time at haying and growing fodder crops. The bunkhouse at the Walrond Ranch on the Oldman River housed six cowboys, three from the United States. From this fragment of evidence one might conclude that the importance of American expertise had decreased during the decade. The same trend was apparent on well-established smaller ranches. Walter Skrine employed a Chinese cook and a Québécoise nurse for his daughter, but his two ranch hands were from Ontario and Nova Scotia. John Norris at Nanton also employed two Canadian-born riders. At the a7 Ranch, Ontarian John Blake was manager and Harry Brown, from England, was foreman. Twenty-six-year-old Andrew Scott was the only member of the community from south of the line.

There were still many American-born ranchers and ranch hands working the range in southern Alberta. They had spread from the large corporate ranches to smaller, individually owned outfits, and many skilled cowboys from the United States had left their positions as employees as soon as they had established a stake. Ed London had brought in a herd from Montana for "Billy" Cochrane's CC Ranch. After staying over the winter to see the stock settled, he moved to Pincher Creek and put down roots and raised a family.[64] George Lane left the Bar U after a bout with typhoid; when he recovered he worked as a buyer for the Winnipeg cattle-shipping firm of Gordon, Ironside and Fares. E.C. Johnson also left the NWCC and worked as a cattle buyer, while W.D. Kerfoot built up a small ranch in Grand Valley, having parted company with the Cochrane Ranch after a court case.[65] Such men remained vital sources of range lore wherever they resided. They had the experience and leadership qualities necessary to plan and manage local roundups and to head up efforts to control mange or to tackle the wolf problem; moreover they were often advocates for local interests to the stockmen's associations and the government. In this manner, the process of the diffusion of technology and know-how went on uninterrupted.

The proportion of those working on the range who had been born in the United States remained much the same as it had been in 1891. Over the same period, the total number of inhabitants of the District of Alberta born in the States rose from 1,251 (5 per cent) to 10,972 (16.7 per cent). One must conclude that the influx of Americans was absorbed in the farm sector and in urban occupations, not on the range.[66]

By 1901, the bunkhouses throughout the foothills sheltered men from a wide range of origins and backgrounds who had honed their range skills on Canadian grasslands. A photograph of the Bar U roundup crew resting in the chuckwagon tent makes this clear. Typical was Charlie McKinnon, who hailed from Ontario and had learned his business on various ranches in the foothills. Milt Thorne was one of three brothers born in New Brunswick but raised in High River. The three Englishmen in the photograph – Hills, Robertson, and Fitzherbert – would be referred to as "New Canadians" today. Hills had arrived in 1885 and had spent all his working life in Canada. Fitzherbert had been apprenticed to a famous Newmarket trainer, but he emigrated when he grew too large to become a top class jockey. Mike Herman came from Montana; his father was Dutch and his mother Irish. A protégé of George Emerson, he was

3.10 Bar U Cowboys in the chuck wagon tent at the general roundup of 1901. Sitting left to right: NWMP Brand Inspector; Lionel Fitzherbert; unidentified; Charlie Lehr the cook with his back to the post; and Milt Thorne. Lying down, left to right: Charlie McKinnon, foreman, boots in foreground and face hidden; F.R. Pike; Ted Hills on his back with his feet on Hugh Robertson; unidentified; and Mike Herman, near the stove with striped handkerchief. Glenbow, NA 1035-6

a wizard with a rope and was destined to spend the rest of his life moving from ranch to ranch in Alberta. Finally, Charlie Lehr, a famed roundup cook, was born in Prussia and moved west by way of the "Pennsylvania Dutch" settlements in the eastern United States. The crew shown in the photograph came from wonderfully eclectic backgrounds and reflected the ethnic makeup of most Alberta ranches of the period. A second picture of the group taken outside shows their faces and a few additional characters mentioned; F.R. Pike, from England, 7U Brown from Ireland, and Jewett Thorne, Milt's brother.

The population of the district round the Bar U Ranch had not changed much since 1891. The Pekisko post office continued to serve some twenty-two families and individuals. There had naturally been some changes. The large Sheppard family had moved from south of the Bar U to a more accessible location on the Highwood River. The Minesinger and Miles families had moved out of the district while the Ward family had settled just to the west of the Bar U cow camp. The completion of the railway from Calgary south to Fort Macleod had already had a major impact on ranchers in the area because it had precipitated the end of the closed leases. However, pressure from farm settlement, starting along the line of the railway and spreading rapidly eastward and westward from it, did not occur until the next decade. For example, the township which included Nanton station had only four people in it in 1901, and the one containing Cayley had only five. However, the population of the eight townships which made up the Pekisko sub-district increased from 70 to 509 between 1901 and 1911. In the townships to the east and north of the Bar U the huge tracts of unused range, which had been so important to the ranch since the cancellation of the closed leases, began to disappear as farmers fenced their quarter sections. The new owners of the Bar U would have to develop new strategies in response to this population pressure after 1902.

A Word about the "Open Range"

The term "open range" is one that is frequently used but seldom defined. Clearly the phrase refers to the days before fences when there was an unrestricted and unmarked "sea of grass" on which to run cattle. By association, the term has also been used to define an extensive method of raising cattle with a variety of characteristics.[67] Many of the techniques and attitudes of this system lingered long after the grasslands were criss-crossed with fences. Terry Jordan has traced the origins of this system to the subtropical grasslands of coastal Texas, and labels it "the Anglo-Texan Ranching System."[68] It involved a bare minimum of inputs. Semi-feral cattle were left alone to multiply and only came into contact with mounted men at the time of the yearly roundup. Indeed, roundups were often called "cattle hunts." Stock fended for themselves, there was no attempt to move herds from one area to another with the passing seasons, let alone to feed stock during the winter or to fatten them for market. The hallmark of the system was pervasive neglect of livestock. Such methods were not viable on the northern ranges and had a limited impact in Canada. Jordan summarizes this in a series of colourful phrases:

3.11 Bar U Ranch Cowboys, general roundup May 1901. Left to right: F.R. Pike, Charlie Lehr (sitting on wagon), Charlie McKinnon, three figures in shadow, Mike Herman (white shirt), 7U Brown, NWMP Brand Inspector, Jewett Thorne, Hugh Robertson, Ted Hills, unknown, Milt Thorne. Notice the two wagons, the cook tent, horse "tack" in the foreground and the "remuda" grazing in the background. Glenbow, NA 1035-5

> Born in a mild subtropical 'lotus land' and based on a breed of cattle that had never known the bite of a true continental mid-latitude winter, the system was singularly unsuited to the far greater part of "Texas extended" to the north … the system died of cold and drought.[69]

However, because the northward diffusion of the cattleman's frontier was so rapid, in the years following the end of the Civil War, there was no opportunity for gradual experimentation and adaptation. It was to be some years before the shortcomings of the system became apparent. For example, during the fall of 1871, Frank Wilkeson was running a bunch of Texas cattle in the semi-arid belt of western Kansas. He describes the lack of knowledge prevalent about winters and grass:

> I had provided no food for my stock. I knew that cattle could and did winter on the plains far north and west of where I was: but I did not know that there was a difference in the nutritious qualities of the different prairie grasses. I did not understand the peculiarities of the climate of the semi-arid belt, nor the effects of rain falling on dead grass (leaching nutrients). Stupid of me, of course, but I had plenty of company.[70]

The results of this ignorance were devastating. Most cattlemen running Texas herds lost more than half of their stock during the winter. In the spring,

> … the creeks were dammed with the decaying carcasses of cattle. The air was heavy with the stench of decaying animals. The cruelties of the business of starving cattle to death were vividly impressed upon me. Every wagon sent from the cattle ranges to the railroad town was loaded with hides.[71]

Even before the onset of the wild investment enthusiasm of the "Beef Bonanza," it was clear that open range methods were risky and therefore uneconomic on the mid-latitude grasslands. Unfortunately, it was a lesson which had to be learned again and again.

In contrast, there were many stock farmers all over the northeast and midwest of the United States and in Canada who had generations of experience of bringing cattle through bitter winters. As these men moved westward onto the Great Plains, their intensive methods were modified but their basic attitudes towards stock rearing were not. From the first, this "Midwestern Ranching system"[72] was marked by efforts to rotate herds to reserve some of the best grass for winter grazing; by constant attempts to "breed up" the herds using purebred bulls; and by a concern for the "calf crop." As early as 1850, the Deer Lodge, Bitterroot, Jocko, and Beaverhead valleys of Montana all supported ranches, and, well before Texans started to drive herds northward, a thriving midwestern-derived ranching industry had occupied western Montana. Success in these mountain valleys depended on putting up large quantities of hay. Soon wild hay was being replaced with tame hay and yields were being increased using irrigation. After he had reviewed all this evidence, Jordan reached one of his most important revisionist conclusions. He wrote:

> The popular image of western cattle-ranching history must be revised to allow a far greater role for these highly successful, if less colorful herders from the American heartland. Their largely British inspired system proved to be the most successful pastoral adaptive strategy in the west and, accordingly, the one that survived.[73]

Cattle raising in the Canadian west drew on both these traditions. From its start in sheltered pockets of the foothills, ranching involved a whole range of strategies, and cattlemen pursued a variety of goals. Several of the first small-scale rancher-farmers round the police forts kept dairy herds as well as small bunches of range cattle. They put up hay and cropped some of their land for oats. Others, like Inspectors Winder and Shurtiffe of the North West Mounted Police, bought cattle and left them almost completely unattended to multiply. The men responsible for establishing the great corporate ranches in western Canada were almost all of them experienced stockmen well versed in the most progressive methods of raising cattle. On their eastern estates they went to great lengths to ensure the superior quality of their bulls; they timed calving carefully, expected a high calf crop, and winter fed all their stock.

These intensive methods were set aside for a time when they came west, because of the lure of endless free grass and the heady mythology of the open range. Less intensive "mass production" of cattle on free grass was wasteful, but yielded good profits most years. As long as immense areas of unstocked or understocked range existed it made good business sense to use it to rear and fatten cattle extensively. But, if those who established commercial ranching along the Alberta foothills borrowed the term "open range" and much of their material culture from the Anglo-Texan cattle raising system, it is clear they also brought with them midwestern attitudes toward cattle rearing. From the very beginning, purebred bulls were imported for foundation herds. This in turn meant that hay had to be put up to feed these valuable animals through the winter. Both Stimson and McEachran were concerned with getting a good hay crop even before their herds arrived in Alberta, and one of the first tasks that Bar U cowboys were faced with during the summer of 1882 was putting up some hay as well as building the first barns.[74]

Nevertheless, John Craig, manager of the Oxley ranch and an experienced stockman, caught the essence of the first decade of corporate ranching when he stated that ranchers simply turned their main herds loose on the prairie, "to wander of their own sweet will, and subsist the year round without any provision for their food and shelter beyond what nature afforded."[75] This meant that a cattle herd paralleled quite closely the dynamics of wildlife populations. The calf crop was low, and losses to predators, fire, drought, and exposure were considerable. However, because the cost inputs per head were kept so low, this extensive system could yield impressive profits.

The risks associated with bitter winters and infrequent chinook winds increased as the most favourably located ranges were occupied and cattlemen were gradually pushed out onto the mixed-grass prairies along the Bow and the Red Deer Rivers toward the turn of the century. Some herds were more vulnerable than others. The greatest losses were consistently experienced in herds that were unfamiliar with the range on which they found themselves, and among immature 'pilgrim' cattle brought in from the farms of Manitoba or Ontario. In spite of the risks, the fact that shrewd men like George Lane, Pat Burns, and Murdo Mackenzie of the Matador maintained large-scale, low-intensity operations well into the 1920s demonstrates that it was a money-making proposition as long as grass could be obtained very cheaply. If returns had not exceeded losses over the years this strategy would have been abandoned. Ranching is a means of turning grass into money, and dependence on natural unimproved range is an important characteristic of ranching. It was neither ignorance nor irresponsibility which encouraged some ranchers to stick with extensive open range methods. It just made good business sense to keep costs to a minimum.

Many small and medium-sized ranches adopted a very different course from the beginning. They invested in purebred horses or cattle and close-herded them to avoid mixing with range herds. These men adopted some fencing almost immediately, often using topographic features to decrease the length of their costly split-pole fences. They put up hay and cropped bottom lands for oats. This more intensive adaptation yielded a viable living for a family in exchange for a lifetime of devotion. Success depended on patient and unremitting attention to a wealth of details pertaining to the welfare of each valuable animal. It was not a formula to "get rich quick," or to earn the sort of returns which would satisfy investors. Not all small outfits were "progressive." A homesteader who ran some stock on adjacent range had neither the capital, the knowledge, or the time, to do much for his cattle. The foundation stock on such places were often of abject quality, and settlers seldom had their own bulls, hoping that their more substantial neighbours would supply this service. There was more to ranching than the ownership of cattle.

Large-scale corporate ranchers were not necessarily reluctant to adopt some measures to intensify production. This almost always meant increased expenditures on labour and materials. The managers and foremen of the big ranches were constantly weighing costs against returns, accepting some innovations while rejecting others. It might be deemed well worth weaning calves and feeding them through the winter, because cows recovered faster and were more likely to bring their next calf to term. On the other hand, experience had shown that range-bred cattle could hold their own through most winters, and it was not economic to attempt to feed the whole herd. Several of the larger outfits compromised by running two or more herds: a more or less intensive cow-calf operation located in the foothills, and a steer herd run on the underused range further east with a minimum of supervision. The later herds were regarded as a short-term gamble against the weather or a downturn in prices. George Lane was justly proud of his breeding herds of Shorthorns at the Bar U and the Willow Creek Ranches, but at the same time he was importing stockers from Manitoba and even Mexico to throw onto the ranges east of the Bow River. This diversification was not confined to the larger corporate ranches. For example, W.E. Cochrane ran a purebred Galloway herd at the CC Ranch as well as a regular range herd,[76] and A.E. Cross maintained a breeding herd at the a7 Ranch and a steer herd along the Red Deer River.[77]

Contemporary observers commented on "the end of the open range" again and again.[78] During the 1890s, those around Pincher Creek were proclaiming the demise of ranching and the triumph of mixed stock farming. Many of the same sentiments were expressed around High River at the turn of the century and in the Maple Creek district a year or two later. But huge acreages of grassland remained unclaimed and unused, for example between the Bow River and the Red Deer, and northward along the South Saskatchewan River from Medicine Hat. These ranges were regarded as being too dry or too far removed from railways to attract farm settlement. Here many of the essential elements of the open range system continued to flourish until the 1920s.[79] It is not surprising that the phrase 'open range' is one that causes some confusion. It is used to refer to a time period – a few brief decades – before fences were built, and also to describe an extensive method of raising cattle. The term is used to introduce techniques used by outfits in Montana and Alberta, but these methods differed markedly from those used in southern Texas where the concept originated. Finally, we should remember that cattlemen are as subject to human failings as the rest of us. The attitudes and methods they had learned as young men were set aside with reluctance. Newfangled ways of "spoiling cattle" were regarded with suspicion if not contempt. Seven U Brown swore that if he could get through the winter on whisky then his stock could survive on snow.[80] Herb Millar was unwilling to feed the herd lest they forget how to rustle.[81] Howell Harris held that you could raise two Mexican steers on the grass which would support a single "improved" Hereford.[82] Thus, some of the philosophy of the open range lingered long in the minds of those who had experienced it.

The Seasonal Round

Cattle ranching follows a seasonal rhythm, although the precise timing of activities may vary according to the weather[83] Spring is a season of calves and foals. Roundups and branding begin in the late spring and used to last well into summer. After branding, cattle are moved to summer pastures and the bulls are turned out with the cows. Midsummer and early fall are devoted to putting up hay for winter feed. In the late fall, mature cattle are shipped to market, and calves are weaned. The cow herd moves to winter pasture. Winter is a time of conserving resources, of feeding calves and needy cattle and preparing for the next season. In the paragraphs which follow, an attempt is made to outline those seasonal activities which would have been particularly important at the Bar U during its first two decades of operation. Others, like haying and cropping, will be described in more detail later.

Stimson's time at the ranch was one of trail-breaking and experimentation. At first the home place, with its corrals and dwelling houses, was like a tiny island in a sea of grass; a few hands looked after a huge herd as best they could. The bulls ran with the cows, and calving was a long, drawn-out process. General roundups were leisurely affairs taking two or three months to complete. Weaning was not generally practised and winter losses were high among cows still suckling big calves through the cold weather. Gradually the life cycle of the cattle was subjected to more and more stringent controls, with the objective of maximizing increase and reducing losses.

March, April, and May: Calving, Fencing, and Horsebreaking

Calves represent the "interest" on a ranching investment: a good calf crop heralds a successful year, while a vicious spring storm could result in losses which wipe out all hopes of increase for that year. For today's ranchers the calving season is a time of worry and incredibly long hours of labour. Each cow that runs into difficulties may expect some attention, and each calf lost is regarded as something of a personal failure. Such was not the case on a large ranch during the nineteenth century where cattle were left to fend for themselves.

At the Bar U during the early days, some cows would calve on the range in January and February, although March and April were the peak months. The cow herd would be loosely held on a more accessible area of winter range where water was available. The cows were encouraged to stay in the locality with salt, and perhaps some hay. Riders would circulate quietly through the herd each day paying particular attention to the places where topography or brush provided cover. Cows which are about to give birth tend to leave the herd and seek solitude, and an experienced cowboy would note where to look for a newborn calf the next day. Intervention was minimal, but the men could play a role by providing some protection from predators; by reuniting cows with lost calves; or by extricating a calf dropped in a tangle of brush.

As the years passed, ranchers in the foothills made efforts to limit the duration of the calving season and to manage its timing. The period of gestation in cattle is about the same as in human beings. If the bulls could be separated from the cow herd until July, most calves would be born in April. By that time much of the snow would have disappeared, the sun would be warm by day and the grass would be greening up in favoured locations. However, the later the calves were born, the shorter was the time before they were weaned in the fall. A late calf may drag a cow down as she enters the winter, while at the same time the calf may be too young to forage effectively. Under open range conditions it was no easy matter to keep the cows and bulls separate. While it was common practice to feed imported bulls through the winter and to close-herd them, the more numerous "range bulls" ran at large.

The first rudimentary steps toward herd management demanded fencing in order to be effective. The first fences erected at the Bar U were corrals for working cattle and training horses, and enclosures where valuable bulls could be fed. Township plans dated from the early 1890s show only a small amount of fencing in the area round the Bar U. Stimson claimed he had two miles of pole fence in his homestead application dated 1884; probably this fence was designed to keep cattle out of the headquarters site. By 1897, Stimson stated that he had three miles of wire fencing worth $80.[84] The next year Fred wrote to A.E. Cross about the possibility of buying barbed wire in bulk:

> Skrine was over here yesterday and he tells me he spoke to you about wire. I understand Lane

> got his wire in St. Paul or Mineapolis for $1.80 per hundred. Could we not get a load delivered at Nanton? I would take 10,000 lbs. Skrine said he would take 4,000 lbs. This would make 7 tons. Could you take enough to make up the car? Sampson believes he wants some....[85]

This correspondence suggests that Stimson was about to embark on an ambitious fencing project; probably his aim was to enclose much of the newly acquired deeded land. The amount mentioned would have been enough to provide for forty miles of three-strand fencing. Accounts of the sale of the Bar U in 1902 mention that most of the deeded land was indeed enclosed. A growing network of fences would have made it much easier to adopt progressive techniques of herd management at the Bar U during the last years of the nineteenth century.

Another sign of spring was the quickening of activity around the horse corrals. While riders used their well-broken ponies for their daily chores, many additional mounts would be required for work during the spring and summer. Each man on the Bar U wagon during the roundup would require a string of seven to ten horses, and the remuda would have totalled some fifty head. In later years specialized horse breakers were usually hired, but during the 1880s this task was handled by Herb Millar. His aim, during the brief period which he spent with each animal, would have been to teach a horse to stand while being handled and saddled, and while a rider mounted and dismounted. He would ride a young horse a couple of times and try to teach it to respond to the reins, but it was incomplete 'schooling' to say the least! The impromptu bucking contests which occurred on the first few mornings of any roundup showed that cowboys were expected to be able to handle a good deal of "mettle" in their mounts.

3.12 A good overview of a roundup camp, showing the rope corral in the foreground, and the two main wagons. The cook is standing beside a side of beef which he had been cutting up; note also the water barrel. Glenbow, NA 2307-55

June, July, and August: Spring Roundup, Branding, Breeding, and Haying

"The feeling of a clammy slicker, sodden boots, and dampish blankets persists to this day...." So starts Fred Ings' nostalgic account of his experiences on roundups during the 1880s.[86] Discomfort, fear, danger, and pain were more than balanced by the euphoria of being part of a big and necessary operation, and riding hard almost every day for a two-month period. The spring roundup was the principal event of the ranching year.[87] It signalled the end of the drudgery of winter, the prospect of new grass and new beginnings. Ranchers had the chance to count their herds, evaluate the increase and establish their ownership. For many cowboys who had been laid off all winter it meant working again and the prospect of a wage packet. For others it meant quitting squalid 'line cabins' and the endless lonely patrols of the winter range. It meant working in company with riders from many ranches;

an intense period of doing what cowboys felt was their real job. There were also opportunities for fun, competition, and the swapping of news and stories. The spring roundup would usually start toward the end of May and continue through the early summer.

One of the largest and most far-reaching general roundups ever held met at Fort Macleod in the late spring of 1885. It comprised fifteen mess wagons, hundreds of riders and more than a thousand saddle horses. Young Ted Hills stated the objectives of the roundup in a letter to his family in Sussex, England. He wrote:

> We shall start probably about the 20th of this month [May], go first to Fort McLeod and then on to the borders of Montana and roundup the country as far as Calgary which will take two months.[88]

This bald and confident statement of intent had the naïveté of the greenhorn, for the general roundup was a formidable operation to organize.

The basic working unit of the roundup was a "wagon." This referred to the men, horses, chuckwagon and other equipment furnished by a particular ranch. For example, the High River district was represented by two wagons, the Bar U and the Mosquito Creek Pool. Each wagon had about fifteen riders besides the cook, night herder, and horse wrangler. Among these riders would be men from smaller neighbouring ranches. Fred Ings explained:

> Our roundups were community affairs, the different ranches in the district sending representatives called "reps" to ride with wagons of that part, I usually rode with the Bar U representing our OH brand. These reps took six to eight horses of their own, their saddles and bedrolls.... Those chosen to go as reps by their different outfits were men who could ride and rope and who were strong and hardy for it was no child's play in which they were taking part.[89]

Wagons would form quite a cavalcade as they moved toward the designated rendezvous. Led by the pilot, the wagons would string out behind, followed in turn by the herd of saddle horses, driven by some of the riders. At day's end each wagon chose its own camp site at some distance from others so as to keep the "remuda" of horses separate and manageable. The noise, dust, and companionship of the outfit on the move contrasted with the quiet of the country through which they moved.

> On the trail down to as far as the cattle might have drifted, we passed through an absolutely unsettled land, no towns, no fences, just one big grass covered range, such grass as we never saw before. The buffalo had been gone for some years, and what cattle there were wandered at will from Sheep Creek almost to the border.[90]

At Fort Macleod the crews from High River were "joined by the Willow Creek, Pincher, Oldman, and Belly River outfits which were all equipped the same way. Lane was elected the Captain which was an important office as he has control of all the different outfits and gives directions as to how the country is to be worked."[91] Being captain of the roundup was indeed a responsible job, but his authority rested on his reputation and on his ability to work with the wagon bosses who knew their local ranges better than he did.[92] From the meeting place at Macleod, some wagons were sent southeastward down Chin Coulee almost to Pakowki Lake before returning along the Milk River. Others scoured the country to the north of the Belly (Oldman) River as far east as the big Bow River.[93]

Each morning teams of circle riders would leave the camp and ride out ten or fifteen miles, from which point they would spread out and slowly drive cattle back toward the new camp site. In this way a broad swath of country could be cleared of stock.

> When all the cattle were rounded up they were put into bunches of about a thousand head each and were held by riders surrounding them. The bunches were kept well apart to permit comfortable working. The calves and cows were separated, and when the cows had been cut out they were driven to a suitable grazing and watering place where a particular outfit was camped. This started the herd, which was moved and added to each day as the roundup moved on. When the herd reached about two thousand head an outfit was sent to drive them onto their range.[94]

Ted Hills was the horse wrangler on the Bar U wagon and gives a vivid description of his job in one of his letters home.

3.13 A group of cowboys from the Bar U relax after a day on the roundup, c. 1902. (Left to right: Unknown, Unknown, Fred Robertson (in tent doorway), Kid Smith (standing), Eddie Moreno (sitting in front), a transient Englishman gathering information, Miles Clink, and A. Melross. Glenbow, NA 285-4

> Since I last wrote to you we have been turning out at 3 am every day, and as I don't take the horses in until 8 pm my hours are rather long. The night herder brings the horses in every morning shortly after 3 am and we have to form a temporary corral with the wagons forming one side and ropes tied to the wheels make a third side, while men stand at the side and rope the saddle horses that are wanted for the morning. At noon I corral them the same way and the afternoon ones are caught.[95]

Perhaps the mists of nostalgia had clouded Fred Ings' memories of those cold, dark, dawn mornings, for he remembered them with affection:

> In looking back now I can see it! hear it! live it! again a roundup day! From the first call to roll out in the dawn till we stamped out the last coal of our camp fire and turned in a little stiff and weary to sleep the dreamless sleep of

> youth. Breakfast at day break was eaten in the mess tent, a hot and substantial meal of meat, potatoes, bread, jam with strong black coffee. Our dishes were tin and we ate sitting around on bed rolls or a box if one were handy, or on the ground. Before we were through, the tinkle of a bell told us the night wrangler was near with the saddle horses.... This bell and the approaching hoof beats was the signal to get our saddles ready and untie our ropes.[96]

Arthur Turner was on the roundup of 1903 and explains why each cowboy needed a "string" of saddle horses.

> We were each given a string of thirteen horse for the roundup and you were supposed to ride them in rotation as they had a very hard life, often doing eight hours hard riding on such grass as they could munch during the short stopping spells. Fat horses were practically unknown in that part of the world, as were also grain fed horses; however, they kept their condition fairly well on their grass diet. We often did 16 to 18 hours riding on the roundup and by now I was able to ride any kind of "razor backed cayuse...."[97]

The objective of the spring roundup was to gather cattle and move them back to their own range. Branding would usually be attended to on the individual ranches. The captain was happiest when a daily routine had been established and there was little to 'spook' the drifting herds. Sometimes, however, a chance encounter led to excitement and a display of range skills. Cross recalled one such occasion.

> On the way ... just west of Tennessee Coulee, we came across some old wild bulls, which we roped and made into stags. There were some lively times when ropes popped and riderless horses ran wildly over the prairie and men on foot dodged furious bulls. In the end, however, no one was much the worse for wear and the work was successfully accomplished.[98]

After some weeks in the country along the international border, the roundup moved northward along Willow Creek. When they reached a point west of the present location of Claresholm, the wagons divided, one group driving their herds northward toward High River, while the other headed westward to search the wooded valleys of the Porcupine Hills and the upper Oldman River. Sometimes, heavy spring rains delayed progress and the crew were forced to remain in the meagre shelter offered by leaky tents, sitting on bedrolls and playing poker or "spinning yarns."[99] Hills reported:

> Since last writing we have moved lower down Willow Creek to within eight miles of McLeod. The creek is very high and for the last four or five days we have been waiting to cross it with the wagons, but we have been unable to. Today was a pretty tough one for the horsemen for three different creeks had to be swum and almost everyone was wet to the waist.[100]

There were many rivers to cross and it required patience and skill to get the cattle into them. Calves presented particular problems:

> These little fellows would break back and had to be roped and dragged into the water. Men would have to herd the cattle as they swam with the current to keep them from milling and turning back to shore, often the horses would be swimming too. It was dangerous work for man and beast, but we made it across eventually and very few cattle were lost. Sometimes new born calves were carried along in the wagon, the mothers followed bawling after them. Many calves were born on the trail. When they were a few days old they could trot along nicely with the others.[101]

It was the end of July before the Highwood River was reached and the Bar U herd was pushed across onto its home range. There were branding corrals built at strategic points close to water and fuel, and weeks of hard work remained to be done before the various herds were all sorted out and branded.

In some respects the methods and objectives of roundups remained much the same from their instigation on the Canadian range in the 1880s at least until the First World War. Indeed, images of a roundup camp, complete with chuckwagon and rope corral photographed in the 1920s could easily be confused with a scene from thirty years before. However, there were ongoing and far-reaching changes in organization and scope. The general roundups of the early years gave way to smaller district roundups before the turn of the century, as the number of outfits multiplied and the ranges became more fully stocked. The districts with their own roundups included Fort Macleod, High River, Mosquito Creek, Bow River, Pincher Creek, Medicine Hat, Red Deer River, Cypress Hills, and the Whitemud River.[102]

Branding is the universally adopted method of demonstrating legal ownership of horses and cattle throughout the west. When the Spanish Conquistador Cortez landed at Veracruz in the New World, he is supposed to have burned three Christian crosses on the hides of the few head of Andalusian cattle he had brought over from Spain.[103] As cattle multiplied around the missions, so a full range of rather ornate Andalusian brand designs were adopted and spread throughout Mexico. By 1537, registration of brands was mandatory in New Spain.[104] Anglo-American settlers pushed into northeastern Texas from Missouri early in the nineteenth century. They brought with them many cultural traits from upland Carolina. In 1840, the Republic of Texas adopted livestock statutes based on English common law and officially endorsed the use of block letters and numerical brands which had originated in Britain.[105] As herds and techniques spread northward over the great plains, this type of brand and its many derivations became more or less universal.[106]

It is easy to see the necessity and efficacy of branding where cattle mixed under open range conditions, but one might have thought that the practice would be outdated now that ranges are for the most part fenced and valuable stock are subject to careful ongoing attention. Times have indeed changed, but branding still plays an important part

3.14 The roper is keeping three groups of men busy in this branding scene at the Bar U Ranch. At the top, one man keeps pressure on the rope while the other prepares to throw the calf; in the middle group, foreman Neils Olsen bends to castrate a calf while Alex Fleming holds a branding iron at the ready. In the foreground, Red Cloud holds down a calf while Tom Weadick kneels at the right. Glenbow, NB 16-264

3.15 Neils Olsen has exerted just the right pressure to create a perfect Bar U brand while two Stoneys, Red Cloud and John Dixon hold the calf. Glenbow, NB 16-263

in ensuring that he who raised a steer gets the benefit. In Alberta, for instance, a private company, "Livestock Identification Services" inspected six million cattle in 2000. High beef prices have encouraged a rise in rustling, but inspection of animals each time they change hands helps reduce these losses to ranchers.[107] The actual methods of branding are under constant review. Dairy cattle have long been individually identified by means of implanted electronic ID. So far, such methods, although technically possible, are not considered cost-effective or "tamper proof" for range cattle.[108] The scale of a branding may be different, and propane may be used to heat the branding irons, but the sights, smells, and noises of a branding in the foothills are strongly reminiscent of the past century.

A key figure in the organized chaos of a branding corral is the roper.[109] He checks that a calf belongs to a particular branded cow, heels it and drags it toward the fire. He calls the brand to the crew so that they know which irons to apply. A team of two wrestlers then immobilize the calf. In short order, it is branded, marked, castrated, and vaccinated. A good roper can keep three teams of wrestlers busy, as is shown in Figure 3.14. At the top left, a man keeps the strain on the rope while his companion flips the calf onto its side. His next job will be to release the rope so that the heeler can pick up another animal; in the middle group, foreman Neils Olsen bends to castrate a calf while Alex Fleming handles the branding irons; in the foreground Red Cloud holds a calf's back legs while Guy Weadick on the right supervises. Bert Sheppard recalled one branding at the Bar U during the 1920s:

> That year the roping was done by that outstanding and expert Stoney Indian cowhand and roper, Jonas Sam Rider ... never again have I seen calves roped so fast. The calves were fairly thick around the fire and he just had to swing the

> horse to get one. Before the first calf was up a second was down and a third in the hands of the wrestlers. There were three sets of husky Indian wrestlers.... Jonas harangued them continually in Stoney, telling them to fly at it and get the rope off.... He roped with a short tied rope and threw a small loop, expertly picking up both hind feet. He could catch with equal dexterity on either side of a calf, using a back-hand throw for the left side. He was right in his prime about then as it was shortly after this that he won the Canadian roping in Calgary.[110]

The "iron men" looked after the fire and made sure that the branding irons were at the right heat: if an iron was too hot it would blotch the brand, but if it is too cool it would not burn deep enough and the hide would not peel and thus prevent the regrowth of hair. Very little pressure is necessary to put on a brand with a hot iron. In Figure 3.15 it is clear that Olsen has applied a near perfect Bar U brand. The "markers" used a sharp knife to cut the loose flesh of a calf's ear, and then to castrate the animal if it was a bull calf. Finally, the calves were tallied as they left the corral and returned to their bawling mothers. Typically, work would start in the cold first light of dawn in late May or early June. Around two hundred calves would be handled in a good day, and the work would be over by mid-morning when it got hotter and the flies became a problem.[111]

Much the same methods were used for handling mature animals except that two heelers were needed to trip and hold a large strong animal. Figure 3.16 shows a branding at the Big Hill corral at Longview in 1892. The two ropers, George Winder on the left and Henry Minesinger on the right, are keeping some tension on their ropes to hold the animal immobile. Jim Byron and George Emerson are applying the brands while Walter Ings and Ben Rankin hold the head and flanks of the animal. The large herds of mature cattle brought in by the founding cattle companies all had to be branded soon after their arrival. Various types

3.16 Branding at the spring roundup, Longview, 1892. Ropers have secured both the front and back legs of this mature animal. This photograph shows several well-known cattlemen and riders mentioned in the text. Mounted or standing left to right: George Winder, Jim Byron, George Emerson, Harry Baines and Henry Minesinger. Kneeling: Walter Ings and Ben Rankin. Glenbow NA 5182-1

into one or the other of two corrals according to sex.[112]

Frank White had a more jaundiced view of the chutes which, he said, kept breaking down and "kept us nearly half the time repairing them."[113] However, it was not long before a maze of corrals, linked by chutes to functional squeezes, were features of most foothills' ranches. Years later they would be used for dehorning and spaying cattle.

During the 1870s, herds of cattle which entered Alberta from the south carried with them their Montana brands. For example, Joe McFarlane's herd had JO on their left hip. After the North West Mounted Police had established themselves at Fort Macleod in 1874, it was not long before they started to take careful note of brands and owners, for one of their roles was to adjudicate between cattlemen and starving Indians.[114] Matters were put on a more formal basis in 1878 when the Territorial Government made provision through an ordinance for brands to be registered at Fort Macleod with C.E.D. Wood, the magistrate and the editor of the Macleod Gazette, acting as the recorder. Fred Stimson registered the Bar U brand on 20 October 1881, even before the North West Cattle Company had obtained its foundation herd.[115] Later, he applied for the double circle brand, which Tom Lynch had

3.17 Squeeze chutes were particularly useful when a large herd of mature animals had to be branded. This picture, sketched for the London Illustrated News, shows one being used to brand newly imported cattle on the Cochrane Ranch in 1882. Glenbow, NA 239-27

of mechanical devices were designed to expedite this process, and one such squeeze chute is shown in Figure 3.17, which shows branding at the Cochrane Ranch in 1882. Frederick Godsal was there and sent the following description to his family:

> … there is a long narrow way leading [out of the corral] of just the width of an animal up which they are driven and at the end is what is called a "squeezer," one side is squeezed against the animal by ropes and a windlass and he is held there while he is branded. As they emerge from the squeezer a swinging gate turns them

used as a temporary "road brand" on the drive from Idaho.

By the mid-1890s some three thousand brands had been registered with three territorial recorders. Many of these brands were too similar to act as effective identification of ownership, while hundreds of small-scale ranchers were using brands which had not been approved. It was time for a thorough overhaul of the system. A new ordinance was passed whereby any person using a mark or brand of any kind had to re-register it, or register it for the first time. This action had to be taken before 1 July 1898. The administrative centre for the new regulations was moved from Fort Macleod to the territorial capital at Regina.[116] Stimson wrote a series of letters to transfer the brands from himself to the North West Cattle Company.[117] When the ranch was purchased by George Lane and Gordon Ironside and Fares in 1902, Lane obtained a note from Sir Montague Allan concerning the sale and sent a request to the recorder of brands that the Bar U brand be transferred to the new owners.[118] At this time Lane already had seven different brands, and this number had risen to nine when a new brand book was published in 1907.[119] After the death of George Lane, the ranch was sold to Pat Burns, and the Bar U cattle and horse brands were transferred to him in April 1928. At this time Burns was divesting himself of his meat-packing empire and was buying a number of ranches.[120] By the early 1930s he owned thirteen ranches, and at his death in 1937, fifteen brands were registered in his name.[121] The Bar U Ranch and brand were sold to J. Allan Baker in 1951.

Today, the brand name "Bar U" is used in popular parlance to identify the famous historic ranch. With the passage of time, the brand, the location, the business, and the history, have become more or less synonymous one with another. This was not always the case. During the period between 1882 and 1902, the location of the ranch was referred to as "Stimson's," or "Pekisko," while the outfit was identified as the

3.18 By the turn of the century a well built squeeze was an essential bit of equipment on most ranches. Here two men sit on a bar which keeps the steer immobilized while another applies the brand through the planks. Glenbow, NA 2307-51

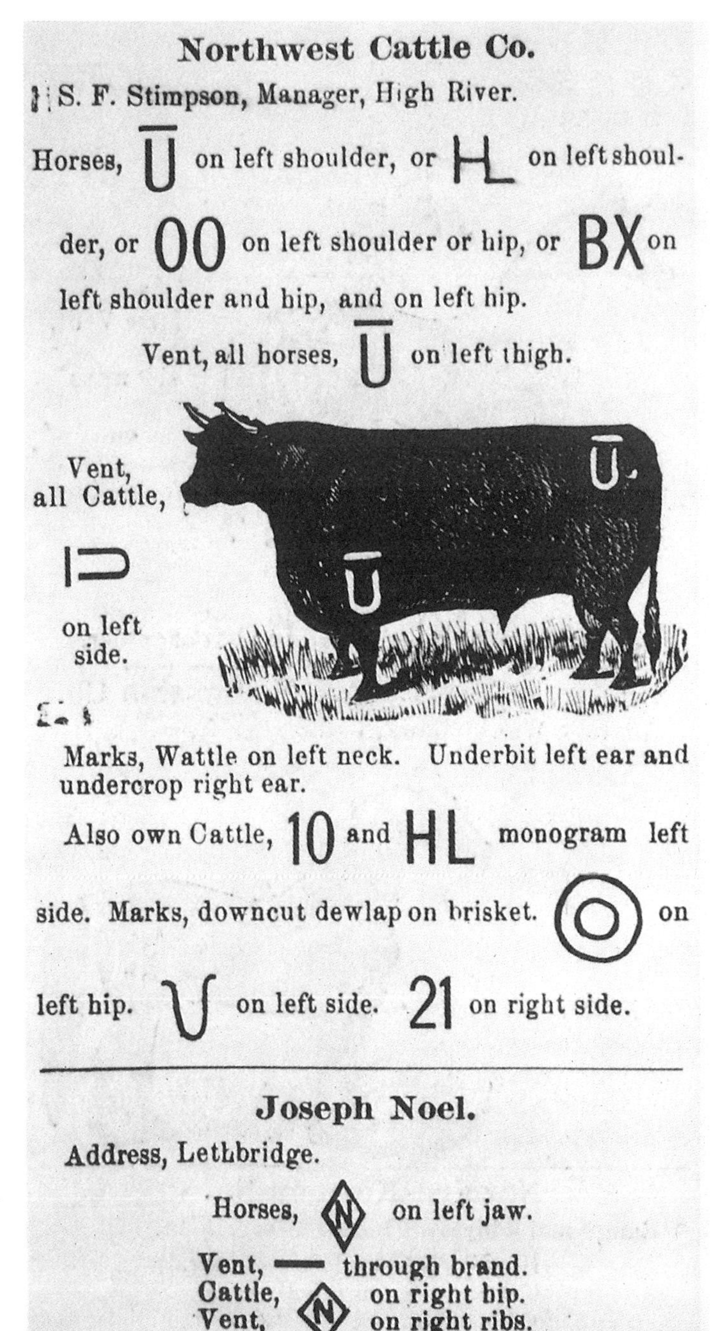

Northwest Cattle Co.

S. F. Stimpson, Manager, High River.

Horses, Ū on left shoulder, or HL on left shoulder, or 00 on left shoulder or hip, or BX on left shoulder and hip, and on left hip.

Vent, all horses, Ū on left thigh.

Vent, all Cattle, on left side.

Marks, Wattle on left neck. Underbit left ear and undercrop right ear.

Also own Cattle, 10 and HL monogram left side. Marks, downcut dewlap on brisket. on left hip. V on left side. 21 on right side.

Joseph Noel.

Address, Lethbridge.

Horses, N on left jaw.

Vent, — through brand.
Cattle, N on right hip.
Vent, on right ribs.

3.19 The brands of the North West Cattle Company shown in Henderson's Northwest Brand Book, 1889. Glenbow, NA 2286-2

North West Cattle Company.[122] The close association between the place and the brand developed during Lane's period of ownership. This happened for two reasons. Firstly, as Lane already owned two ranches before he bought the Bar U and was to acquire others, the brand Bar U became a useful way of referring to a specific unit in a complex multi-unit ranching empire. Secondly, during the first decade of the twentieth century, Lane emerged as the surviving giant in a ranching industry which was changing fast and was increasingly dominated by a large number of medium-sized family ranches. While the Cochrane, the Oxley and the Walrond had disappeared, the Bar U brand was burned into more cattle than ever before. The brand came to symbolize a grand regional tradition.[123]

After the spring roundup and branding, when the cow herd was once again grazing on its summer range, it was time to turn out the bulls so that the reproductive cycle could start once more. A mature bull might serve forty to fifty cows in a given year, but a widely scattered herd needed at least five bulls to a hundred cows. The quality of the bulls was as important as their number, for although they make up only 5 per cent of the herd they account for half of its genetic makeup.[124] Under range conditions it was almost impossible to ensure that all the cows received the attentions of a bull. Bulls have a tendency to cluster together, and it was the job of cowboys to keep them dispersed among the herd and to gather cows that had wandered too far. A measure of calving success is the number of live births obtained as a percentage of the number of cows in the herd. Today in Alberta, where a large proportion of cattle still graze on open and unsupervised range land, an 89 per cent calf crop is typical, while in Ontario the figure is 95 per cent, the highest in Canada.[125] In the late nineteenth century a calf crop of 65 per cent was considered good.[126]

Haying was part of the seasonal round at the Bar U from the very beginning. As we have seen, Stimson was emphatic that mowers and rakes bought from T.C. Power and Company should be sent up the Whoop-Up Trail to the headquarters of the North West Cattle Company even before the herd arrived. The first task of the small crew on Pekisko Creek in 1882 was to put up wild hay to feed to the imported bulls during the first winter at the ranch. Almost all the early ranchers cut enough hay to sustain their working horses, and to act as an emergency reserve to feed to their most vulnerable stock during a bad winter. In addition, many outfits had a small fenced field where they grew oats.[127] Fred Godsal described putting up hay on his ranch during 1882 and expressed the prevailing attitude among ranchers very well. He reported to his family at home in Iscoyd Park in Shropshire close to the Welsh border:

> We have been hay making today.... We cut it straight off the prairie.... We use it in winter to feed the horses when they come in off a journey perhaps or if one is being left in the stable to catch the others and so forth, otherwise they forage for themselves like cattle. Cattle are never housed in any way (how could you house 1000s?) or fed at all, but a store of hay is always kept in case of emergency as it is easily got and an extra hard winter might come when it would be useful. But if it can possibly be avoided cattle should not take hay in the winter or they will not hunt for themselves but hang around the rick and die.[128]

3.20 This haying scene taken in the 1890s shows wild hay being gathered using a minimum of equip-ment: a mower, a rake, and a hay box mounted on a wagon. Glenbow, NA237-14

While small ranchers like Godsal and Ings might aim at putting up thirty to fifty tons of hay, the larger outfits required substantial amounts. Frank White, treasurer for the Cochrane Ranch, noted in November 1882:

> Mr. Cochrane thinks that at least 1,000 tons of hay should be cut next summer so that there may be plenty for calves and foals, and that all calves should be taken from cows and fed through the winter.[129]

By 1884, this objective had been partially realized on the Cochrane Ranch along the Belly River, where contracts had been arranged to provide five hundred tons of hay.[130] It seems likely that the severe

3.21 Edward F. Hagell, "Handout better than Willow Brush." From the first, wild hay was put up and fed to valuable imported bulls, saddle horses, and weak cows and calves. In the picture a six-horse team pulls a sledge to distribute hay in deep snow. Glenbow, Art Collection 4549

winter of 1886–87, and particularly the publicity and commentary which followed it, encouraged an expansion of hay production.[131] As we have seen, there was a move toward controlling the time of calving and the weaning of calves so that the mothers could recover before the coldest part of the winter. This required much more hay to feed the calves and weak cows. The size and number of haystacks became a measure of modernity and progressiveness among foothills ranches. Before the turn of the century the Bar U was already contracting with enterprising homesteaders to come and put up wild hay on the ranch.[132] When the Bar U was sold to Lane in 1902, there were a thousand tons of hay on hand left over from the previous season.[133] This reserve could only have been built up with a considerable deployment of labour and capital. Haying was already an important part of the seasonal round at the Bar U.

September, October and November: Fall Roundup and Weaning

The fall roundup on the big foothills ranches during the last decade of the nineteenth century was a more local affair than the general roundup of the spring. During the summer riders would have kept an eye on the company's herds and would have had a good idea as to exactly where the bulk of the herd was located. Starting in late September or early October, the herd was gathered and worked to achieve a number of objectives. One of these was to cut out three- and four-year-old steers and drive them to a convenient shipping point on the railway to begin their journey to market. Another reason for the fall roundup was to brand calves that had not been so identified in the spring. After branding all the calves were separated from their mothers and moved to a holding ground where they could be winter fed.[134] The Bar U calf camp was at the site of the old Mount Head Ranch and was run for many years by Henry Minesinger. The cows would be herded onto their winter range, and the open fall weather would give them a chance to recover and to put on weight and condition before the full onslaught of winter.

December, January and February: Winter Feeding

The Bar U fed its imported bulls during the winter of 1882–83, but Professor Sheldon, visiting in 1884, expressed the opinion that it would be to their advantage to feed calves and weak cows too.[135] William Pearce, who spent so much of his time touring the range and talking to ranchers, was of the opinion that the winter of 1886–87 encouraged ranchers to put up more hay and to wean and feed calves.[136] In 1888, the Bar U fed seven hundred calves and two hundred weak cows and reduced their winter losses to almost nothing.[137] The aim of winter feeding was to bring the vulnerable elements of the herd through the winter with minimal losses. No attempt was made at this time to feed mature cattle so that they could be marketed in the spring. When the ranch was sold to Lane in 1902, it was reported to be feeding eighteen hundred head of cattle and calves at three camps, and expected to finish the winter with a thousand tons of hay on hand.[138] For more than a century the rolling lease lands to the north and west of the Bar U headquarters have been regarded as some of the best winter range anywhere. Sheltered valleys and brush provide shelter, and chinook winds clear the rounded hills and expose cured rough fescue grass for grazing. Allen Baker, who owned the Bar U from 1950 to 1977, explained that for twenty-three consecutive years he never had to feed the majority of his herd.[139] Given this extraordinary natural endowment it did not make economic sense to try and feed the whole herd. The number of riders at the Bar U was reduced during the winter. Those retained would be the most experienced and responsible men. Winter riding was unpleasant and could be dangerous, but it was vital to keep springs open and to identify, cut out, and bring in weak cattle that desperately needed feeding. Fred Godsal described both the purpose of winter riding and the manner in which he prepared himself for the ordeal:

> Tomorrow I shall probably be going out again on what I call Cow Relief Expeditions in which most of my time is spent viz. riding out to look for any cows that require feeding, or calves weaning or any other help such as driving them out of bad places and so forth. I have a good buffalo skin cap, made to my own order, which covers my ears and has a strip across the nose, then I have a woollen comforter round my neck, then a buffalo coat, buckskin mitts with knitted mitts inside them on my hands, or else more generally lined sheepskin mitts with a pair of silk gloves

> in my pocket in case of necessity, then buffalo trousers, and on my feet a pair of woollen socks, over them a pair of thick lined German socks and over these a pair of buffalo moccasins, hair side in and wrapped over these and the bottoms of my buffalo trousers to keep the wind from going up the legs, I wrap horse bandages which I got from home.[140]

Many of the hopes and dreams shared by our group of eastern Canadian entrepreneurs as they made their way up the Missouri River in 1881 had been realized by the turn of the century. In particular, Fred Stimson had acted with determination and a sense of urgency to make sure that the North West Cattle Company was among the first large cattle companies to stock the Canadian range. He was able to obtain a share of rich government supply contracts, and this meant that investors got a timely and generous return on their risk capital. Although the arithmetic of those who promoted the "beef bonanza" was creative rather than realistic, the foundation of their case was solid. Consistent profits could be achieved as long as grass was unlimited and almost all capital was tied up in cattle and not in improvements or labour. In spite of harsh winters, fires, wolves and rustlers, cow herds grew by half each year.

An attractive and functional headquarters had been chosen and developed on Pekisko Creek. It housed a community of more than a dozen people and was already a regional centre for both the police and the post office. Shrewd choices had been made in the appointment of senior personnel, and the growing reputation of the ranch meant that it was possible to attract and keep the best riders, cooks, and chore boys available. Fred Stimson had become a man of some importance both in the ranching community and in the growing city of Calgary. He represented his peers on his business trips to Ottawa and Montreal, and his counsel was sought by men like A.M. Burgess, the Deputy Minister of the Interior, and William Pearce, whose responsibility it was to shape grazing policy. Stimson relished his membership of the Ranchmen's Club and was a friend of judges and legislators. With Fred as host, the Bar U had already assumed its role as a showplace, and received a steady flow of prestigious visitors. The termination of the closed leases in 1893 might have been a severe blow, but the government had been considerate both in the timing of the changes and the compensations they offered. The North West Cattle Company had acquired a substantial package of deeded land and this had been fenced. This, in turn, enabled progressive innovations in herd management to be introduced. There were some looming issues. The Bar U still exerted de facto control over its vital winter range to the west of the headquarters, but it had no legal basis for so doing. To the east, too, the custom of pushing cattle across the Highwood and holding the herd there all summer was being challenged by settlers. It was clear in the long run that farm settlement would occur along the line of the railway between Calgary and Fort Macleod, and that this would limit the Bar U's access to its summer range. But in 1900 this threat seemed far removed.

The sale of the Bar U Ranch in 1902 should in no way be taken as a sign of failure or as an indication that the ranch was not profitable. It was purely a business decision which reflected changing attitudes and personalities on the board in Montreal rather than any problems at the ranch. To establish a new business enterprise and to run it for twenty years was a remarkable achievement. To recoup nearly a quarter of a million dollars from a company which had an authorized capital of $150,000, and which had paid handsome dividends over the years, was surely a triumph rather than a sign of failure.[141] The question facing the directors was not whether the ranch would continue to be a safe and profitable venture, but rather how did it fit into their larger investment plans? Was it the *most* profitable investment available in a growing and diversifying economy? The board chose to "cash in their chips"; they did not slink from the table as losers. Their decision meant that the fate of the Bar U would be guided by a very different owner for the next two decades of the twentieth century.

4

George Lane Buys a Ranch

"History humanizes us by giving names, faces and texture to our physical and mental landscapes."[1] – James P. Ronda

The story of how George Lane travelled to Montreal in 1902 to confront the Allans and buy the Bar U has a central place in the mythology of the Canadian range. The tall, gangling Lane in his rumpled suit is pictured as a naive man from the boondocks. Across the dining table in the St. James Club were some of the most urbane and best established financiers in the country. Rankin's version of the story goes like this:

> The Allans asked $250,000, Lane offered $220,000.... They fenced and sparred and bluffed for a good half hour, but eventually the Allans, upon repeated offers from their guest, looked at one another, laughed, and said, "All right George, you can have it at that, but put $50,000 down and the balance upon signing the deeds." They sat back in their chairs and motioned the waiter to fill up the glasses; they smiled and nodded at one another good humouredly, they did not think Lane had the money. But they reckoned without their guest; they forgot what manner of man he was. Then and there, he "just dug down into his jeans" and came up with the $50,000.[2]

The author then invents an imagined report on the event by Lane, which is rendered in Rankin's version of "western speak":

> "It has allus been my rule in life," said Mr. Lane afterwards, "to be prepared for any emergency, so I went to this intervu' with my rope unslung 'case some steer w'd stampede. I was prepared to do business. When I suggested so much down and the balance on signing the deeds, I know'd it w'd take two weeks to get the deeds ready, and in that time, with the ranch as my hitching post, I could easily raise the money. You can't bluff a cowboy – men trained to dominate all living things around 'em."[3]

This invention is at odds with my vision of Lane, for Lane was a naturally taciturn man and would have been especially reluctant to discuss the details of business matters. Moreover, self-promotion, or telling stories which put the teller in a good light, was too close to boasting to have been acceptable behaviour to him. In fact, it was more than a decade since George Lane had been "only a cowboy," and he had spent those years shrewdly ploughing his profits as a cattle buyer and "speculator" into ownership of two not inconsiderable ranches, both well stocked with cattle and horses.[4] Moreover, he knew all the leading figures in the cattle business of western Canada and had earned their trust and respect. Since the early 1890s he had been working closely with James T. Gordon, and had become one of the principal buyers for the Winnipeg firm of Gordon, Ironside and Fares. He combined a thorough knowledge of the cattle business from the bottom

up with "a striking air of cosmopolitanism...."[5] When he headed east to Montreal, Lane knew that Gordon, Ironside and Fares were seeking to integrate their cattle shipping operations by acquiring ranches. In 1896, they had established the Two Bar Ranch in the Wintering Hills, and not long afterward they had bought the SC Ranch near the forks of the Red Deer and the South Saskatchewan Rivers. He thought that they might be prepared to come in as silent partners on his bid to buy the Bar U.

Lane also knew that the death of Andrew Allan in 1901 would, in all probability, mean the sale of the Bar U and the winding up of the North West Cattle Company. It had been Sir Hugh Allan who had seen the potential of the trade in live cattle with Great Britain during the 1870s. As his ships began to carry increasing numbers of Ontario stock down the St. Lawrence and across the Atlantic, so the potential profits of running cattle on the open range in western Canada encouraged him to back the North West Cattle Company. Although this was a modest investment in the context of his large and diverse portfolio, he

4.1 The roundup camp on Willow Creek, 1895. George Lane sits at the entrance to the tent on the right conferring with Walter Wake, the foreman of the Oxley Ranch. At this time Lane owned the Flying E Ranch on Willow Creek. Several of the cowboys in the photograph worked for Lane after he bought the Bar U in 1902. From left to right: David Baird, George Winder, Mike Herman, Charlie Haines, Jim Johnson, George McDonald, Duncan MacIntosh (Captain of the roundup), George Lane, Walter Wake, the last two unknown. Glenbow, NA 118-3

was personally interested in the venture and visited western Canada just a few months before his death in 1882.[6] He was succeeded as chairman of the board by his younger brother Andrew Allan, who was an "unabashed admirer" of his brother's business acumen, and would have been extremely reluctant to abandon any enterprise initiated by Hugh.[7] So in the early 1890s, when the government withdrew the closed leases and the future of the NWCC was discussed, Andrew Allan had been prepared to overrule the naysayers. He had sided with Stimson and authorized the purchase of the full amount of land to which the company was entitled.[8] Now, in 1902, the economic landscape had changed. The Allans were firmly entrenched in the growth sectors of the Canadian economy; transportation, banking, communications, manufacturing, and insurance. A modest holding in primary production must have seemed increasingly anachronistic. Thus it was not long after Andrew Allan's death, in June 1901, that the new chairman, Hugh Montagu Allan, whose primary interests were in banking and the deployment of capital, considered once again disposing of the NWCC.[9]

It seems extremely unlikely that Lane would have proceeded to Montreal to make an offer to the Allans without stopping in Winnipeg en route to confer with Gordon, Ironside and Fares. The Winnipeg exporting firm was riding a tide of prosperity as trade in live cattle to the United Kingdom reached unprecedented volumes and their share of that trade increased. Their gross annual turnover in 1902 was about $6 million.[10] The three partners had incorporated in 1901, with a capital of $1 million. The newly established company's purpose, outlined in letters patent, reflected its expanding interests. It was involved in the importing and exporting, shipping and raising of stock; the manufacturing, producing, and packing of livestock; cold storage and related business.[11] The company was also building an abattoir in Winnipeg adjacent to the CPR yards. Their willingness to join Lane in the purchase of the largest and most prestigious ranch in the Alberta foothills was just another facet of an expansion program which was redeploying their profits into long-term investments. In 1902, they also purchased a two-hundred-thousand-acre ranch in Chihuahua, Mexico.[12] They were poised to consider Lane's scheme with interest.[13]

Available evidence suggests that Lane cut an extremely favourable deal with the Allans when he purchased the Bar U Ranch.[14] There were two major components to the value of the ranch: the land and the stock. Of these the stock accounted for more than 80 per cent of the total. This would have been true for any of the major outfits on the Canadian range during this period. The idea of tying up capital in land was contrary to the basic tenets of this method of stock rearing.[15] Indeed, Stimson had only been forced by circumstances to purchase some deeded land during the last years of the nineteenth century.[16] The cattle herd numbered about eight thousand head, and at the heart of the herd were the three thousand breeding cows. They were serviced by three hundred bulls and produced between fifteen hundred and eighteen hundred calves each season. Some of the best young heifers were retained to replace old or dry cows, and there were some three thousand head of two-, three- and four-year-old steers.[17] Each component of the herd would have had its own value, from the purebred

4.2 George Lane in his forties, chopping wood at a camp on the range. Parks Canada, Mildred Lane Day Estate Col-lection, 95.01.01.09

bulls, purchased for thousands of dollars, to the calves which were 'thrown in' for free in most deals. Similarly, a four-year-old steer might be valued at $70, while a yearling might only be worth $20. An overall average price of $30 a head seems a realistic estimate, and yields a total value of $240,000 for the herd.[18] The five hundred head of horses also varied in quality and value, from well-known "blood" animals like Fallback Roan and Circle Bar Sorrel, to working cow ponies. An average of $40 per head provides a total of $20,000. The eighteen thousand acres of deeded land had been acquired only a few years before at $1.25 to $1.50 an acre from the Calgary and Edmonton Railway Company. By 1902, the demand for land was rising, but the Bar U lands were at some distance from the railway. A conservative estimate of the land value would incorporate a modest appreciation to $2 an acre, giving a total value of $36,000.[19] Later inventories suggest that "improvements," whether buildings or fences, were considered of very little additional value. Taken together, these estimates give a total value of $296,000 for the ranch.[20]

The press reported that the price "involved a cash turnover of over $200,000."[21] Rankin, from his interviews with Lane, mentions the figure of $220,000, a sum also mentioned in court documents.[22] The total assets of the North West Cattle Company in April 1902 were estimated by the treasurer, E.W. Riley, as $233,000.[23] This figure would have included cash and office property on the books as well as the ranch. Stimson, the resident manager of the ranch, who was perhaps in the best place to evaluate its worth, thought that the ranch had been "given away" to Lane. In his opinion it was worth $100,000 more than Lane had paid for it.[24] It seems that George Lane secured a bargain. He achieved this by taking negotiations to the Allans in Montreal, and offering them a means by which they could immediately and cleanly withdraw from an investment which they had held for twenty years, but which did not fit with their vision of the future.[25] Lane's partners, Gordon, Ironside and Fares, must have congratulated him.

Rankin portrayed Lane as a folksy American cowboy, complete with "unwavering steel-blue eyes," but Lane's actions suggest a quite different persona. In particular, Lane exhibited an unusual ability to step backward from the consuming details of the immediate situation in order to evaluate the big picture and divine the likely course of events. Secondly, Lane was ambitious in the sense that he was driven to explore the next avenue of development, and the next, and the next.... He was not content, as many would have been, to enjoy running his Willow Creek Ranch and raising his young family. Instead, he took the huge risk of buying the Bar U. A letter from the Union Bank makes it clear that he was already heavily in debt. It is worth quoting:

> Union Bank of Canada, Quebec,
> January 18, 1902.
>
> Dear Sir, As requested, I hand you herewith a letter of guarantee to the President of the North West Cattle Company Limited, undertaking to pay the sum of $85,000 on your account, the first of June next, and may mention that your present indebtedness of $62,500 is to be reduced by you before the 31st May next by the sum of $25,000.
>
> In addition to the payment of $85,000 on the first of June next, we will advance you the sum of $12,500 on that date to pay your indebtedness to Gordon and Ironside. Your account will then show advances from us amounting to $135,000 after these payments have been made. Your share of the expenses of operating the ranch between now and the first of October next we understand will amount to $8,000.00, and that you undertake to reduce your indebtedness during the summer by the sum of $68,000 or a net reduction after providing for expenses and management of $60,000, bringing your debt to us on the first of October next to the sum of $75,000.
>
> Yours truly,
> E.E. Webb,
> General Manager[26]

Lane had the capacity to see opportunities and then the drive and determination to exploit them. He was not motivated by a desire for personal wealth but rather by his vision of how things ought to unfold; he and his family seem to have adopted a modest, middle-class way of living, He was a booster of western Canada in the best sense of the phrase. He had a gift for charismatic leadership coupled with an uncanny touch for the practical. Finally, if Lane was much more than a cowboy, he was also much more than an American. His wife noted that he was "very enthusiastic about the country," like many new Canadians.[27] He certainly put down roots in Canada in short order. Only a year after his arrival, he married Elizabeth Sexsmith, whose family was one of the first to settle near High River. Soon after that he became a Canadian citizen, and thereafter he was careful to buy Canadian goods and to insist that his achievements be attribute to the boundless opportunities available in the Dominion of Canada.[28] This identification with his country of adoption did not go unnoticed. One of his obituaries noted, "George Lane was a good citizen too. Born out of the country, he came into it, lived his life and died as true a Canadian as the native born. He believed in Alberta and Canada and stood by both whenever they were under criticism or misrepresentation."[29] But Lane was a mature man of twenty-eight when he came to Alberta, and his upbringing had provided him with a wealth of skills and experience which would continue to shape his attitudes and actions in the years ahead.

Who Was This Man Lane?

George Lane was born in Boonville, Indiana, in 1856.[30] His parents were Joseph William Lane and Julia Pidgeon Lane. His mother's people were originally from Pennsylvania and were of Quaker stock, while the Lanes originated in Kentucky and Virginia. George was the oldest of five children, but two of his brothers suffered accidental deaths. The family drifted westward, first to Kansas, and then, following the Overland Trail, to Nebraska, Wyoming and finally Montana. The discovery of gold at Alder Gulch in 1863 initiated the largest placer gold mining boom in Montana's history and attracted some ten thousand people to Virginia City and the surrounding area.[31] Joseph Lane was one of this hopeful crowd, and he was joined later by his sixteen-year-old son, George. There is no mention in the family memoirs of them actually filing a claim or working the gravels, and it seems much more likely that father and son worked to provide services to the miners in a store or livery stable.

In 1874, when he was eighteen, George Lane began working for the U S Army. He was probably recruited as a stable boy, but he soon graduated to dispatch riding and scouting. In a press interview later in his life, Lane recalled this phase of his young adulthood. He remarked:

> Do you know what was the best thing I ever did? Well, it was when I gave my mother $1,400 which I had earned carrying messages for General Miles. I was only twenty then and $1,400 was a heap of money.[32]

It certainly was a considerable sum of money and must have represented the reward of several years of service. Lane was therefore involved in the last chapter of the conflict between the US Army and the Indians of the northern plains. It had started when General George Armstrong Custer led a punitive expedition into the Black Hills in 1874; climaxed with the death of Custer, along with about 260 of his 7th Cavalry, at the Little Big Horn in June 1876; and ended with Colonel Nelson A. Miles' merciless winter campaign which resulted in the surrender of Crazy Horse and the flight of Sitting Bull to Canada in 1877.[33]

When Lane left the military he served a seven-year apprenticeship on some of the best-run ranches in Montana. The attitudes and perceptions with which he came into contact during this period affected his behaviour later in life. As we have already seen, the methods of these "Midwestern" cattlemen were much more careful and intensive than those employed on the southern plains.[34] Lane would have been exposed to these techniques when he worked for Conrad Kohrs on his ranch in the Deer Lodge Valley. Kohrs was born to a middle-class German family and had run away to sea at the age of fifteen. Having tried his luck in the Californian gold fields, he was drawn to Alder Gulch in 1863. However, it soon became clear to him that there was more money to be made from feeding the miners than by wielding a shovel. He set up a butcher's shop, and started ranching in a small way to provide a steady flow of beef to his shops.[35] In 1866, Kohrs bought the well-established ranch of a Canadian, Johnny Grant, who had decided to retire to the St. Lawrence River. This spread became his home place, and he brought in registered Shorthorns and Herefords to upgrade the western Durhams which made up the ranch herds.[36]

It was not long before Kohrs' ranching interests expanded well beyond the Deer Lodge area. Soon he was running herds on the open range in four states and even across the Canadian border. Thousands of head of his fat cattle, surplus to local needs, were driven to railheads, first at Ogallala and later at Cheyenne. Lane would have moved from outfit to outfit, and gained experience not only with the intensive and careful methods used at the home place, but also of the more extensive techniques used on the open range and on long cattle drives.[37] In fact Kohrs' ranching operations during the 1870s and 1880s appear to have been replicated by George Lane in Canada around the turn of the century. Both men had breeding herds in well watered and highly productive locations in the foothills, and range herds of much more mixed quality, and subject to far greater risks, fattening on the open range to the east. Moreover, Kohrs not only maintained his butcher's shops, he also diversified his operations by investing in mining, real estate, and water rights. Lane, as we shall see, did much the same.

A number of factors conspired to draw Lane's attention northward to the Canadian range during the early 1880s. Even while enthusiasts were promoting the "beef bonanza" and emphasizing that there was no limit to potential expansion, those already working the ranges were becoming apprehensive about overcrowding and overgrazing. As early as 1883, the editor of the *Rocky Mountain Husbandman* pointed out that, "The increase of herds and continual driving in of herds ... begins to tell but too plainly that the overcrowding of our ranges is only a question of a very few years, unless some plan is hit upon to prevent it."[38] The speed and scale of the transformation from abundant underused free grass to competition, overcrowding, and overgrazing made a deep impression on men like Granville Stuart who had been in the Territory for many years.[39] At the same time the potential of the Canadian range was being discussed. Several major cattle companies, like Kohrs', were shipping beef north for the Indians and the North West Mounted Police; others were making spectacular deals with Canadian entrepreneurs to provide foundation herds for new Canadian ranches. For example, Frank Strong drove the Cochrane herd to the Bow Valley in 1882. On his return he told Lane that he was impressed with what he had seen and that he intended to return to Canada to live.[40] The following year, Lane was responsible for gathering and driving eighteen hundred head of beef cattle from the Sun River, Montana, up to the Belly River in Canada. This gave him a chance to evaluate both the state of the grass to the north of the border and the political and economic climate. The excited talk in Fort Macleod was all of the new lease legislation and the coming flood of capital investment from eastern Canada and Great Britain. Lane returned to the Sun River in late summer, but the prospects in Canada were in the forefront of his mind.

In the spring of 1884, Senator Cochrane and Sir Andrew Allan wrote to the Sun River Stock Association of Montana asking for help in finding experienced stockmen to serve as foremen on their newly established ranches.[41] Robert Ford was president of the Association and lost no time in contacting Lane.

> "George, you have been up in that Canadian country twice. You know the situation. Now, would you like to go back as foreman for a company?" This seemed an opportunity which suited George down to the ground. "What's the wages?" he asked guardedly. "If you make good, my boy, you won't have to worry about wages," was the quick reply. He then showed his correspondence and telegrams, which indicated that he wanted two more men to go up, and told him who he had in view. To cut a long story short, George Lane accepted and thus commenced his Canadian career.[42]

Lane headed north with Jim Dunlap, who was to join the Cochrane outfit. He met the 1884 roundup close to the present site of Lethbridge. Fred Stimson, the manager of the Bar U, had accompanied his crew southward on this occasion with the express purpose of meeting Lane and engaging him as foreman.

We know more of George Lane's life from his actions and achievements than we know of the man himself. He was, as far as we know, not a diarist. If he was a reflective man, he kept his ruminations to himself. His wife recounts only the stark details of his birth and upbringing. His contemporaries and peers wrote tantalizingly little about the man, until his passing produced any number of eulogies. Lane's physical presence was striking and earned him the undivided attention of unruly cowpuncher and self-important politician alike. William Sexsmith catches this quality well when he writes:

> A man considerably over six feet tall … he moved in a loose jointed yet positive manner which left no doubt as to the virility and aggressiveness of his nature. Brilliant blue penetrating eyes which concentrated on who[m]ever he at the moment was talking to, left an impression of searching behind the words being said to ascertain any deviousness or hidden meaning.[43]

Fred Ings echoes several of these images when he described his friend and colleague: "He was a tall gaunt man with long arms. His leonine head was covered with a shock of hair. His face was clean and clever with quiet grey-blue eyes."[44] Ritchie claimed that Lane possessed "The long, lean enduring physique of a mustang."[45] Journalist Norman Rankin, who visited Lane on several occasions and had some knowledge of the cattle business, presents a vision of Lane which could have graced the pages of a dime western! But it helps us to look at Lane's face in static photographs in a new way, and suggests some possible reasons why he had such an impact on those he came in contact with:

> He is a picturesque figure; and attractive personality; a double-barreled, back action, high pressure, electrical dynamo at full speed; a living example of perpetual motion, mental, physical and corporal; a six-foot giant with tow-coloured hair and the smile of a Sister of Charity.[46]

Lane was "a strict autocrat with all the boys. His word was law and no one disputed it," wrote A.E. Cross as he recalled the general roundup of 1887, at which Lane was overall captain.[47] He knew exactly what he wanted and expected his employees and his colleagues to share his vision and drive. Higinbotham puckishly labelled George "the Mussolini of cowdom."[48] He was writing in 1933, before all the qualities of the Fascist dictator had become apparent, and no doubt he wanted to emphasize Lane's no-nonsense drive and efficiency.

In spite of his somewhat imperious manner, Lane earned a reputation as an excellent employer. "His men simply swear by him, one reason for this unusual loyalty may be, that he never swears at them."[49] Unfortunately, those who worked for him have not left a record. However, it is true that a cadre of key men "remained with him for 10 and 20 years, and upward. That in itself told quite a story."[50] One of the qualities which endeared Lane to his employees was the way in which he delegated responsibility. Once you had earned his trust, he would give you almost unlimited freedom of action. The story is told that Lane sent a young man, who had only been with him a year or two, to France to purchase some Percheron stallions.

> The instructions were, accompanied by a $25,000 letter of credit, "Treat yourself well. Tip well. Go over to London and get acquainted. You will have to spend money to do it. When you have closed the deal take a couple of weeks for sightseeing in France. Enjoy yourself and don't take things too hard."[51]

Lane never took his employees for granted and was aware of the debt he owed them. For example, in 1912 Lane organized a reunion on the Bar U for all the men who had worked for him on his various outfits. He advertised in the local paper and got a good response.[52]

A letter, written from the roundup in 1885 by a young Englishman to his parents in Sussex, England, presents a rare glimpse of Lane. Ted Hills had had little difficulty getting a job on the Bar U wagon because so many men were engaged in military action to the north.[53] There were few of the regular riders left to choose from. On 1 September, after two months on the roundup, Hills picked up his mail again and hastened to reply:

> I was very glad to get letters from you and Elf today at Fish Creek post office dated August 9th. While I remember I will answer the questions you asked me in your letter re the Round-up. The first part of the time I used to sleep with Lane [Stimson's foreman] and Paul. Paul was a very nice sort of fellow but I can scarcely say the same for Lane. The tent was a wretchedly leaky one overhead and had no fourth side to it at all and as I had no second waterproof sheet to

> throw over my blankets, they used to get wet of a bad night. After the first fortnight or so I used to sleep in a covered wagon by myself which was more comfortable.[54]

Why did Ted Hills express such an aversion to Lane? The probable answer is that the son of a British Admiral was still carrying a lot of "class baggage." In an earlier letter written just before the roundup began, he wrote: "I know I shall like the work; of course there are a few gentlemen but a great many of the men are from Montana and a pretty rough lot, but I think I can look after myself pretty well among them."[55] Hills had been working for a man named Wilberforce who was also British and would have shared his unconscious sensibilities. He had met a number of young men from his old public school, Uppingham, and had remarked on the number of Englishmen in the west. Mr. Barter, the manager of the Quorn, whom he had approached for a job, had been suitably deferential, even though Hills was a greenhorn. Similarly, Hills, because of his background, was on friendly terms of equality with ranchers like the Skrine brothers, "Billy" Cochrane and A.E. Cross. Lane was a different case entirely. He valued a man according to his demonstrated ability and experience, and ignored completely both how he spoke and his background. He was "devoid of artificiality, brusque sometimes to the point of rudeness."[56] As a contemporary remembered, "there was absolutely no side or swank to Lane, he was always natural, and breezy even to the point of brusqueness, but beneath his rough exterior he had a kind heart."[57] There was a gulf of culture and experience dividing the foreman from the young Englishman, which made communication difficult. It should be noted, however, that Hills learned to appreciate Lane's leadership qualities and was a regular hand on the Bar U or Willow Creek wagon until he left the district about 1906.

I have been at some pains to suggest that we should remember George Lane as much more than a cowboy. Nevertheless, Lane did possess in abundance all the varied skills of the range rider. The fact that he could out-ride, out-rope and out-shoot most of his employees did much to establish his effortless authority.[58] He was in the saddle working cattle from about 1877, when he left the military, until 1889, when a bout of typhoid fever changed the course of his career. Fred Ings told a revealing story about Lane's arrival at the roundup in 1884.

> The previous year the Bar U bought a bunch of BX horses in BC. It was 7U Brown who brought them into this country. Among these horses was one called Fallback Roan, an outstandingly handsome horse. They had another horse, Circle Bar Sorrel, for some time. These were the first horses that Lane rode on roundup. They were by no means gentle, in fact they were only half broke. It was customary when a new man came into an outfit to give him something tricky in the way of horseflesh. Nigger John and I went to give Lane a hand, but he didn't need any help. He was quite equal to the task.[59]

Lane's long legs gripped a horse like a clothespin, while his upper body strength imposed his will on his mount. He also had a reputation as a first-class shot, both with handgun, rifle and shotgun. This talent was illustrated by an incident that occurred in the fall of 1886.

Lane's first child, Lillian, was born on 7 October 1886, at the home of Mr. Nickle in Calgary. It was riding back to the Bar U from a visit with his new daughter that Lane had a run-in with a pack of wolves. The encounter became the subject of a painting by the American cowboy artist, Charles M. Russell. It confirmed the perception of Lane as the quintessential cowboy with blazing six-guns! [Plate 3, colour section] Lillian Lane summarized the family recollections concerning this incident in a footnote to her mother's reminiscences. She explained:

> In October 1886 Father rode to Calgary where my mother was expecting me. After I was born he rode back. There had been a heavy snowfall and it was cold. He [George] had a fine bay horse and wore chaps and a moose hide shirt. About sun down, 12 miles from the Bar U, while riding upwind, he came upon a group of six wolves attacking a weak cow. He was able to ride up to them and shoot some of them with his revolver, one jumped for his leg but missed it and was killed. He shot three or four more and returned the next day with the wagon to pick up their hides. This incident was commemorated by a water colour painting by his friend Charles Russell in 1912 [*sic* 1914]. This painting was given to father and subsequently willed by him to his son Roy.[60]

Roy Lane hung the picture in a place of honour at his ranch on the Bow River.

So many of his contemporaries remembered Lane as a stern man with firm convictions that the repetition made me wonder if he was entirely bereft of a sense of humour. It was with some relief that I came across this brief reference to that essential ingredient in the makeup of a well-rounded individual.

> ... on occasion he could be light hearted and display his cowboy sense of humour. During the summers he enjoyed fishing and picnicking along the creek with his family. Once, when he suspected his daughters' admirers were smoking his cigars, he planted an explosive one to cure the culprit. Unfortunately, it was not one of the admirers, but Lane's brother-in-law Marston Sexsmith, who got the surprise.[61]

It is customary when writing a eulogy to laud the qualities of the departed, but it is also usual to try and get to the essence of the man; to focus on his inner qualities rather than the froth of his temporal achievements. Lane was remembered as a man who had made a real difference, and for his frankness and honesty. He was a man who never forgot a friend or remembered the insults of an enemy.[62] Who could wish for a better epitaph?

George Lane and the Young Napoleon[63]

GEORGE LANE AND HIS PARTNERS, GORDON, IRONSIDE AND FARES, BOUGHT THE BAR U RANCH DURING A PERIOD OF optimism. The dreams of a western-based stock-rearing industry furnishing beef to the burgeoning industrial population of Great Britain had, in large measure, come to fruition. Beef prices were high, and live cattle had risen to become a major Canadian trade staple.[64] Nevertheless, the excellent short-term prospects did little to mask looming uncertainties. One of the first things Lane had to do having acquired the Bar U was to try and ensure its territorial integrity. He had eight thousand head of cattle to provide for and a legal hold on only eighteen thousand acres of deeded land. The grazing lease system, which had done so much to give a semblance of legality and security of tenure to the ranching industry, was in disarray. The original closed leases had been cancelled in 1896. Ranchers had been given the option to take out new leases which were open to homestead entry and subject to cancellation, but most of them saw little value in paying the government for the privilege of grazing cattle on range which could be invaded by settlers at any time. Instead they joined the growing numbers of small ranchers who grazed their herds on the open range without let or hindrance and without paying the government a cent.

Burgess, the Deputy Minister of the Interior, regarded this development with concern, not only because the government was for forfeiting a worthwhile source of income, but also because an important and powerful industry would be operating on Crown lands without any legislative controls. This could, he felt, lead to many of the abuses evident in the United States, especially overgrazing, illegal fencing, and 'range wars.' He went on to suggest an entirely new system based on a per capita charge for every head of stock grazed on the public domain.[65] This proposal was one of Burgess's last acts as Deputy Minister. He was demoted, and his place was taken by James S. Smart, a close political friend of the new minister, Clifford Sifton.[66]

Sifton has earned a vaunted position in western Canadian history as the man who transformed the lethargic bureaucracy of the Department of the Interior and opened the doors to an unprecedented influx of farm settlers.[67] However, Sifton proved indecisive with respect to the administration of grazing lands.[68] For almost a decade the Department of the Interior wavered from one response to another: new initiatives were suggested, uncertainties were deprecated, committees of inquiry were established, but no decisive action was taken. On the one hand, Sifton was not unsympathetic to the ranchers who had built a successful industry on the western grasslands, but, on the other hand, he was reluctant to allow any perceived obstacle to the march of farm settlement. Harassed by petitions from the Western Cattle Growers' Association, by letters from small ranchers, and from homesteaders in grazing country, Sifton was also needled by his fellow Liberal, Frank Oliver, who for years had waged a bitter campaign against the cattlemen. The minister reacted to the uncomfortable position in which he found himself on this front by doing nothing, giving ground first to one party and then to the other.

The ranchers had hoped to maintain their *de facto* control over the range through the continuation and extension of "stock watering reserves." These reserves were designed to prevent continuous settlement along surface streams and rivers, or around important springs. This policy was justified on the grounds that the reserves prevented a minority from controlling the vital water resource, and thus made grazing available to all users. This position was articulated in a letter from Fred Stimson to his newly elected Member of Parliament, Frank Oliver. He referred to a particular spring situated on the headwaters of Mosquito Creek, and continued:

> ... this is one of the most important springs on the winter range. Several thousand cattle wintering and ranging round it in winter, it is situated in a valley which affords shelter and is also a driftway, if this spring is allowed to be settled upon, it will mean the loss of a very large area of the country as it practically controls a great stretch of the range. There are a great many miles of Mosquito Creek, High River, and Little Bow open for settlement that will not in any way interfere with the cattle industry, but the settlement of important springs will deal a very great blow to the cattlemen ... I trust the stockmen will have your support in this matter, it seems an injustice to those who have invested their capital here, that these important watering places should be taken up to the detriment of the stock industry.[69]

As long as Burgess and Pearce were handling the affairs of the Department of the Interior, men like Stimson could expect a sympathetic hearing. After the Liberals were elected in 1896, the power of both these stalwart supporters of the ranchers was severely curtailed, and the stock-watering reserves were auctioned off. In Frank Oliver's view, the bewildered incoming settlers were victimized because they were unable to find out precisely what was reserved and what was not.[70] He felt that the big ranchers had the country as "buttoned up" under the water reserve system as they had formerly under the closed leases, and he was determined to break their hold on the range.[71]

The location of the Bar U in the foothills, some miles to the west of the main axis of north-south movement along the "Whoop-Up Trail" and later of the railway, meant that it had not been a primary focus for friction between the ranchers and incoming settlers. The Cochrane leases along the Bow Valley west of Calgary had witnessed angry clashes during the 1880s, while the inflexible defence by Duncan McEachran of his Walrond leases precipitated violence on the Upper Oldman River during the 1890s.[72] However, as settlement increased westward from High River along the Highwood River, so the customary behaviour of the ranchers who made up the "Mosquito Creek Roundup District" was challenged. Sixty-six settlers from the region signed Robert Findlay's petition to Frank Oliver in which Findlay detailed the several ways that cattlemen were undertaking concerted action to drive settlers from their newly acquired lands.[73] Findlay explained that it had been:

> ... the practice of the cattle companies and large individual cattle owners living south of High River: that it has been their custom annually for the last ten years or more to gather their cattle, some 12,000 head (this is a small estimate) and drive them from the south to the north side of High River and to hold them there all summer

> by placing riders along High River to keep them from going south.[74]

This meant that the grass close to the settlers' holdings was eaten out and they had no chance to cut wild hay. If they attempted to fence their hay fields, the wild range herds would destroy the fences and lure the settlers' cattle away. If these beasts were unbranded and were caught up by the roundup, they were likely to be sold by the Stock Association as mavericks. In the short term, these grievances seem to have resulted in no response from the government; indeed the ranchers taunted the settlers by asking what had happened to the "Findlay Bill."[75] However, the balance of power was changing and the status quo of the previous twenty years was about to be altered.

The pressure of incoming settlers on the country westward from High River was part of a much wider movement, for during the opening years of the twentieth century the range cattle industry was challenged for the first time with competition for space from commercial agriculture which was sustained by eastern capital and spurred by rapidly evolving technology.[76] The rise in the price of wheat, the extension of railway mileage, the wet cycle of rainfall, advances in dry-farming technology and agricultural mechanization, as well as the energetic policy of the Department of the Interior, all contributed to a positive surge of settlement.[77]

To the ranching industry, the expansion of settlement along the railway between Calgary and Fort Macleod was the most critical. Mere stopping places on the railway, marked only by corrals in 1902, grew rapidly into thriving small towns like Claresholm, Nanton, and Stavely. Superintendent Primrose reported: "The increase in this district has been enormous, to the north from Macleod to Nanton, to the east to Kipp, to the south to the boundary, and west to the Crow's Nest Pass, nearly every available section of land has either been taken up or purchased."[78] A continuous broad band of settlement cut ranchers off from their customary summer grazing pastures. At the same time they had to compete with settlers for the continued use of winter range in the Porcupine Hills and in the valleys of the eastern slopes of the Rocky Mountains.

The dilemma facing the cattlemen was trenchantly summarized by George Lane in a letter to the Commissioner of Dominion Lands, which is worth quoting in full.[79]

> As one of the cattle men of Southern Alberta, I would like to call your attention to the uncertain position we are in at the present time with our herds. For the past two seasons settlers were crowding into the country, and what used to be our summer range is practically all gone now and our cattle are confined to the hills and rough country along the eastern slope of the Rockies.
>
> You can easily understand a man, who has some hundreds or thousands of cattle, and who can neither lease nor purchase land for grazing, being in a very uneasy state of mind as to what the final outcome of his business will be if settlement still continues to crowd in upon him.
>
> There may be good reasons for refusing to lease or sell large tracts of land for grazing purposes where it can reasonably be contended the land

is fit for farming, but the same reasons cannot apply to the high, rough, broken country known as the foot-hills, and I would strongly urge that the Government reserve this latter country as a permanent grazing district. The map enclosed will indicate roughly the country along the east slope of the mountains to which I refer. This country is very rough and broken and stands at an altitude of 4,000 feet and upwards, and it is quite safe to say that in the average season it is subject to frost and snow ten months in the year.

The only arable land in this district is found in valleys lying between high hills and along river bottoms, and it would all be required for growing green feed for wintering cattle. Though settlers might wish to take up locations in the hills I think it would be undesirable to allow them to do so until the men who are now there with their cattle are allowed to acquire a reasonable amount of land, for over-crowding would mean scarcity of grass with the necessary result that every man's cattle would be thin, poor and unfit for even the home beef-market, to say nothing of the export trade which has hitherto been a very important business in that part of the country.

In the past cattle owned in the foot-hills, drifted out on to the plains for the summer and the grass in the hills was thus saved for the winter months, but the large settlement along the Calgary and Edmonton Railway has cut off the summer range and the cattle are now practically confined to the hills all the year round.

Further, homesteading in the hills, the only range that is left now, will simply mean the destruction of cattle ranching in that part of the country.

I am enclosing a fairly accurate list of the cattlemen in the district to which I refer, and approximately the number of cattle and amount of land owned by them individually, and from this list you can easily see very few of these men are in any kind of position to reasonably protect their stock. It might be of interest to you to know that during the past two seasons nearly one third of all the beef produced in the Territories was grown in the district to which I refer. Now, for the protection of this important industry I would suggest that a reservation from homestead entry or further settlement for the present time, be made along the east slope of the mountains in such a way as to include in the reservation all that portion of the country which I have described as rough and broken and unfit for farming purposes, and roughly indicated on the enclosed map, and that stockmen who are now bona fide settlers within that district be allowed to secure by closed lease, for a period of twenty-one years at a reasonable rental, a quantity of land proportionate to the number of cattle now owned by them.

In order to prevent the old cry of "The big man squeezing out the small man" I would suggest that small ranchers in this district be allowed first to select the land to which they may be entitled, such land to be taken as nearly as possible in the immediate vicinity in their present locations, and that after this, the larger stockmen make their selection out of the remaining lands. If this

4.3 George Lane in his fifties, dressed in his habitual dark three-piece suit. Glenbow, NA 2520-68

> principle be adopted I do not see how any hardships can be done to any one either in or out of the reserved district.
>
> I hope the above pro-position will meet with your early and favourable consideration.[80]

Surprisingly, Lane's suggestions were heartily endorsed by Frank Oliver, who was usually a vociferous opponent of the cattlemen. His letter to Sifton makes it clear why, on this occasion, he felt impelled to support the suggestions of "my friend Mr. Lane of Calgary."[81] He was consistent in his support of those actually in occupation of the land, whether they be ranchers, as in this case, or farm settlers. He pointed out that the southern country had been occupied for over twenty years and that it was evident that some parts were suitable for grain growing while other parts certainly were not. Lane's proposals encompassed land which "is unquestionably beyond the limits of successful agriculture. It is grazing country, and only valuable as such." But Oliver stressed more than once that he could only support the proposals as a matter of general policy, not as a special case. On this important matter of principle he was to be disappointed.

Lane's letter, with its complete lack of self-serving polemic and its willingness to present both sides of the question, must have impressed Sifton. He had prevaricated long enough. His own department had completed a comprehensive review of how grazing lands were administered around the world.[82] The conclusion reached was that the only way to avoid abuse of the grasslands of western Canada was to turn them over to individuals who would have proprietary rights and would therefore look after them. A three-tier lease system was proposed which went a long way to meet the demands of the cattlemen. Long-term leases would be available for land which was proven to be unfit for agriculture.[83] However, the ranchers had little time to celebrate. By an odd quirk of fate, Clifford Sifton tendered his resignation to Prime Minister Sir Wilfrid Laurier on 27 February 1905, the very same day that the detailed provisions of the new grazing lease regulations were published.[84] He was succeeded as Minister of the Interior by Frank Oliver, and, by the end of the summer, Sifton's regulations had been replaced by provisions much less favourable to the cattlemen.[85]

In June 1905, G.W. Ryley, the secretary of the Department of the Interior, wrote a memorandum to acquaint the new minister with the situation with regard to grazing leases. In it he reviewed the history of the lease system and detailed the recent changes. In particular he drew attention to the difference between the open lease, which allowed for homestead entry, and the closed leases, which did not. At the very end of his letter – almost as a postscript – he wrote:

> I may add that the following closed leases have been issued which do not provide for the withdrawal of lands for homestead entry or for cancellation upon giving two year's notice:
>
> | | |
> |---|---|
> | Milk River Cattle Company | 60,000 acres |
> | Grand Forks Cattle Company | 47,218 acres |
> | Grand Forks Cattle Company | 47,615 acres |
> | Messrs. Brown, Bedingfield et al. | 55,747 acres |
> | George Lane | 43,736 acres |
> | Glengarry Ranche Company | 13,794 acres[86] |

In this manner the Bar U Ranch received one of a handful of highly prized "closed leases." Nearly forty-four thousand acres of grazing country comprising all the available land in two townships to the north and west of the home place.[87] This lease secured vital winter grazing lands and was essential to the continued functioning of the big ranch; it was to remain in force for the next forty-five years.

The "paper trail" in the archives of the Department of the Interior is not sufficiently detailed to allow us to reconstruct exactly what happened. Clifford Sifton, the outgoing minister, was a master of patronage, and the two major shareholders of the Grand Forks Cattle Company were his close political friends, J.H. Ross and J.D. McGregor.[88] The assumption has been that Sifton indulged in a "last fling" to benefit his friends before leaving the Department. However, it is not easy to establish clear links between the other recipients of closed leases and Sifton. Certainly, Lane and A.B. Macdonald of the Glengarry were Liberals, but 7U Brown and Frank Bedingfeld were largely apolitical.[89] It seems likely that both Sifton and his successor Oliver were convinced that the lands applied for in the foothills were totally unsuited to cropping, and that these leases at least were awarded on their own merits. The precise timing of events adds to the mystery. A handwritten note entitled "Closed Grazing Leases," gives the dates on which the leases were granted. George Lane's was the first, dated 26 April 1905; others followed over a period up to 9 May 1905. Did Frank Oliver in fact approve these leases or did Sifton set things in motion late in 1904, when the Order-in-Council detailing the details of the new regulations was framed? We shall never know. The hard fact remains that George Lane and the Bar U Ranch emerged from a period of bureaucratic turmoil with a firm grasp on their land base, to their deeded holdings they could add nearly two townships of leased land.[90]

4.4 Teams of Percherons taking out winter feed on the Bar U, 1917. Although tractors were used for some tasks, large numbers of horses were employed on the ranch through the 1940s. Glenbow, NA 5652-75

Lane ran the Bar U and the Willow Creek ranches as "cow-calf" operations. The objective was to turn out young cattle that would put on weight and bulk rapidly and, after two years on the range, earn premium prices in the beef market. In spite of the large scale of operations, production techniques on these ranches incorporated many of the advances that had proved successful during the 1890s. Hay was put up in large quantities and fed to calves and weak cows. Every effort was made to upgrade the quality of the stock by using purebred bulls, and the "calf crop" was improved by ensuring that there were enough bulls to service the cow herd. Fences were used to protect hay and crop land during the summer, and to reserve particular areas of the range for winter grazing. The different elements of the herd – cows, heifers, yearlings and steers – were, to a considerable extent, segregated and used different areas of the huge ranch at different periods of the year. For example, weaned calves were moved to the southern extremity of the ranch in the fall and were fed from haystacks put up on the "Bar U Flats" during the winter. Cows in calf were kept in the sheltered field along Pekisko Creek just to the north of the ranch buildings. Steers and dry cows were wintered in the rolling hills between the ranch and the Highwood River, where chinook winds cleared much of the snow. After the acquisition of the new closed lease in 1905, increasing use was made of summer pasture for cows and calves across the Highwood and up into the aspen breaks and alpine meadows of the Bull Hills.[91] It was from his foothills ranches that Lane drew the stock which would so impress buyers in Chicago or Liverpool, but the yearlings were fattened on the mixed-grass prairies along the Bow River.

Through the two decades spanning the turn of the century, Lane did his best to retain or even extend his hold over as much unpatented Crown land as he could. He used his YT Ranch on the Little Bow, which he had purchased in 1898, as a base from which to pasture both horses and cattle on the extensive open range reaching eastward to the Bow River. The 1901 census shows that sixteen people were living at the YT, the largest ranch community identified in the grazing country.[92] Herb Millar was foreman, and his wife Ella was the only woman on the place. Adrian Jarvis was the cook and all the other men described themselves as ranch hands. The crew was far larger than would have been required for local operations. These cowboys scoured the ranges between the Little Bow and the Bow, cutting out fat cattle belonging to Lane and his partners and herding them to shipping points on the railway. Interestingly enough, eight of the riders had been born in the United States. The demands of this particular job, which involved selection, cutting out, roping, and trailing, were just the kind of work that required top-line cowboys, and it was work that traditional riders enjoyed. At the YT they could continue to spend all their working days on horseback. They did not have to undertake 'demeaning' work like haying or fencing.[93] It is probably not a coincidence that two of the cowboys from the United States were over forty years old, despite the fact that 'cowboying' was usually regarded as a young man's vocation.

To the north, Lane's partners, Gordon, Ironside and Fares, controlled the Wintering Hills area from the Two Bar Ranch, and the lower Red Deer River from the SC Ranch. As farm settlement spread rapidly eastward from the Calgary to Fort Macleod railway during the early years of the new century, the YT was abandoned and Lane moved his herds across the Bow River to the semi-arid mixed-grass prairie between the Bow River and the Canadian Pacific main line, from Bassano to Medicine Hat. Here he was able to use land that had been assigned to the railway company but was unsuitable for farm settlement until irrigation was provided.[94]

The herds Lane turned loose to rustle on these extensive ranges were "steer herds," young cattle being fattened on grass for two years before marketing. One source of yearlings was Lane's foothills ranches. An observer in High River commented on this connection:

> About 1,600 head of young cattle were driven from the Bar U Ranch last week and passed High River on Saturday on their way to the Bar U grazing lease away east of the Big Bow. About ten cowboys were in charge and they presented a novel sight as they drove the enormous herd along the road. The Bar U Ranch west of High River is utilized as a breeding station from which place thousands of head of well bred stock are raised each year, and later driven to the grazing limits near Bassano.[95]

The Flying E Ranch, or the "Willow Creek Bar U" as it was often called, George Lane's original home place, played the same role. However, because their cow-calf operations in the foothills provided but a fraction of the young cattle that Lane and his associates required, they also bought large lots of "stockers" or "dogies" wherever they were cheap and available. Sometimes they obtained young cattle from the farms of Manitoba or Ontario, while other young cattle were shipped from Mexico.[96] Profits were made on these ranches by keeping investments to a minimum and by maintaining a high volume and rapid turnover. The larger the herds, the lower were the labour inputs per head. Steers were expected to rustle for their feed and 3 to 5 per cent winter losses were accepted as normal. These were the classic techniques of the open range and were highly speculative. Ranchers were betting

4.5 George Lane's headed note paper, used in 1910, shows 13 cattle and horse brands. Glenbow, M652

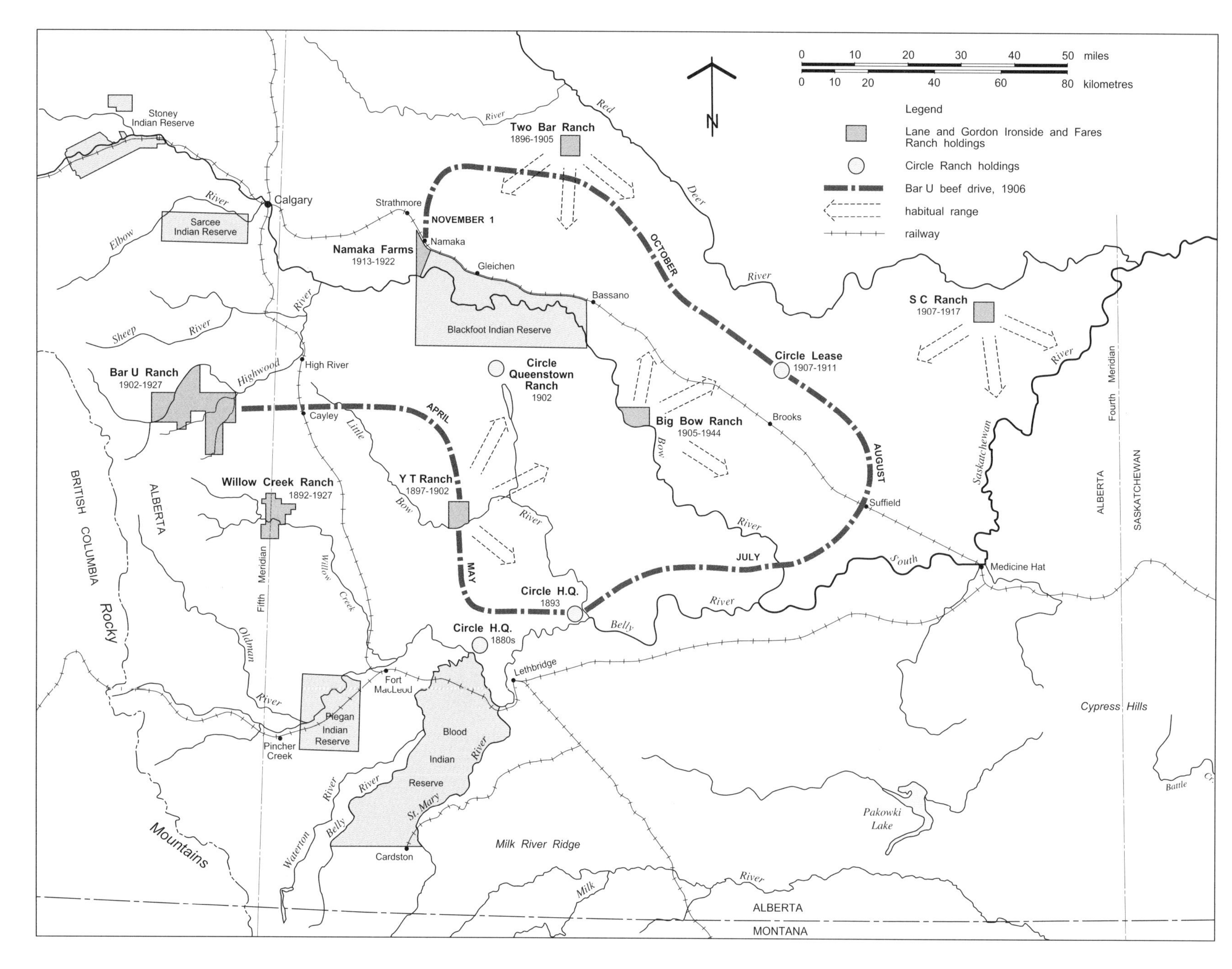

Map 4.1 The ranches and ranges of George Lane and his partners, Gordon, Ironside and Fares, c. 1906, showing the route of a beef drive and the locations of the Circle outfit. See also colour section.

not only that herds would survive the depredations of winter, fires, and wolves, but were also gambling on the markets at local, national, and international scales. The westward flow of stockers to the western ranges increased from 18,500 head in 1896 to 63,235 in 1902.

Lane's far-flung operations on the mixed-grass prairie were supervised from a base on the north shore of the Bow River near the present site of Bow City [Map 4.1]. Lane's son Roy explained:

> The Bar U come over here in 1905, they crossed over from the other side of the river, they used to run these cattle over around that Vulcan country on the Little Bow, and they moved over on this side of the river in 1905.[97]

The company proceeded to buy out the small ranchers who had squatted or homesteaded along the river, and absorbed their small lots of cattle.[98] The Millie homestead, near the present site of Bow City, was developed as the headquarters. Bill Henry was living there at the time and was offered a job. His first task was to expand the buildings and corrals to provide suitable facilities for a company running thousands of head of cattle.[99] Roy Lane joined the crew in 1908 when he was sixteen. It was to be his home until 1944, except for a period spent with the Canadian forces during the First World War.[100]

An account of a drive of beef cattle from the Bar U to the Bow River in 1906 illustrates the way in which Lane and his partners integrated their holdings and utilized the open range.[101] The outfit, under the leadership of "Slippery" Bill McCombe, consisted of two wagons each pulled by a four-horse team. One was the chuckwagon, the other the "bedwagon" which hauled the rope corral, two tents, a thirty-foot pole, and various other equipment. The cavvy had about a hundred horse in it and each man had a horse for every day of the week.[102] The crew consisted of the boss, the cook, eight cowboys, the horse wrangler, who looked after the horses during the day, and the night wrangler, who herded the horses at night and drove the bedwagon.

The objective of the drive was to gather three- and four-year-old steers of various brands owned by the partners and deliver them to the shipping point at Namaka. Experienced riders would ride out ten to fifteen miles on either side of the wagons, cutting out mature steers and driving them back to join the growing herd. The men worked slowly and carefully so as to keep the disturbance to the grazing cattle to a minimum. The herd itself would drift along at a walking pace for a few hours in the morning and again in the late afternoon. The aim was to put weight on the herd even as they moved. At strategic locations the wagons would halt for two or three days while riders explored even further afield. North of Fort Macleod the outfit stopped for a week; then, as they crossed the Little Bow they were joined by four cowboys from the Circle outfit. These men escorted the Bar U herd across their range, pushing aside Circle cattle so that they would not get swept up with the moving herd. By July, they had crossed the Bow River and were within about twenty miles of Medicine Hat; there they turned northwestward and paralleled for a time the Canadian Pacific main line. After a broad detour northward toward the Red Deer River and the Wintering Hills,

4.6 A camp on the drive to Brooks showing the "cavvy" or herd of saddle horses from which the cowboys selected their mounts. Parks Canada, Florence Fleming Collection, 95.07.01.26

4.7 Bar U cattle stopping for a drink en route to Brooks stockyard. Parks Canada, Hugh Paulin Collection, 96.11.01.31

they ended up at the shipping pens at Namaka with about two thousand head of beef steers. George Lane met them, and most of the prime steers were shipped to Chicago, while others moved to Winnipeg for export. Another Bar U beef wagon pulled into the shipping point a week or two later.

Much the same strategy for using available open range areas, and one employing traditional methods, was used at least until the 1920s. An account of a roundup in 1918 makes clear the continuing presence of the Bar U and Gordon, Ironside and Fares in this area:

> There were four roundup crews and wagons on the 1918 roundup between Suffield and Bassano and the Red Deer and Bow rivers. They were the Bar U, Hill and Butler, The Circle and the Anchor P. The wagon bosses were Ernie Lane, Harry Herbets, George Crooks and Sam Howe. We pulled out with the Bar U wagon about the twentieth of September, and finished loading out beef steers on the eighth of November at Brooks. We loaded out a trainload, 12 head to a car, for Mr. Fares of Gordon, Ironside and Fares. The weather turned bad and there was quite a lot of snow before we finished after dark. We were on CPR land all that fall and [it was] all open land, and a hundred brands to contend with. All of the stock were fat on the 1915–1916–1917 grass, and all of it free grass ... there were no fences or anyone living there ... it took one or sometimes two men to clear stuff [other cattle] out of the way.[103]

Thus, the essential elements of 'open range' ranching were still very much in evidence more than twenty years after the first reports of the demise of the system had been confidently predicted.[104]

At the Bar U Ranch on Pekisko Creek, methods and techniques were refined and intensified during Lane's long tenure of the ranch. Gradually, higher yielding "tame hay" was grown and put up alongside massive quantities of wild hay. From 1907 onward, efforts were made to divert water from the creek onto the bench, where it was used to irrigate an ever-increasing acreage of land reaching eastward to the Bar U farm.[105] The small area of the ranch that was ploughed and sown to oats to provide winter feed for saddle horses and work horses was expanded to meet the demands of the growing herd of Percheron horses. Surplus or frosted grain was available for the cattle herd. As early as 1908 selected steers were being winter fed and shipped for export in April.[106] In 1911, fifteen hundred acres of the Bar U flats were ploughed and sown to winter wheat. The editor of the *High River Times* commented gleefully:

> When one of the biggest ranches in the country becomes one of the biggest wheat fields in the country and the ancient red man hires out as a farm labourer, the modernization of the west appears to be complete.[107]

At this time Lane envisioned running his home place like a huge stock farm. He explained:

> … from now on I shall feed all my cattle every winter. I began feeding this year [1911/12] simply because feed was cheap, and I have figured out now that it is much more economical and businesslike than the old system of wintering cattle on the range. It was different a few years ago when cattle were cheap and feed was dear. It then paid better to leave the cattle on the range and stand the inevitable loss. Even in the most favourable years this amounted to 5% of the

Please write for an explanation of anything unsatisfactory

ROSENBAUM BROS. & CO.
(INCORPORATED)
LIVE STOCK COMMISSION MERCHANTS.

UNION STOCK YARDS, CHICAGO, ILL. 10/23 1917

SOLD FOR ACCOUNT OF Geo Lane

SALE No. 66110 SHIPPED BY FROM

DATE OF SALE.	PURCHASER.	CATTLE.	HOGS.	SHEEP.	WEIGHT.	OFF	PRICE.	AMOUNT.	TOTAL.
	Wilson	49 Strs			67900		1200	7548 00	
	Anglo	159 "			180870		1080	19533 56	
	"	145 "			167000		1080	18036 00	
	Wilson	36 "			39530		950	3755 35	
		389			450750				48867 91
	U̅								

R.R. Soo Line CHARGES—FREIGHT (INCLUDING FEED ON THE ROAD) 3920 26
STATION Brooks Alta YARDAGE. 97 75
CAR NOS. WEIGHT RATE 7100 POUNDS HAY. 122 39
22 cars 22000 each 22 BUSHELS CORN.
FIRE INSURANCE 10 CTS. PER CAR 2 20
TERMINAL CHARGES $2.00 PER CAR 44 00
UNLOADING 25 CTS PER CAR 5 50 INSPECTION. 1 10
Cleaning 16 50 SHIPPING CHARGES.
Sand 22 00 COMMISSION. 258 20 4451 40
Clearance 16 78 NET PROCEEDS. 44416 51
Postal 250 20 50 00
T.C. yds 220 00 44366 51
Weighing 4 50
Less frt 9 cattle 94 02
Eg. A. Fleming
Dominion Bank Calgary Alta Can.

E & O E.

4.8 An invoice for 22 rail cars of cattle from Chicago based agents, Rosenbaum Brothers. The gross return on 389 head of cattle was $48,867, and costs for cleaning, sand to give the cattle a good footing, hay, etc amounted to $4,451. Parks Canada, Florence Fleming Collection, 95.07.01.30

4.9 Cattle in the stockyards at Brooks awaiting loading for shipment to Chicago. Parks Canada, Florence Fleming Collection, 95.07.01.27

herd. With cattle worth $50 a head it pays to feed the whole herd and obviate this loss. Besides you have your beef cattle in shape for market at anytime, and especially at the season when they command the highest price.[108]

At the time Lane explained that he was feeding thirty-five hundred beef cattle on frosted wheat for which he had paid only $5 a ton. He was on his way to Idaho to consult a well-known authority on the growth of alfalfa.[109] The incident was described as follows:

> ... with a twinkle in his clear blue eyes, and a broad smile of contentment, Mr. Lane tells why he fell in love with Alfalfa. It was because he found this plant would fatten three times as many horses, steers, hogs, etc, as anything else he could grow.[110]

Later in the year, three hundred head of this bunch of cattle were shipped to Chicago, where they were sold for the highest prices ever paid for beef steers.[111] Lane's hopes and dreams of "wrapping his grain inside the hide of his well bred steers"[112] were disrupted in the longer term, first by the wartime rise in the price of grain, and later by the droughts which started in 1919.

The scale of operations pursued by Lane and his partners was huge. With a breeding herd of more than three thousand cows, little attention could be paid to individual beasts. The calf crop was rather lower and losses rather higher than on smaller, family-operated ranches in the foothills. However, economies of scale ensured higher returns. Lane stayed with Shorthorns during his long career and emphasized the importance of running cattle with "range smarts." He explained that he had tried breeding from stockers, but they were such poor mothers that calf losses were unacceptably high. Cows raised on the range had a remarkable ability to look after their calves even in harsh spring weather. Similarly, Lane was suspicious of purebred bulls brought in from the east, for often he said, "[T]hey are in full flesh and when turned out to rustle for themselves they go backwards rather than forwards."[113] He preferred purebred bulls raised on the range like those obtained from local ranchers like

"Billy" Cochrane of the CC Ranch.[114] No lesser cattleman than Pat Burns appreciated Lane's efforts to maintain the quality of his herds:

> He saw the need to improve the grade of beef cattle and he imported from Mexico large numbers of hardy range cows and purchased some of the best thoroughbred bulls obtainable. The result was an extra fine herd of hardy range cattle, the fame of which was only second to that of his Percherons.[115]

A word about numbers: enough has been said to convey something of the variety and complexity of the operations run by Lane and his associates. This means that there is much confusion about the size of the herds actually run at the Bar U, and the number of cattle owned by Lane. Cattle on all the Lane ranches – the Bar U proper, the Willow Creek ranch, and the leases along the Bow – were referred to as "Bar U cattle." In addition, it would have been impossible to separate the overlapping interests of Gordon, Ironside and Fares and Lane; they shared individual short-term enterprises as well as their co-ownership of the Bar U Ranch. When a trainload of stockers came from the GIF ranch in Mexico, some would be run onto the Two Bar Ranch, while others would be fattened with Lane's herds. In his letter to Greenway in 1904, Lane claims that he had twenty thousand head of cattle.[116] But only a fraction of this total would have been run along Pekisko Creek. A possible deployment of his herds might have involved: eight thousand head at the Bar U; four thousand head at Willow Creek; four thousand head along the Bow; and some share of the herds at the Two Bar and the SC Ranches, perhaps totalling another four thousand head. As the number of horses at the Bar U increased during the first decade of Lane's ownership, so the cattle herd was reduced to four to five thousand head. The press of settlement around the Bar U meant that there was no longer the ability to throw more and more cattle onto the range. Rather, the total number

4.10 Bar U cattle loaded for their journey to Chicago, October 1917. Parks Canada, Florence Fleming Collection, 95. 07.01.28

of stock grazing had to be strictly controlled. The ranch comprised sixty-four thousand acres, and each steer required about sixteen acres to support it year round. This would have meant an optimum herd size of four thousand head. However, the introduction of irrigation, tame hay, and fodder crops increased the carrying capacity of the ranch.[117] The sustainability of this, and all of Lane's ranching enterprises, was about to be tested by an environmental cataclysm.

The Killing Winter of 1906–7

The long and viciously cold winter of 1906–7 has gained mythic status among historians of the Canadian range. Observers at the time blamed the end of the open range era on the losses it inflicted on cattlemen. However, many viewed the ranchers' losses as but a necessary prelude to a new and progressive future for the west based on farming grain.[118] Perhaps the most famous evaluation of the winter was that offered by Wallace Stegner in his book *Wolf Willow*. After a narrative about an outfit battling the snow and cold along the Frenchman River, south of the Cypress Hills, he summarized the results of the winter:

> The net effect of the winter of 1906–07 was to make stock farmers out of ranchers. Almost as suddenly as the disappearance of the buffalo, it changed the way of life of a region.[119]

More recently, Hugh Dempsey uses his carefully crafted description of the heroism of individual cowboys and ranchers in the face of implacable nature as the closing chapter of his study of *The Golden Age of the Canadian Cowboy*.[120] These are but examples of a considerable literature which depicts the bad winter as a turning point in the agrarian history of western Canada. [Plate 4, colour section]

Joy Oetelaar has investigated the winter of 1906–7 from a number of innovative points of view.[121] She asks questions like: was it really "the worst" winter on record? To what extent is it possible to measure the degree of severity of a winter objectively? Who were the observers who reported on the experiences of the winter, and the sights and smells of the carrion spring? Surely, she argues, their perception was coloured by their background and experience. We should also be skeptical of "official" witnesses, because they all had their own visions of development and agendas to pursue.[122]

Oetelaar does us a service by demanding that we take another look at the killing winter of 1906–7, and she points us in new directions. However, she would be the first to admit that her ideas could be extended. To carry out a definitive evaluation of the climatology of the event it would be necessary to consider all the meteorological stations covering the prairies, and to use the diurnal maximums and minimums rather than merely the monthly averages.[123] A good model for such a study would be Chakravarti's work on the changing patterns of drought in the prairies during the 1930s.[124]

Although we still lack a definitive climatological study of the bad winter, few would dispute that the intensity of the winter varied widely from place to place. Toward the end of February 1907, William Roper Hull, a veteran Calgary-based cattleman, returned from his ranch in the Porcupine Hills and reported that "cattle loss was certainly exaggerated in that portion of the country." He was careful to explain that he could only give an opinion based on what he had seen in the foothills. Here at least, he was sure that ranchers would not suffer any greater loss than they experienced in a ordinary year.[125] On the other hand, the full fury of the winter fell on the region from Sounding Lake southward to the Red Deer River and through the Hand Hills to the Bow River from Gleichen to Medicine Hat. A series of letters from Charlie Douglas to A.E. Cross chart the course of the winter and his despairing efforts to bring a bunch of cattle through it. The cattle sheltered in the breaks along the Red Deer River south of Dorothy and had to be driven up onto the prairie to be fed. Douglas wrote:

4.11 Edward F. Hagell, "Counting the Storm's toll." Cattle drifted blindly into coulees where they froze to death or smothered in the drifting snow. Glenbow, Art Collection 33

> The hardest part is hauling [feed], I have got two teams going every day, two loads each, and there is hardly a day passes without a blizzard or wind to fill up the trails. One can hardly imagine the drifts, stacks are buried in snow and a crust on the level ground, all the flats have been belly deep

> to a horse since November, so that the cattle have been living on brush. It has been so cold with winds when one breaks trail and puts them out on the banks to patches of sage or grass the poor brutes fight you right back to shelter.[126]

A week or two later Cross was suggesting putting the weak cattle out of their misery.[127] In these extreme conditions even those who had hay could not bring it to where it was needed, nor could the stock be kept from drifting away from the home place. The loss "was not always lightest where the rancher was well prepared with abundance of winter food, nor heaviest where he was unprepared. Stock drifted before the storms to places where it was impossible to get feed to them, and often where feed could be had it was impossible to supply water and it is a question if many animals did not die from lack of water as well as the absence of food."[128] At one point some thirty-five thousand cattle were congregated in the great bend of the Belly River just west of Lethbridge, but this huge herd was broken up and dispersed to other areas during a lull.[129] Others were not as fortunate: they either starved or suffocated in the drifted coulees. Harry Otterson, the foreman of the T Bar Down on the Whitemud River south of the Cypress Hills, rode from the ranch through the wintering grounds toward the end of March. He described what he saw:

> It surely was a gruesome ride. The cattle were in all stages of dying. The brush was simply lined with dead cattle. The live ones, at night, would lie down on the dead and many would not be able to get up again: consequently they were literally piled up.[130]

Nor was the desperation of winter relieved by an early and sudden arrival of spring. T.B. Long, another Cypress Hills rancher, stressed that "the late spring was what finally dropped the axe on those cattlemen."[131] Douglas, Otterson and Long, and others like them, were experienced ranchers and not given to fanciful story-telling. Although, sometimes, it was years before they committed their impressions to paper, their collective witness sketches a range tragedy of epic proportions. Gradually the toll was reckoned up. Otterson judged his own losses and those of the other major herds in the Cypress Hills area at between 60 and 65 per cent. The Conrad Price outfit lost 60 per cent of their longhorns.[132] Even the canny Matador Cattle Company, which always put up plenty of hay, lost 40 to 50 per cent of their two-year-olds.[133] While these estimates may be accurate for specific herds in particular districts, they should not be interpreted as typical for the whole range country.

Undoubtedly the bad winter had a devastating effect on many cattlemen and reduced the number of cattle on the range, but it is important to evaluate its impact in the context of other far-reaching changes which were taking place. In an article published nearly twenty years ago, I tried to set the effects of the winter into this broader context.[134] I suggested that the dominance of ranchers over the grazing lands of the North West rested on three foundations: the support of the government of Canada, the widely held belief that the short grass prairies of Palliser's Triangle were too arid for grain farming, and the insatiable demand of the British market for beef on the hoof. All three of these foundations, I argued, were being eroded even before the killing winter.

The Census of Canada provides data on livestock numbers which can help us set the effects of the bad winter in a rather different light, and in the absence of any authoritative estimates of the losses incurred, it is a good place to start. In Alberta, the total number of "Other Horned Cattle" (not milch cows) increased by 158 per cent between 1901 and 1906; it then dropped by 30 per cent between 1906 and 1911. These figures indicate that a major, if temporary, dislocation took place in the buildup of cattle on the prairies.[135] If we consider those census districts which cover "The Grazing District" of southern Alberta and southwestern Saskatchewan, the changes are more marked: a decline of 304,273 head, or a loss of 45 per cent, was experienced between 1906 and 1911.[136] We must surely conclude that some important changes were taking place on the range in the five years after the winter of 1906–7.

It is certainly not my intention to suggest that the losses in cattle numbers indicated by the census were solely, or even primarily, the direct result of the bad winter. The period from the turn of the century until the First World War was one of the most tempestuous in the history of western Canada.[137] The population doubled and doubled again as immigrants flowed in at a rate of thirty thousand a year. The lure was farmland, and homestead entries and crop acreages increased exponentially. The rapid expansion of farm settlement was accompanied by a massive programme of railway building [see Maps 4.2 and 4.3]. The huge areas of untouched open range, which stretched both northward and southward from the main line of the Canadian Pacific Railway in 1906, were reduced to a few scattered pockets by 1912, as branch lines were built and elevators constructed.[138] Thus, the main reason for the decline in numbers of cattle was the increase in cropped acreage: wheat had replaced meat as 'the way to get rich on the plains,' and this time it was not a handful of investors who sought to profit, but rather thousands of farmers. Many of these newcomers were experienced cash grain farmers. Some had sold farms in the United States for $15–20 an acre, and were purchasing land in Canada for $4 or $5 an acre. They had capital to invest in machinery, and they hastened to break up their holdings and to extend the scale of their agricultural operations.[139] Advancing farm settlement was spurred by the tide of enthusiasm for dry farming which was generated by the work of Hardy Webster Campbell.[140] Contemporary reports reflect the excitement the pace of change engendered. There were some eighteen steam ploughs working in the Lethbridge area in 1908, and the acreage in winter wheat increased tenfold.[141] Along the Milk River newly broken land was yielding thirty to forty bushels to the acre.[142] Around Fort Macleod cattle buyers were finding it hard to find enough cattle to meet their needs. Superintendent Begin reported from Maple Creek:

> Ranching in this part of the province will soon be a thing of the past. Ranchers are going out of business. Most of the land has been opened for homesteaders. Old ranch grounds are gradually being cut up by farmers. Stock cannot anymore roam over the country as hitherto.... Maple

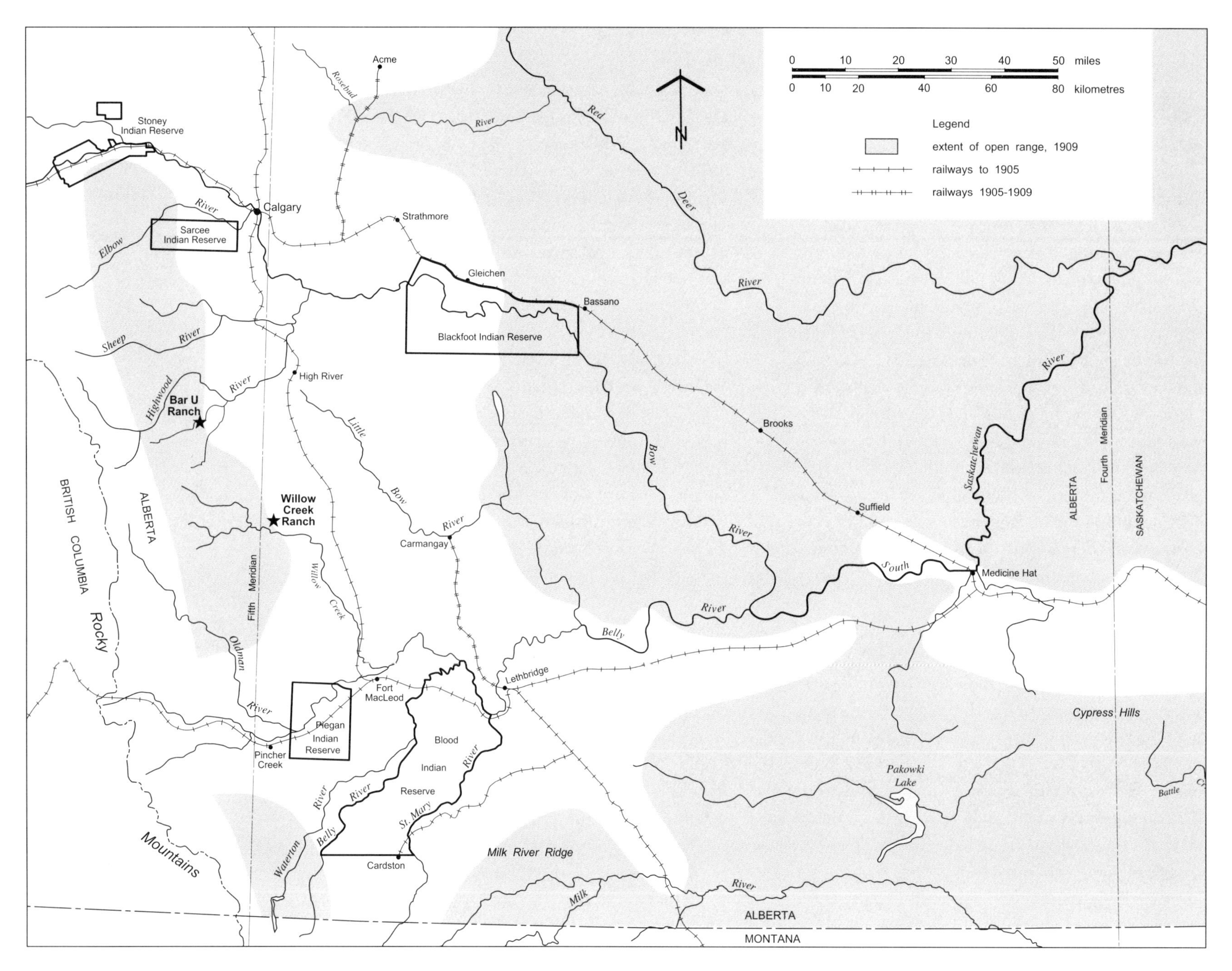

Map 4.2 The extent of open range, 1909. Generalized from Department of Interior Cereal Map, 1910. Shading shows townships with less than 1000 acres of crops planted.

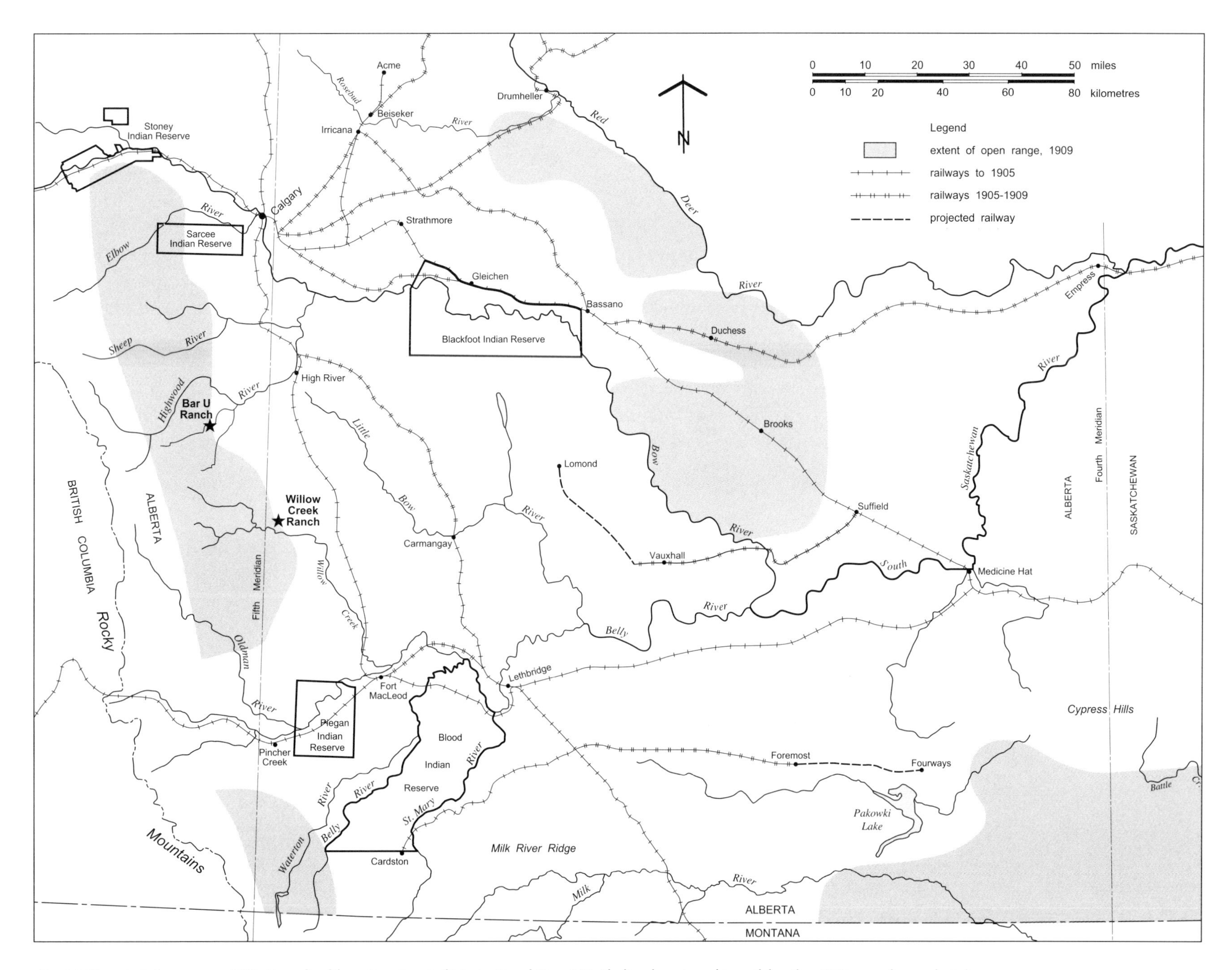

Map 4.3 The extent of open range, 1913. Generalized from Department of Interior Cereal Map, 1914. Shading shows townships with less than 1000 acres of crops planted.

> Creek is supposed to be ranching country with a great number of cattle, but the price of beef in the town of Maple Creek is higher than any other place in the province.[143]

This was a time of rapid change in the rural economy and land use patterns of the prairies. The causes of these changes were deep seated. They involved conditions in eastern Europe, improvements in land and sea transportation, and industrialization and urbanization in both Europe and North America. Together these forces comprised an irresistible juggernaut on an intercontinental scale. The occurrence of an unusually hard winter in western Canada was an event of a much more local and limited importance.

Nevertheless, the losses caused by the winter disrupted the status quo and added momentum to the pace of changes which were already underway. Several large cattle companies from the United States that had moved onto Canadian range in the early years of the new century either went bankrupt or retreated back across the line into Montana. By 1909, the Turkey Track, the T Bar Down and the Conrad Price Cattle Company had all closed down their Canadian range interests.[144] Wilkinson and McCord, the pioneer ranchers of the Sounding Lake area, were forced out of business.[145] Smith and Musett sold out to the Massingill brothers.[146] Veteran Canadian cattlemen like George Emerson went bankrupt, as did the Prince and Kerr Ranch and the High River Trading Company.[147] Some ranch companies, like the Medicine Hat Ranch Company and the "CY" Ranch north of Taber, sold out to land development companies.[148]

The small stockman was also profoundly affected by the severe winter. For years the homestead farmer, who was characteristically short of both capital and labour, had looked upon a small herd of beef cattle as a sheet-anchor for his farming operations. The heavy losses of the winter, taken in conjunction with low cattle prices and rising wheat prices, forced farm families to reassess their position. A great many smaller stockmen sold the cattle that had survived and devoted their entire attention to growing grain. The recorder of brands noted in 1907 that there had been a decrease of nine hundred cattle brands in Alberta and Saskatchewan as well as a long list of transfers.[149] One observer remarked: "This year men who have lived in the west for years as ranchers raised their first crop, and will raise more next year."[150] This transformation from rancher to farmer had started as early as 1903, because it had become increasingly difficult to dispose of cattle raised on farms. Export buyers were not impressed with small lots of inferior cattle, and local markets were oversupplied. Moreover, the incoming flood of settlers quickly reduced the available unused land adjacent to farms that had been used for grazing. Far from forcing ranchers to become stock farmers, the bad winter disrupted the growth of mixed farming and the gradual transfer from extensive to more intensive methods.[151] Pat Burns commented on this precipitate change: "Dazzled by dollar wheat, many farmers are selling their stock to make room for more wheat. Keep your cows for breeding. We are now shipping cattle which should be kept at home."[152] Stock shipments eastward from Alberta to Moose Jaw and Winnipeg reached a peak of 115,000 head during 1906 – before the winter – and a large number of stock presented for shipment in the fall were turned back onto the ranges because there were not enough railway cars available.[153] This contributed to the winter losses as cattle started the winter off their accustomed range and with no provision for feed. The process of "beefing out," or spaying cows and selling off breeding herds for beef, started before the bad winter and continued for several years after it.[154]

Thus, one thing compounded another to decrease the number of cattle on the range. The killing winter resulted in severe losses and was followed by a very low calf crop. Stock shipments out of the grazing region were exceptionally high even before the winter and picked up again in 1908, 1909, and 1910. George Lane addressed the National Livestock Convention in Ottawa in 1912 and summarized several of these trends, he started off:

> Now, gentlemen, I have come down here from Alberta to protect the interests of the poor old cow, the animal which has driven the mortgage away from the farm, the animal which has made so many homes happy, and in spite of this, has been and is now being destroyed in the west in such a shameful manner. Last year 30,000 calves were killed [slaughtered for veal] in our western country. All these calves were still sucking their mothers and were of the very best beef strains that could be produced. In addition to this I want to tell you that at a very conservative estimate at least 65% of all the cattle slaughtered during 1911 in Alberta and British Columbia were she stock.... The wholesale destruction has been going on for a number of years until we have dropped down to scarcely cattle enough for our own home use.[155]

Lane went on to point out that exports for 1911 were less than 16 per cent of exports for 1906. He enumerated the pressures from land agents, machinery salesmen, gasoline salesmen, and banks which encouraged farmers in the perception that "wheat is king." But Lane urged stockmen to resist these blandishments for, he said, over a career which spanned thirty years he had seen a great many ranchers who had, by diligence and careful planning, done very well in the cattle business.

Lane Takes a Hit

So far we have reviewed the context of the bad winter. But what factors explain the fact that some herds were decimated, while others came through virtually unscathed? And how did Lane's various ranches fare? Here, we will examine some of the variables which influenced the effects of the winter at a larger scale, and thereby construct a model to estimate the losses among Lane's cattle herds.

The severity of the winter was not uniform. While every region experienced frigid periods of Arctic air, some places, particularly the Bow Valley and other areas close to the mountains, enjoyed some respites brought by chinook winds. However, even these reprieves were not without their problems. In some transitional locations, snow melted and then refroze into an impenetrable crust which prevented stock from rustling and cut the legs of cattle as they broke through and then fought their way out. Nevertheless, it is clear that the mixed-grass prairies in the plains to the east suffered the worst conditions.

During the summer of 1906 there were more stock on the ranges than there had ever been before. Along the Bow and the Little Bow Rivers and in the valleys of the foothills, incoming homesteaders had taken up and fenced bottom lands which had once been reserved for winter grazing. Further east along the Frenchman River, Harry Otterson admitted that the Turkey Track and the T Bar Down had allowed their herds to graze grass during the summer and fall which they had usually set apart for winter use.[156] Throughout the grazing country the "security net" provided by huge acreages of unused grassland had been severely reduced by settlement.[157]

The age, sex, breed, and origin of cattle affected their capacity to survive the winter. Cows that had calved late or were still nursing large demanding calves were particularly vulnerable. Three- or four-year-old steers, with larger body mass and longer legs, could move with the wind through drifted snow and had much better chances of survival. Range-savvy herds raised on western grasslands did much better than stockers. Dairy strains and young cattle reared and fed in barns in Ontario and Manitoba were ill-adapted to face the rigours of even a normal winter, and died by the thousands.

Mange contributed much to the high death rate of cattle during the winter of 1906–7.[158] This debilitating disease had been particularly widespread during 1904 and 1905, but had been contained by a programme of compulsory dipping. The consensus among cattlemen was that this expensive and time-consuming process need not be continued in 1906. However, it is characteristic of mange to disappear almost entirely during the summer months only to reappear when temperatures drop. Billy Cochrane wrote to his friend and neighbour A.E. Cross in the fall of 1906:

> I saw George Lane on Saturday last at Cayley.... He told me to tell you that your cattle on the Red Deer were badly affected with mange. If you think it necessary to do any dipping, Lane said he had a good dipping plant at Bassano ... your cattle were clean last fall when we put them through the Mosquito Creek vat, with the exception of a few old bulls which we boiled plenty. They must have become reinfected by mixing with cattle from High River which were driven to the Red Deer last winter.[159]

Later, Cross admitted that his losses, amounting to some $25,000, had been so high because his mange-infected cattle "went into the winter in a bad state and the hard winter completed the destruction."[160] The disease was more widespread than ever in 1907 after infected stock had drifted far and wide. It is clear that many cattle were carrying mange in a latent state in the fall of 1906.[161] The disease has the effect of reducing the thickness of the animal's coat, which may become, in severe cases, an almost hairless mass of scabs.

George Lane survived the losses of the winter and expanded his cattle interests in its aftermath. His greatest successes and his best years in the ranching business were still to come. However, one should not underestimate the consequences of the winter on this account. Mrs. Lane remembered, "In 1907 it looked

4.12 Around the turn of the century, tremendous efforts were made to control mange. As the photograph shows dipping vats were substantial structures and expensive to construct. Cattle from all over a district would be driven to the nearest dip. However, the circumstances of the open range meant that "clean" cattle might get reinfected by mixing with those which carried the disease. Glenbow, NB (H) 16-460

as though all the big cattlemen were broke, George Lane included, but the situation was saved with grim work and trying."[162] During the middle of May 1907, Lane had had an opportunity to tour his ranches and make an initial evaluation of the damage. He was receiving reports of stray cattle turning up, many in Montana. An exact accounting would have to wait another month or six weeks until the various roundups had been completed, but for the present he was quite encouraged by what he had learned of the situation. A correspondent interviewed Lane and wrote:

> The reports of his [Lane's] sustaining heavy losses is wrong. For a great many years his losses averaged from 3 to 4 per cent, which is an exceedingly low average. This year his losses will average about 15%. Mr. Lane is buying cattle now which is very wise for cattle are bound to increase very much in value ere the end of the year.[163]

If we take this "15% loss" at its face value, what can we infer concerning the number of cattle deaths and financial losses? The cow-calf operations at the Bar U and Willow Creek Ranches would have escaped the worst depredations of the winter. Not only were the conditions in the foothills less severe, but also the broken topography and vegetation provided good shelter. Large quantities of hay had been put up to feed calves and weak cows, although there was no provision made to feed the whole herd. Moreover, conditions permitted access to the stacks and allowed the distribution of feed during lulls between the storms. Furthermore, the Bar U, with its deeded land base and extensive closed leases, had not suffered, as had many foothills ranchers, from homestead settlers taking up the bottom lands which had in the past been used for winter pasture. Losses on these two ranches were probably around 10 per cent of a combined herd of ten thousand head. Deaths would have been greatest among calves and nursing cows, and less among older dry cattle. The one thousand head lost might have had a combined value of $23,000.[164]

The range herds along the Bow River south of Bassano would have been hit harder. The winter was most severe here, and natural shelter was less available. Little or no provision had been made for feeding these herds.[165] However, the huge area of unfenced range between the Bow River and the CPR main line had hardly been used prior to the arrival of Lane's herds in 1905. There was plenty of grass available and the cattle entered the winter in good condition and free from mange. Again, some elements of the herd were more vulnerable than others. First-year 'dogies' shipped westward and thrown on the range during the summer would have suffered worst, while robust, range-bred young cattle from the Bar U and Willow Creek Ranches would have fared much better. Losses here might have amounted to about two thousand head, valued at more than $80,000.[166]

Altogether, our calculations suggest that Lane lost three thousand head of cattle to the ravages of the winter, and had to write off more than $100,000 in livestock assets. Moreover, the calf crop in 1907 would have been small and would have delayed natural recovery until 1908. This was a heavy blow, but it should be balanced by remembering the scale of Lane's operations. In a normal year he sold five thousand head of grass-fed three- and four-year-old steers at prices which fluctuated from a low of $50 to a high of more than $70 per head, for a gross return of between $250,000 to $350,000.[167] To use the language of the stock exchange, Lane had suffered a "market correction" rather than a wholesale crash. The Fordney-McCumber Tariff imposed by the United States in 1922 was far more destructive to Lane's interests than the winter because it cut deeply into his profit margin year after year. Lane still had a well-established breeding herd and the capital resources to put the effects of the winter of 1906–7 behind him. In this regard he was unusual but by no means unique. While Lane extended his landholdings in the mixed-grass country by acquiring the SC Ranch, his partners in the Bar U, Gordon, Ironside and Fares, bought up the remnants of the herds belonging to the Turkey Track, the T Bar Down and the "76," and ran the consolidated herd along the Frenchman River. The great Circle outfit came through the winter intact but with considerably reduced sales. Their cheque from Burns and Company, their main buyer, was for $78,000 in 1907 instead of the average of more than $100,000.[168] After the winter they acquired the stock and the leases of Lord Beresford's Mexico Ranch. These examples were well-managed outfits with "deep pockets," and they understood the "effects which erratic markets and capricious weather might have on a year's balance sheet."[169]

We have examined the mythic winter of 1906–7 from both a general and a particular point of view. A few conclusions emerge. The winter deserves its reputation as a "killing winter." More cattle died on the Canadian range during it than at any other time. The winter of 1886–87 may have been as severe and as long, but the country was so lightly stocked that total losses were relatively small, especially when compared with the devastation suffered on the overstocked western ranges of the United States.[170] The winter of 1919–20 was equally severe, but stock losses were nowhere near as numerous as in 1906–7. However, Oetelaar explains that the long-term financial implications of this post-war winter were huge because ranchers spent so much on feed, and the droughts of the next few years prevented them from recouping their losses.[171]

The bitter winter of 1906–7 caused widespread suffering to stock, and inhabitants of small towns in the grazing country saw some grisly evidence of this. In the late spring the losses were all too obvious: coulees and fences were lined with the bodies of cattle. These real and gruesome images left an indelible impression on those who witnessed them. Nevertheless, many of the estimates of the numbers of cattle that died were greatly exaggerated. In the best of times, correspondents credited large outfits with thousands of head of cattle which they never really had. After the winter, these inflated totals were cut in half to provide estimates of the losses.[172] As we have seen, the pattern of losses was anything but uniform and was explained by a number of variables. The Livestock Commissioner estimated that 50 per cent of the cattle on the range were lost.[173] He had access to statistical information and jumped to the conclusion that the drop in numbers of cattle in the grazing country was linked solely to the effects of the winter. But we have established that "beefing out" herds started before the winter and continued after it.

The losses incurred during the winter affected all those who had stock on the range to a greater or lesser degree. But their reactions to those losses were as varied as was the pattern of intensity of the winter itself. Wallace Stegner suggested that the winter "made stock farmers out of ranchers."[174] This was certainly the case with some individuals in some locations, but we must resist the temptation to accept this change in attitude, scale, and technique as a "universal" response.[175] Many small ranchers were already displaying the attributes of stock farmers before the bad winter. This was a period which saw a marked growth in the number of stockmen who concentrated on purebred herds of cattle or horses, animals which were so valuable as to demand intensive methods to raise them. At the other end of the scale, some big operations were swept away, while others expanded and continued to use the last isolated fragments of the mixed-grass prairies much as they always had. Many who had aspired to develop a ranch based on a homestead entry had their hopes dashed not so much by their winter losses but by the press of settlement. They needed a section or two of unused range adjacent to their holdings to pasture stock while they cropped their homestead. As neighbours moved in and took up this vacant land, stock rearing ceased to be a viable option. Dry farming wheat offered many times the return from a limited land base than did stock rearing in areas where a single animal required at least thirty acres to sustain it. In 1908, odd-numbered sections were offered for sale in the form of pre-emptions or purchased homesteads at $3 an acre.[176] It was estimated that some twenty-eight million acres would be covered by the provisions of this act. This immense reserve of land which had been available for grazing was soon taken up by farmers. One aspiring rancher remembered:

> In our district (Ghost Pine Creek north-east of Drumheller) the little fellows just getting a start at ranching like myself had but two choices. They could sell out entirely and leave the country, or they could turn farmer. The later course meant pre-empting another quarter section, and selling off all but a few cows and possibly the few three year old steers that they happened to have. The proceeds from the sale of cattle would have to be invested in another team of horses and in farm machinery.[177]

In the park belt between Red Deer and Edmonton things were different again. More precipitation and less evaporation allowed more intensive mixed farming to flourish. The risk to the grain farmer in this region came as much from the short growing season as from moisture deficiencies, and the frozen grain crops of 1906 were marketed on the hoof.[178] Oats rather than wheat remained the principal crop in this area. By 1910, cattle shipments from the area north of Calgary more than balanced those from the old ranching district to the south, and Stettler, Vegreville, and Camrose vied with Bassano, Medicine Hat, and Maple Creek as shipping centres on the railway.[179]

The winter of 1906–7 occurred at a time of rapid change in western Canada. The losses inflicted by the winter were assessed by a diverse multitude of decision makers. To some long-time ranchers it was a fatal blow, to others it was an opportunity. Many small stockmen had expected to continue to make their living from selling beef, but the winter's losses forced them to take a new look at their futures. Some had more options than others. Those whose deeded land lay adjacent to hills, forest reserves or broken coulee country still had access to range lands and could continue to raise cattle, while others who had taken up land in the parched mixed-grass prairie were more likely to adopt the higher (but inconsistent) returns offered by dry farming. The sharp decline in numbers of cattle in the grazing region of Alberta and southwestern Saskatchewan between 1906 and 1911 was the result of an infinite number of individual choices made by families and individuals who had different experiences, abilities, and aspirations.

5.1 Percheron mares and colts on pasture at the Bar U, c. 1916, Parks Canada, Florence Fleming Collection, 95.07.01.21

5
Building an Internationally Famous Percheron Stud

"In no other department of human knowledge has there been such a universal and persistent habit of misrepresenting the truth of history as in matters relating to the horse."[1] – John H. Wallace

George Lane was a cowboy turned businessman. Since leaving the Bar U in 1889 he had made his living by buying and selling cattle, and his purchase of the Willow Creek Ranch and the YT were natural outgrowths and rewards of his commercial ability. Starting in the 1890s, Lane developed a new business enterprise: the breeding and sale of purebred Percheron horses. This interest played a more and more important role in his life after the purchase of the Bar U Ranch in 1902. For the next twenty years he invested heavily in the best available stock, and made this ranch the showplace for the biggest Percheron stud in the world.

What were Lane's motives for embarking on this new venture? What messages was he receiving from the economic environment that prompted him to diversify his operations and to enter the specialized field of horse breeding? Unfortunately, George Lane preferred to keep his thoughts to himself, except for speeches reluctantly made "in the line of duty." We shall never know for sure what motives lay behind his decision making. On the other hand we have the benefit of hindsight to help us reconstruct the ideas spurring his actions. Lane must have been acutely aware that the open range methods which he had perfected were under threat, and that the status quo was no longer an option. Even on his foothills ranches, where so much had already been accomplished, more had to be done to intensify production. Experiments with tame hay and fodder crops had to be initiated, and the slow process of improving herds through selective breeding had to be pursued. Such changes involved both capital costs and ongoing expenditures on a larger labour force. It must have seemed unlikely that such investments would have earned profitable returns given the prevailing prices for cattle. Breeding and rearing purebred horses, although it required all the expensive innovations outlined, produced a product of much higher value. By producing superb draft horses for incoming farmers, Lane turned a threat into an asset; it was a brilliant and courageous example of business foresight.[2]

The infrastructure of the Canadian west was put in place during the two decades preceding the outbreak of the First World War. While railways provided vital links to and from continental and international markets, it was the countless teams of draft horses which provided the power that underwrote the transformation of town and country. Every mile of railway grade levelled and every road improvement required horse power. Circulation and distribution within the growing towns, and out into their service hinterlands, were all handled by horse-drawn vehicles. The number of horses in Canada rose by a million in each of the intercensial periods, 1901–1911, and 1911–1921. Closer to home, in the new province of Alberta, there were 226,000 horses in 1906, and 828,000 by 1921.[3] This was a market undergoing extremely rapid expansion, and the Annual Report of the Department of Agriculture of the Province of Alberta for 1906 explained excitedly:

> During the year the demand for horses, especially of the heavier classes, has been exceedingly brisk. It is altogether likely that this demand will continue for years, at any rate as long as the tide of settlement continues as it has been for the last few years. There is especially good demand for horses suitable for dray purposes from all cities and towns. The heavy work incidental to pioneer life makes great drains upon the horse flesh of the country, consequently the demand arising from this, together with the demand for stock by new settlers, gives a very bright outlook to the horse breeding industry of the province.[4]

The following year the department reported that demand for heavy horses remained good, "not only on account of the large acreage brought under cultivation but for railway construction as well."[5] This importunate demand was to continue, for in 1911 the Alberta Horse Breeders noted with satisfaction:

> ... with the continuous marvelous development of this province, and the large amount of railway construction, a good horse market is assured for Alberta horse breeders for years to come.[6]

In his vivid memoir of life on a ranch during the first decade of the twentieth century, J. Angus McKinnon remembered that it was hard to hold onto work horses:

> ... farmers were continually coming along desperate for horses. It was not unusual to sell several animals and unhitch them to be led away by the new owners. This required a breaking crew to be constantly driving out young horses and keeping the depleted ranks filled up.[7]

Moreover, it was not only farmers who were after draft horses, McKinnon continued:

> There was also a terrific demand for horses building railroad grades, digging irrigation ditches and dams, as well as road building in the local improvement districts. Freighting supplies

> and materials to these jobs was also done by horses.[8]

McKinnon tells the story of a contractor who came to the ranch when nobody was home. He took a saddle horse from the barn, chased a bunch of horses, cut out ten of them and took them away to work on the Carsland dam. He came back later to pay for them, without being overly concerned about the price.

Experts predicted that the outbreak of war in Europe would intensify the demand for horses of all kinds. J.G. Rutherford explained that "more than one million horses were engaged in the war, and that the average life of a horse on active service at the front was estimated at only seven days for cavalry and thirty days for artillery horses ... he felt that twelve million horses would be needed to sustain the war effort."[9] The Livestock Commissioner looked forward to the termination of the conflict:

> When the war is over and the various European countries restock their devastated farms, and when enormous tracts of open prairie in the North West come under cultivation, the demand for good draft horses will be unprecedentedly great and the prices probably higher than we have ever known.[10]

Such was not to be the case. The last year of the war ushered in a period of much more fickle demand and lower prices. "Conditions prevailing in the horse raising industry ... have not been such as to cause farmers and ranchers to increase their stock of brood mares."[11] Continuing drought, lower grain prices, and the impact of the severe winter of 1919–20, combined to reduce demand. "Horses have perhaps never been lower in price."[12]

It was not until 1924 that there was a "stronger tone in the horse market," and "a revival of interest in horse breeding."[13] By 1926, a shortage in horse power was forecast as stock presently on farms was considered to be aging.[14] Continuing strong demand and good prices through 1927 and 1928 led to a revival:

> ... a feeling of optimism among breeders of stock which has not existed since 1919 ... it is unfortunate that there are not the number of good pure-bred horses and cattle in the province that there were ten years ago. Never was there a time in the history of the province when there was such a dearth of good sires, and it is to be hoped that with the higher price ... this condition will soon be remedied.[15]

As Alberta entered the Depression the demand for draft animals remained strong. Many farmers weighed the low price of wheat against the high price of gasoline and continued to opt for horse traction. In 1932 there was "great demand for farm chunks weighing 1,400–1,500 lbs., though there is also a demand for the real good big draft horses, but only a few of them are to be found."[16] In the light of these reports it would be simplistic to suggest that the decline in the fortunes in Lane's Percheron breeding operation in the early 1920s was a direct response to mechanization on farms.

Horse breeding was a specialized business. The original Alberta horse ranches, like the Quorn and the Bow Valley Horse Ranch, aimed at producing riding horses and teams to pull light conveyances. Of course, the Quorn was named after the famous Leicestershire hunt, and the lofty aspiration of the shareholders was to produce hunters for the British shires and remounts for the imperial cavalry.[17] Similar expectations were widespread. Harry Sharpe, writing on "The Future of Alberta Horse Breeding" in 1894, started his piece with an advertisement for Canadian horses in *Field* magazine. He stressed the demand for good horseflesh in Britain, and suggested that Alberta could share in the lucrative market for first-rate hunters.[18] The outbreak of hostilities in South Africa produced an upward surge in prices of Canadian light horses, and the visit of Colonel C.H. Bridge, C.B., C.M.G., Assistant Inspector of Remounts for the British War Office, was reported with enthusiastic expectation. Some 116 head of horses were indeed shipped to South Africa; however, even the exigencies of wartime did not lead to a large-scale export trade in light horses. Economies of scale and low production costs could not be realized in horse breeding as they could be in the cattle business, and price differentials between Canada and the United Kingdom were not sufficient to offset the international freight rates.

This class of light horses was the first to suffer from the onset of mechanization. Bicycles, electric trams, and motorcars replaced horse power in the cities and towns, and their influence soon spread to rural areas. George Lane obtained his first car in 1910, and the famous Bar U cowboy, Charlie Millar, made history the same year:

> "I'm the first cow puncher from the Bar U that ever made the trip from Pekisko to High River and back in three and a half hours," exclaimed Charles Millar on Sunday as he motored back to Pekisko in a Maclaughlin Buick machine in clouds of dust. The auto is certainly revolutionizing things in this country and is becoming a popular mode of traveling and it reduces distances very materially; the trip to Pekisko being made in about one hour and a half.[19]

By 1916, the fate of light horses had been settled:

> All light horses, including coaches, hackneys, and roadsters … have gone by the board and like the Othello of old, their occupation is gone, probably never to return.[20]

In contrast, the mechanization of farm operations was a much more gradual process drawn out over two decades.[21] The slow, heavy, and cumbersome steam tractors were particularly valuable for breaking prairie sod for the first time. They were operated on a custom basis, for they were far too expensive and difficult to operate for individual farmers. It was not until the First World War that light, fast, and manoeuvrable tractors like the Fordson became widely available at a price that was competitive with that of a four-horse team. The Government of Canada played a role in diffusing these machines to meet the wartime demand for greater food production and the shortage of farm labour. The individual decision to switch from horses to tractors was always a complex one.[22] It depended on the size of the farm and the mix of enterprises undertaken, as well as on physical factors like soil quality and rainfall. On most farms, during a long transition period, tractors and draft horses coexisted, each being employed for the tasks at which they gave the best return. Nor was the diffusion process uninterrupted. Changes in the balance of costs and returns produced a swing back toward horse power during the early 1930s. The Livestock Commissioner explained:

> The farmers of late have been relying on motor power, and therefore have not been raising many colts. In the past year a great many of them were forced back into using horses and this has created considerable demand.[23]

Nevertheless, the total number of horses in Canada was down by half a million in 1931 from the peak number recorded in 1921. In the west, horses lingered in large numbers even after their working role had been usurped by mechanization. It was not until 1944 that a determined effort was made to reduce the number of idle animals eating valuable grass.[24]

The timing of Lane's foray into horse breeding could hardly have been better. He enjoyed a decade of booming markets and rising prices between 1908 and 1918. Of course, the purebred Percherons from the Bar U did not pass directly to incoming farm settlers; most of them were sold to other breeders. Nevertheless, it was the insatiable demand for horse power in general which supported the highly specialized sub-sector Lane had chosen to invest in. It is clear that the "bullish" market suffered a severe setback during the years after the First World War. This contributed to Lane's financial difficulties during the early 1920s.

Introducing the Percheron

Historian and horse enthusiast Grant MacEwan provided a striking introduction as to the origins of the Percheron breed, he wrote:

> A search for Percheron roots leads to three distinct and powerful areas of influence in the homeland: first the ancient and unrefined Flemish horses of western Europe; second, the transforming force of the North African and Arabian horses left behind by an army in flight; and, third, and comparatively recently, the united pursuit of clearly defined type and performance ideals by French horsemen in the lovely old district of le Perche, southwest of Paris.[25]

Le Perche lies on the southern margins of Normandy. It is a transitional region between the open fields of the Paris Basin and the Bocage (mixed woodland and pasture) of the west. Its steeper relief distinguishes it from its neighbours, and its wooded hills straddle the watershed between the English Channel and the Loire valley [Map 5.1]. Le Perche is a vernacular region, recognized by inhabitants and neighbours alike, but having no precise legal or geographical definition. The heart of the region associated with the development and promotion of the Percheron horse is the area within thirty or forty kilometres of the town of Nogent-le-Rotrou. In spite of its centrality and relative proximity to the capital, this is a region of rolling hills and valleys where traditional farming and old-fashioned values linger. [Plate 5, colour section]

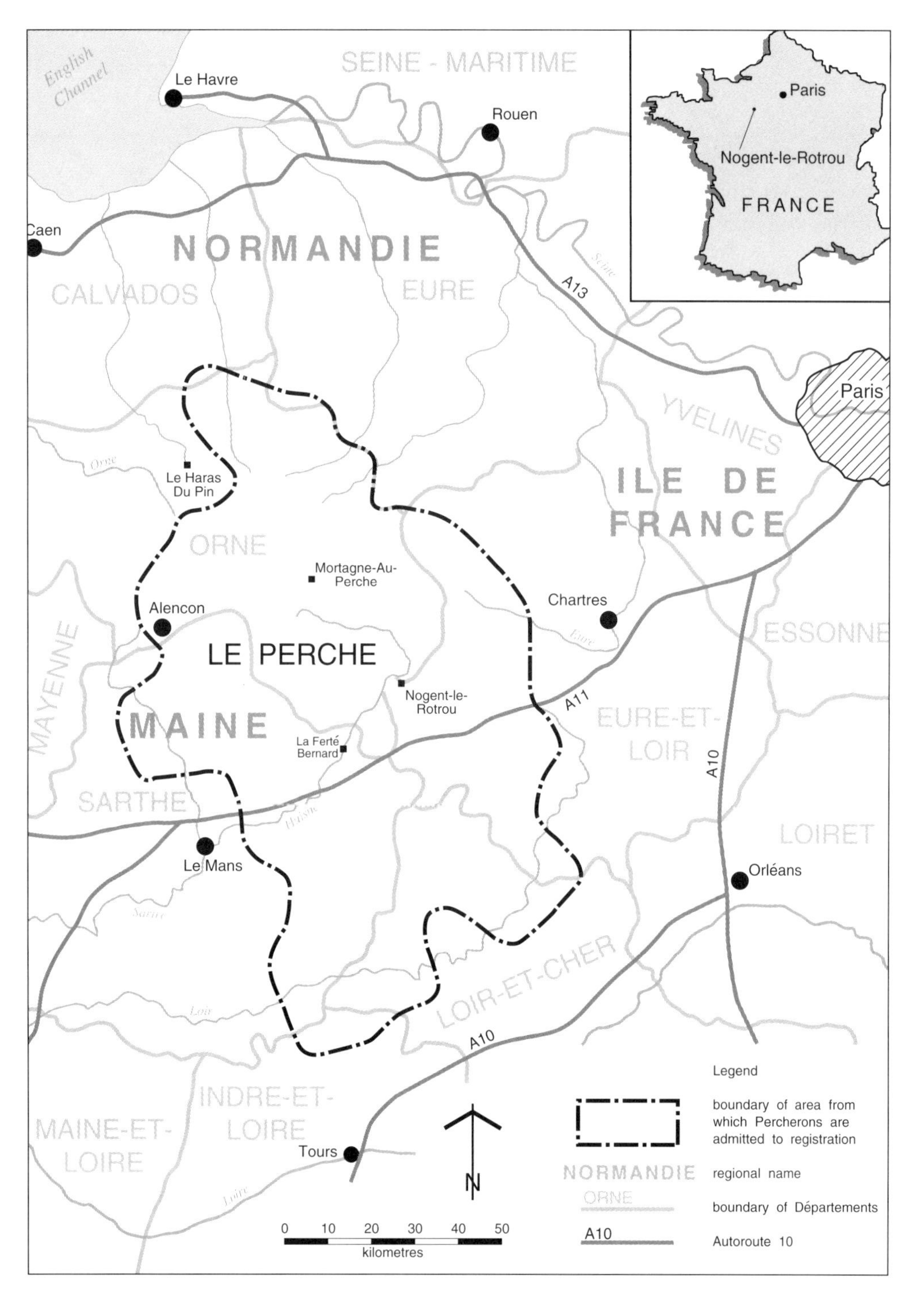

The Percheron was developed as a general purpose or *diligence* type of horse; medium-sized, active, and tireless, and a specialist at pulling the cumbersome stagecoaches over rough and often muddy roads. Indeed, the French favoured grey or white horses because they stood out well at night. As railways began to displace carriages around the middle of the nineteenth century, so larger and larger animals were bred to satisfy the growing demand for heavier animals for both agriculture and street use. The credit for this work of converting the old-time trotting Percheron into a heavy horse was due to the breeders of the district themselves. "Centuries of evolution in a small country where the soil, the climate, the forage, and the very air itself conduced in the highest degree to the production of good horses, have accomplished the result so admired today."[26]

The hardiness and versatility of the breed impressed all who came into contact with Percherons during the South African War and the First World War. The *Live Stock Journal* of London claimed that, "the Percheron type has made many friends in England and … is firmly established in the hearts and minds of the responsible officers of the British army. The Percheron fills the bill best of the many types brought from all the world over since the outbreak of war."[27]

Map 5.1 The location of Le Perche, a vernacular region which sprawls across the boundaries of several Departments. See also colour section.

Percherons were regarded as "easy keepers," and thrived when confronted with inclement weather, poor housing, and hard work. They could be taught to rustle and were adaptable and robust. By the turn of the century, after several decades of breeding for size, Percheron conformation conveyed a sense of real power. Sires like the Brilliants weighed over a ton each, while Pink, a famous stallion brought to Alberta by R.W. Bradshaw in 1910, weighed 2,100 lbs., and stood seventeen hands. In general Percherons outweighed their rivals the Clydesdales by some 300 lbs. Those who worked with Percherons in the fields or streets commented on their staying power and their capacity to work consistently for long periods. They were intelligent and yet docile and biddable. The fact that most of the horses used to pull omnibuses in London during the early 1900s were Percherons is a testament to the breed's sang-froid.

The Percheron stallion possessed another quality that was less obvious, but that paid even greater dividends. He crossed well with almost any kind of mare. "No matter what he is bred to, the progeny of a Percheron stallion will always take after the sire in all practical essentials, and for this reason will be a good salable horse whether dammed by a large or small mare."[28] A common mare bred to a Percheron would produce large strong colts which could be sold for $30–40 per head more than the progeny of a common sire. Moreover, half-blood Percheron mares made great brood mares, and might well produce large colts which might mature to weigh 1,600 to 1,900 lbs., and be sold for more than double their smaller kin. For a farmer or small rancher with some capital and plenty of patience, the possession of a Percheron stallion could be a high-yielding investment. Lane was at pains to stress this point in a promotional album sent to a valued customer in New York during 1917. A series of pictures of handsome grade Percheron geldings were accompanied by the following captions:

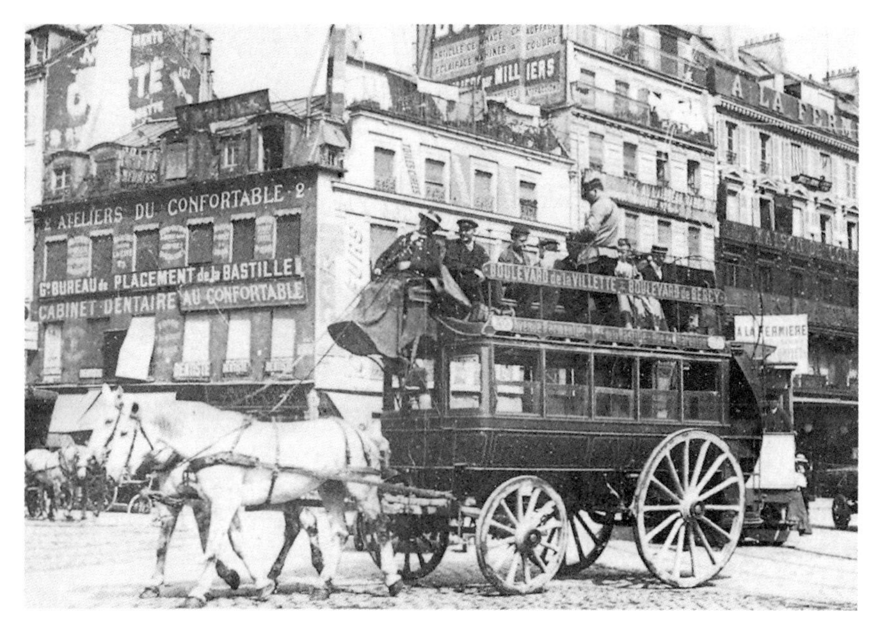

5.2 Percherons pull an omnibus over the cobbled stones of the Place de la Bastille, in Paris. Association des Amis du Perche, *The Percheron Horse Past and Present*

> It is easily within the power of any farmer with small but sound mares and the exercise of common sense to produce these valuable types from such mares. Work horses on Namaka farms bred up from very small stock to as high as 1,800 lbs. This magnificent type of grade Percheron weighing over 1,800 lbs., perfectly sound, with splendid conformation exhibits the possibilities of sound breeding. Usefulness, pride, character and reputation are gained from the ownership of such animals.[29]

5.3 Three famous Percheron stallions at the Bar U, left to right, Garou, Halifax, and Americain, 1916. Parks Canada, Hugh Paulin Collection, 96.11.01.16

During the second half of the nineteenth century, the Percheron was introduced to North America. Individual American visitors to France were struck by the attributes of horses from le Perche. One saw the Percheron team that pulled the wagon advertising the "Magasin de la Paix" department store in Paris, and just had to have it! Others noticed the gait and vigour of the Percherons that were pulling their carriages and bought them on the spot.[30] Between 1853 and 1870 some ninety stallions and twenty-one mares were exported to the U.S.A. During the next decades this trickle became a veritable flood as breeders like Mark W. Dunham of Wayne, Illinois, I.L. Ellwood of Dekalb, Illinois, and James Harvey Sanders, the founder of the American stud-book, returned to France again and again and bought hundreds of Percherons. The period from 1870 to 1880 saw the export of 1,250 horses, while in the next decade this figure rose to 7,552.[31] Pioneer visitors like Dunham were joined by hundreds of other American breeders who wanted to see for themselves the homeland of the Percheron and the depth of breeding available. One regional history explains:

> Until the approach of the war in 1914, the arrival of Americans in the Perche must indeed have seemed like the coming of the swallows, the harbinger of spring. They used to arrive at Nogent and La Ferte Bernard early in April, creating a proper 'home away from home' in their hotels; they gave concerts for their friends and let their families go off on trips around the Perche.... Small buyers who could not make the journey individually, formed groups entrusting the buying to one of their members; local breeders gave them three, four, or five years credit and were never let down.[32]

This export trade earned considerable wealth for the region of le Perche. It is hard to put a dollar value on it because the price paid for individual horses varied so greatly. However, George Lane paid an average of between \$400 and \$500 for his purchases during the first decade of the twentieth century, and total exports averaged about 2,500 horses per year; this leads to the conclusion that sums of over \$1 million flowed into the region each year.[33] As the lion's share of this influx of capital was retained by a handful of top breeders, it is not hard to see where the money to extend and beautify their homes and stables came from.

Indeed, the "American connection" played a significant role in transforming le Perche. The present cultural landscape of this beautiful region of France reflects the glories of its 'Golden Age,' when horse enthusiasts from all over the world made their way to the spring sales at Nogent-le-Rotrou. Several of the great stud farms still exist although they no longer concentrate on turning out purebred Percherons. When I visited the region in 1995, I took with me the illustrations from Alvin H. Sanders book *A History of the Percheron Horse*, which was published in 1917.[34] Even after the passage of some eighty years and two World Wars, I felt a definite sense of déjà vu as I negotiated the narrow lanes and viewed the wide flood plain of the Huisine. It was amazing how countryside and townscape had survived, modified of course, but still easily recognizable.

The most striking and easily accessible example of 'the past in the present' is la Touche, until recently the home of the Aveline family. Located on a hill just outside Nogent, the estate was used as a showplace to entertain prospective buyers and to show them the stock available. The huge flagged courtyard is flanked on three sides by horse stalls and barns and overlooked by a gracious residence. Passing through an archway one finds oneself in a formal garden sloping gently down toward the river. Looking back from the bottom of the garden, the magnificent western facade of the house is framed with overarching chestnut trees. Horses were brought here from the Aveline family's outlying farms to be fitted and shown. Around the doorway of each horse stall there is a mosaic of prize plaques, weathered now to a variety of blue colours, and bearing witness to past triumphs. For a hundred years visitors from all over the world came here looking for their dream horses. George Lane and Alex Fleming probably spent days here in 1907 and 1909 assessing the relative merits of various stallions in their loose boxes and the mares and foals grazing in the pastures below the house. They must have been confused by the bewildering range of talent paraded for their consideration. In the same tradition, Japanese buyers visited as recently as the fall of 1980 on an equally successful buying mission.[35]

We found another historic farm in the tiny village of Dorceau. This was the home of Joseph Aveline, a friendly competitor and no relation of his namesakes, Charles and Louis Aveline. The U-shaped layout, with the house looking out over the parade courtyard, was similar to la Touche, but here the most obvious visual connection with the glories of the past was a magnificent avenue of Sequoia trees. These trees had grown from seedlings brought over as gifts for the family by American breeders. They are about a hundred years old and tower over the barns and outbuildings of the farm. [Plate 6, colour section]

The medieval town of Nogent-le-Rotrou (population thirteen thousand) was the centre where most American and Canadian visitors stayed when they visited le Perche. It was exciting to follow in their footsteps up the narrow Rue Saint Hilaire to turn into the Place Saint Pol. I felt a real sense of familiarity as I looked down into the square from the balcony of the town hall. It looked much the same as it did a hundred years ago when cavalry officers inspected the horse stock of the district there. In particular the roofline of the houses is unchanged although the commercial enterprises which they shelter have gone through many transformations. Crossing the Rue Villette Gate, we spotted the Dauphin Hotel, the home away from home for American and Canadian visitors. A little further down the street we came across the evocative sign of the "Societe Hippique Percheronne de France." This is the headquarters from which the breed is promoted and the storehouse where records and photographs have been accumulated since the society was founded in 1883.

One of the most delightful things about exploring in le Perche is that one may be surprised around any corner by a herd of Percherons grazing in an apple orchard. Everywhere there is a consciousness of the Percheron heritage. The local rugby football club is called les Percherons; the bakery makes Percheron pastries; above all the present generation of breeders have inherited a passion for the breed from their illustrious ancestors. In spite of changing times, they still have the Percheron in their blood, and will explore ways of maintaining and extending the breed's profile. They are hoping to establish a historic site celebrating the Percheron horse at la Touche. The cultural landscapes which developed in le Perche before the First World War – sustained by a massive influx of capital from North America – are of considerable interest and deserve to be more widely known. The texture of the French countryside, in its infinite variety, always repays careful examination. An explanation of the history of the Percheron horse and a tour of the famous old stud farms would provide a means of encouraging visitors to the region to look beneath the surface.

The breed was established in the United States during the early 1880s. Mark Dunham's Oaklawn Stud, at Wayne, Illinois, became the pre-eminent Percheron breeding centre in the new world. Two of his imports, the black stallions, Brilliant and Brilliant III, established the bloodline of most North American born Percherons.[36] Year after year, the best young horses bred by the Perriot brothers in le Perche, found their way to Oaklawn. Later, Dunham estimated that he had spent some $350,000 to obtain the top bloodlines for his stud.[37] Since almost all North American Percherons trace their ancestral lines to one or other of the two Brilliants (sire and son), the Canadian Percheron heritage is so interwoven with that of the United States that it would be extremely difficult to draw firm lines of separation between them.

The expansion of the Percheron breed in western Canada was handicapped by the entrenched position of the Clydesdale.

> The obstacle, briefly stated, lay in the fact that early settlers in Canada, outside of Quebec, came very largely from Scotland, brought their own livestock with them and were being financed by stockmen in the old country who were naturally anxious to sell the surplus from Scottish herds, studs and flocks.[38]

As Percherons began to attract more and more supporters during the early years of the twentieth century, so rivalry between them and Clydesdale owners became more marked.

> ... in some districts, where breed prejudices seemed to get out of hand, Clydesdale supporters sat defiantly at one end of the judging ring bleachers and Percheron supporters with the same serious scowls sat at the other.[39]

During the First World War this rivalry became political. The Canadian Percheron Association felt that their horses were being discriminated against with respect to public works, in competitions, and especially at the College of Agriculture at the University of Saskatchewan, where Dean W.J. Rutherford held sway. However, the dispute was diffused when the government matched its purchase of a first-rate Clydesdale stallion in Scotland, with the acquisition of the Reserve Grand Champion Percheron from the Chicago Show in 1919.[40]

Early Percheron Purchases

Lane was impressed with the contribution which Percheron stallions had made toward improving horse stock in Montana. He once explained why he was drawn into breeding Percherons in the following words:

> When I first came to Canada from Montana in 1883, to take charge of the cattle on what was then the newly started Bar U Ranch, I was particularly impressed by the absense of horses such as we had been accustomed to ... by this I mean horses that were able to do ordinary hauling and farm work, and that also had the endurance and speed necessary for making long trips to the railroad ... I came to the conclusion that it was Percheron blood that was lacking in the horses of western Canada.[41]

To redress this lack, Lane visited the Mauldin Horse ranch at Dillon, Montana, and bought the entire band of "Diamond O" purebred Percherons, thirty-five head in all.[42] Many of these young horses had been sired by the stallion Americo, which James Mauldin had imported from France, and their quality was high. Because of a general economic downturn and a sudden panic in horse-raising circles caused by the popularity of the bicycle and the adoption of the electric tram, Lane was able to obtain these purebred animals at $20 a head, which was a fraction of their "book price." Lane also bought more than a thousand head of the best grade Percherons available in Montana. He acquired the pick of the Flurry bunch, a lot from Poindexter and Orr, and all of the ST Bar horses from the Samms Cattle Company. These horses were purchased at prices between $5 and $15 a head.[43] Next, Lane visited the Riverside Ranch in North Dakota in order to enlarge his stock of pedigree horses, and a short time later, Lane made his first pilgrimage to the "shrine" of Percheron breeding in North America, the Dunham establishment at Wayne, Illinois. He was able to obtain the black stallion Presbourg, imported from France, and the American-born Paris. These two studs appear with the registration numbers 1 and 2 respectively in Volume 1 of the *Canadian Percheron Stud Book*.[44]

In 1907, Lane made the first of several trips to the home of the Percheron breed. During this visit cordial

5.4 Pete Hardy, Bar U horseman, trotting out a Percheron, c. 1920. Parks Canada, Florence Fleming Collection, 95.07.01.16

relations were established with Charles Aveline, one of the pre-eminent Percheron breeders in France. This contact bore fruit in 1909 when seventy-two mares and two stallions were imported to Alberta.[45] Scarcely had these animals settled in, than Alex Fleming left for France again in company with another notable Alberta Percheron breeder, W.B. Thorne.[46] In January 1910, fifteen stallions arrived in High River. They were two-year-olds, and the lightest of them weighed over 1,600 lbs. They were, "in the estimation of professional horsemen, the best bunch that ever entered Alberta."[47] In the spring, the *High River Times* reported:

> George Lane left last week for France to purchase 100 more head of Percheron mares for his Pekisko horse ranch. It is rather difficult to obtain good Percheron mares in France and the price has gone way up, but Mr. Lane is determined to supply the demand if possible.

Coincidentally, on the same day, an advance guard of sixteen mares arrived in High River. Twelve were three-year-olds and two only two-year-olds. The heaviest mare, a beautiful white Percheron, with the typical lines of a brood mare, weighed 1,700 lbs., while the lightest turned the scales at 1,400 lbs. The shipment came via La Havre, Southampton, and Montreal, and were under the care of Allen Williams, the journey taking only thirty-five days.[48]

Lane returned home in July, and announced with some satisfaction that he had been able to obtain a bunch of "the very cream of French horses." The scene where the transaction had been concluded was described in a French newspaper:

> The transaction took place at the show of the Percheron Horse Society of France, which was held on July 1–3, at Nogent-le-Rotrou. This quaint old town was, in the middle ages, the capital of the ancient feudal duchy of Perche, whence the Percheron breed of horses took their name.... A Canadian gentleman [George Lane] was conspicuous among the purchasers. A lot of 75 mares and 25 stallions was due to leave for Antwerp under the guidance and supervision of Mr Louis Aveline, son of the President of the Percheron Society. All of these animals are registered in the French Percheron studbook and are of the purest origin. With this important lot of choice animals, Mr Lane contemplates establishing in Canada 'un elevage' of Percherons of the very first class, such as never hitherto existed on the continent of America.[49]

Many French breeders were reported to view with regret this loss of so much of their best horseflesh.

Lane's energy and his collector's enthusiasm were close to becoming obsessional during this period. In December, 1911, he attended the Chicago show, and was impressed with the stallion Imprecation, which took the grand championship. However, the price proved to be prohibitive, and so Lane determined to pursue Imprecation's sire. He cabled his friends in France to check the stallion's breeding, and found that the sire, Pinson, had been bred by a French widow, and sent to the United States three years previously. After an anxious search, the stallion was traced to Fort Wayne, Indiana. Lane hurried there, checked out the horse and some of his "get," and purchased him from a consortium of owners who had no idea that he was the sire of the famous Imprecation.[50]

> Pinson is a beautiful gray, stripe, left hind pastern white, trace of white on right front pastern, head clean cut, size proportionate, neck crest arched and muscular. The symmetry of the animal was declared by those who know, as being perfect, with the legs squarely placed and well set under the body. Its temperament might be called energetic and kindly.[51]

This was the end of the period of wholesale imports to build up the Percheron stud. Thereafter, individual horses were purchased as required. By 1916, the Percheron herd at the Bar U was at its largest, with seven hundred registered horses and four hundred brood mares. It was described as the largest Percheron breeding establishment in the world.[52] When Louis Aveline visited the Bar U from le Perche in 1915, he recognized many of the French imports and declared that Bar U Percherons were the equal to any found in his home region. According to Lane:

> Among the earliest and best sires we used were 'Paris' and 'Presbourg.' Both gave us good results, and 'Presbourg' particularly proved to be an extremely potent sire of high class brood mares. Experienced judges, in going over the band of brood mares in 1914, with a view to selecting the brood mares of the best type, picked on more 'Presbourg' mares than mares of any other breeding, and these mares have been noted as among the best producers in our stud.[53]

However, the greatest breeding stallion used by Lane was Halifax. He was acquired for $3,000 after winning the title of 'supreme champion' at Winnipeg in 1909, and established an enviable record. Halifax stood seventeen hands, weighed over a ton, and had an unusually good character. He was regarded by both French and American judges as one of the ten best Percheron sires in the world.[54] His outstanding youngsters were mothered by daughters of Presbourg, thereby laying the foundation on which the Percheron breed was to rest in western Canada for the next sixty years. A son of Halifax, Lord Nelson, was exported to North Dakota and contributed significantly to the breed in the United States. A second son moved to Great Britain and assisted in founding the breed there.[55]

Methods and Techniques

Lane credited the unique physical environment of the foothills for the success with which his breeding program was blessed. In 1915, ninety purebred foals were raised, and by 1916, when the operation reached its full potential, no fewer than 117 young horses were born. Horses were raised as far as possible under natural conditions, and there was no pampering. A regime of fresh air and exercise was lauded by experts:

> Warm stables, which in turn are responsible for short coats, impure air, lack of sunshine, develop

weak constitutions and impair digestion, while the very opposite is the case when horses run out of doors and get plenty of wholesome food and shelter from storms.[56]

In the fall of each year the mares and foals were rounded up, driven into a vast network of corrals and separated. At this time each youngster was branded under the mane with an individual number, the same number being placed in a memorandum book, and eventually on the individual's pedigree papers. Neatly applied with a small set of copper irons, it was possible to distinguish any animal in pasture, barn or corral, and to determine its name, registration number and breeding. No foals were registered until they were yearlings, for all too often a black foal would turn grey when it shed its winter hair. Each year's crop was identified by a code letter at the start of its name. From the crop of 1916 came Oyama, Osler, Olive, and Olympic Maid; while 1917 saw Perfection, Pride of Pekisko, Paragon, and so on.[57]

In the spring, each youngster was carefully inspected, those showing potential being retained as stallion prospects, while the culls were altered and developed as geldings. Entire horse colts were grain fed even while at pasture during the summer. They ran in large pastures surrounded by fences of woven wire. Feed bunks were installed in each pasture, in which colts received their daily ration of grain. The weaning fillies were well cared for their first winter. Then they were turned out on native pasture, receiving no grain from that time on. The choicest young horses were registered, and the cream of the crop were fitted for the show ring. Brood mares at the Bar U were never pampered. They ranged the hills west of the ranch in summer and were moved to the Bar U flats for the winter. Grazing on the native "prairie wool," these mares rustled their own feed, never receiving hay or grain. While snow did get deep on occasion, the chinook winds would periodically expose large areas of native grass.

The leading stallions were housed in a complex of barns separated from the rest of the stabling area. Several stallions were bred by hand, some were ranged with the mares each spring, while others were used by way of artificial insemination, which was still at the experimental stage. For example, the huge white stallion Garou was turned out each spring with a selected harem of mares. Percy Gardner recalled, "a collar that would fit old Garou would fit an elephant." The black stallion Americain was a perfect gentleman about

5.5 Two-year old Percheron colts at feed bunks at the Bar U. Note the massive fence posts and page wire which was much safer for the horses than barbed wire. Parks Canada, Hugh Paulin Collection, 96.11.01.22

the stable, but turned into a devil when taken to the breeding corral. He would grab the saddle horse of the mounted groom by the neck, and then stamp on it! A strap iron nose guard was devised to prevent such incidents.

Bert Sheppard worked at the Bar U in 1922. He remembered that the five hundred mares were run in two bunches during the breeding season. The mares and colts ran in the wet mare field just to the north of the ranch, while the dry mares and the three-year-old fillies ran in what was referred to as the dry mare field, adjacent to and south of the ranch. Bert Pierson was the man in charge of the breeding program at this time. He corralled all the horses in one field in the morning and did likewise in the other field in the afternoon. Two men that looked after the stallions saddled up in the morning and led out to the corrals the studs which Bert had specified.

> Bert was an old and experienced horseman and was as active as a cat. He used the same horse as a teaser right through the summer. He would lead this stud to the middle of the corral, give him six, eight or ten feet of rein and let him strut his stuff: there would be mares squealing and hoofs flying, with Bert dancing just out of range. Looking back at it now, his feat of horsemanship was just as exciting and just as dangerous as the show put on by the rodeo bull riders of today. Only once did I hear of him getting into trouble. A cranky old mare took a run at him and knocked him down and trampled him; it was a mare which had been brought up from the Namaka farm and Bert did not know what she was like. The teaser had a rough time of it too, and was pretty well beat up by fall and usually died the following winter.[58]

Rewards and Returns

The excellence of Lane's Percheron purebreds, coupled with the scale of his operations, meant that Bar U horses dominated the show rings of the North West during the decade from 1909 to 1919. In that time the Bar U Percherons collected prize ribbons too numerous to count and, according to one writer, sufficient trophies "to stock a galleon."[59] Headlines drawing attention to each new triumph were commonplace in the *High River Times*. Typical would be "Pekisko Percherons Sweep Everything," or "Bar U Makes Clean Up."[60] The show Percherons were first exhibited close to home. For example, at the Provincial Horse Show held at Calgary in April, 1909, "George Lane's Percherons were the dandiest of them all. He captured several prizes and the horses merited the prominence given them."[61] The Bar U gained first and second prizes for Percheron mares of two years and over, and stallions Garou and Epatant also won cups. Later the same year, Lane decided to exhibit at the Seattle Fair, a much more competitive forum. On 2 October 1909, he wired his home-town newspaper:

> Sixteen firsts; first with six horse team; first reserve champion and grand champion with mares. First reserve champion and grand champion with stallion. Going Some![62]

The classic matron Bichette was champion mare in Seattle, Spokane, Calgary, Saskatoon, Regina, and Brandon, for the 1909, 1910, and 1911 seasons. The ranch house at the Bar U became the depository for a magnificent collection of cups, shields, and medals. Among the most treasured awards was a picture of President Taft handing George Lane the shield for the finest exhibit of livestock entered by one concern at the Yukon Pacific Exhibition in Seattle, 1912. Lane was also particularly proud of a gold medal presented to him at the Dominion Exhibition at Regina, for he felt that it was evidence of the appreciation of the French government and the Percheron Society for his efforts to build the reputation of the Percheron in Canada.[63] Another more utilitarian trophy was the set of silverware awarded to the Bar U by the Percheron Horse Association of America. The knives, forks, and spoons were all crested with the engraved head of a Percheron.

George Lane played a key role in establishing an independent Canadian Percheron Horse Breeders' Association. He and his neighbour and friend, W.B. Thorne, had spent a year or two laying the ground work, and their planning came together at a meeting in Regina in August 1907. Six western horsemen, and a representative of the Dominion Live Stock Commissioner, elected W.B. Thorne President, R.P. Stanley, Moosomin, Vice-President, and F.R. Pike, part of the management team at the Bar U, Secretary-Treasurer. Lane, who always tried to avoid conspicuous roles for himself, agreed to be a director. The Association met annually in the following years and worked hard to produce the Canadian Percheron Stud Book in 1912. This was the official register, showing names, birthdates, colours, breeders, sires, dams, owners, and registration numbers for more than two thousand stallions and a similar number of fillies and mares. Where animals were already registered in the United States or France, their names would be followed by registration numbers enclosed in various types of brackets. Canadian numbers were written with square brackets, French with rounded brackets, and the United States without any brackets. Lane's overwhelming contribution could be judged by a count showing that his name appeared for 431 entries, roughly 10 per cent of all the entries in the stud book.

While the cups and trophies amassed at the Bar U provide one measure of success of the Percheron stud, it is far harder to chart the extent to which the venture was a commercial success. Lane had certain fixed ideas concerning sales, and these may have had an effect on turnover. He explained:

> I want to sell my horses, of course, but I want to sell them at a fair price to the purchaser, and not at one third or one half more cost to him, into the pocket of the middleman. It is true that by sticking to this principle I have lost many a sale, but what of it? Why should the purchaser pay $2000 for a mare when it can be bought for $1,000 or $1,200, or $3,000 for a stallion when it can be purchased for $1,700 or $2000? I made enemies and lost sales at first, but eventually, and as soon as dealers knew my principle, the sales increased.[64]

Buyers who did visit the ranch were granted every opportunity to select superior breeding stock at prices suited to their means. The stock shows at which Percherons from the Bar U were such consistent winners were also an opportunity to affect sales. After one such show, the *Farm and Ranch Review* commented:

> There must have been about $50,000 worth of horses sold at the Calgary Spring Show: George Lane had at least $10,000 worth booked before the end of the week.[65]

Bert Sheppard remembers young studs being sold for about $1,000 a head at two years of age. They were shipped out of High River by the carload.[66] Buyers from the United States were frequent visitors at the ranch, and at least one stallion, Oyama, was sold to Japan. However, it was probably the shipment of Percherons back to Europe which gave Lane the greatest satisfaction. Twenty-six mares and one stallion were shipped to England in 1918. A larger shipment followed a year later. This consisted of fifty-three head, including the stallions, Perfection and Paragon. After their arrival these studs won first and second in their class at the Royal Show.[67] These exports provided an important foundation for the developing Percheron breed in Britain. They were distributed among the top breeders in England, like Lord Minto, Sir Alexander Parker, Sir Henry Hoare, and Robert Parker. Colonel Henderson, A.D.C. to the Governor General, purchased two mares when he visited the Bar U in connection with the visit of the Prince of Wales in 1919.[68]

For all the successes recorded in the show ring, and notwithstanding the spectacular sales which made headlines from time to time, Lane's Percheron-breeding enterprise was extremely vulnerable to changes in the economic climate. In general terms, as long as farm settlers flooded in to the Canadian west, so the demand for heavy horses to provide traction remained strong. However, few if any struggling farm families could afford purebred stock. From each annual foal crop at the Bar U, the best colts were often the last and the most difficult to sell.[69] The unique qualities and lineage of a particular horse ensured its value in France, and this established the "book price" for the animal. However, the value of that animal and its offspring, when translated to western Canada, depended entirely on how much stockmen were prepared to pay for it. The actual figure would reflect the general state of the economy, competition from other breeders, demand, and other intangibles like snobbery, fashion, and reputation.[70] Any downturn in confidence would mean a growing gap between the book price and the returns actually achieved. In these circumstances the more stock which was sold, the greater the "book losses" to the enterprise.

Breeding purebred horses is not a task for the faint of heart. The risks are considerable and the returns modest. A stud book in the Clifford/Alwood Collection allows us to follow the "family history" of mares born in 1908.[71] There were fifty-one mares in this group, and all their names started with the letter I, from Incuse to Iquique. They were first bred when they were three-year-olds in 1911. Most mares were serviced twice, while some were serviced three or four times. Only two stallions were used, Epatant and Intitule. Thirteen foals were born, two of which died soon after birth. Illettree was imported in foal and gave birth to Alberta Boy, and Invasion was never bred.

Thus the "yield" was eleven live foals from fifty mares (22 per cent). The following year, Epatant remained the dominant sire but Pinson played a subsidiary role. There were seventeen live births, but two foals died the same month and there was one abortion. The yield was fifteen live foals from forty-six mares (33 per cent). The last year for which information was recorded was 1913. Pinson was the only stallion used, and only one service is recorded for each mare. There were thirteen live births and no deaths from forty-two mares (31 per cent). Nine mares had died by the end of 1913. In this generation of mares 32 per cent did not foal in the first three years of their reproductive life; 52 per cent had one foal, and 16 per cent had two.

Another item in the collection – Frank Pike's small leather-bound notebook – provides a detailed picture of the1915 foaling season.[72] In it he recorded when a mare foaled, the sire, the gender of the offspring, and its fate. The spring foaling season must have been one of excitement and stress. It started slowly at the end of April, reached a crescendo in May and tailed off in June. There were nineteen new arrivals during the fourth week of May, and a total of eighty-four births during the season. Halifax and Americain were the dominant sires, accounting for 70 per cent of the foals. Inkerman and Garou, made smaller contributions. There were a few more fillies than colts born this particular season. The triumphs and joys of foaling must have been somewhat muted by the high rate of losses among the newborns. Seventeen foals died at birth or soon afterward. Thus, one in five of the new arrivals failed to make it. The laconic entries in the diary indicate that it was the delivery itself which caused most losses. Typical entries include: "... coming twisted, had to take it from her dead; colt choked had to pull it away; had to take it from her with tackle head deformed; twins both dead." Accidents added to the death toll: two foals drowned in the slough in the south field, another mare gave birth in the north field and the colt was never found. An infection of the navel killed another. This relatively high mortality rate must have had a marked effect on returns from the operation. Did the unusually large scale of the Bar U Percheron stud mean that these losses were unusually high? This was certainly the case with the extensive cow-calf operation, but with the Percherons things were different. The prestigious stud attracted a group of experienced men who specialized in horse care, and the best veterinary help was available. Mares foaled in ideal conditions with optimum care. Although the Bar U stud was a large-scale operation it could offer more expert care than could any farmer or small rancher. Nonetheless, even with experienced and specialized help and first-rate living conditions, the risks of breeding horses were considerable. Many young mares proved to be barren and the fertility rate in general was low. Mortality among newborn foals further reduced returns.

The enormous scale of Lane's operations presented another problem. The most obvious market for a top-rate stud was among other livestock breeders seeking to upgrade their herds. However, with close to a hundred youngsters being foaled each year, this specialized market was soon in danger of being glutted. Thus the majority of sales were made direct to farmers, and the prosperity of the purebred enterprise depended on the willingness of these men to pay premium prices for superior teams. In good times they

was owned by George Lane and his partners Gordon, Ironside and Fares. Lane's consistent efforts to diversify his activities, first by developing a Percheron stud and later by growing grain on parts of the ranch, led to the building of a series of barns to house the horses and the development of a subsidiary farm complex a mile or so to the east. These innovations in turn meant a larger labour force, especially during the growing season. Moreover, Lane sought out and hired experienced men to run the various departments of the ranch. Several of these specialists were middle-aged family men and required suitable on-site accommodation. As the community increased in size, so further diversification took place as efforts were made to make the ranch as self-sufficient as possible. Two large gardens provided vegetables; modern pig barns housed hundreds of pigs for pork, bacon, and sausages; and a score of dairy cows supplied milk, cream, and butter. Lane was fascinated by new ideas and technical developments and sought to incorporate them into his operations wherever they were relevant. The Bar U was linked by telephone to the outside world,[73] had its own electricity generator,[74] and experimented with irrigation not only to make water available to pastures but also to supply domestic requirements and to provide fire protection.[75] Lane had spent all his working life in Alberta travelling from range to range, and from one enterprise to another, by rail

5.6 The community center with Percheron barns in the background at the eastern end of the site. Early spring 1919. Glenbow, NA 1075-1

might be prepared to purchase a "Rolls Royce" team to enhance their prestige, but they could always get by with a "Chevrolet." Lane had built an internationally famous stud, but the economics of the undertaking were to be found wanting in the years ahead.

Making Room for Horses on the Cattle Ranch

The headquarters site of the Bar U Ranch was transformed during the twenty-year period it

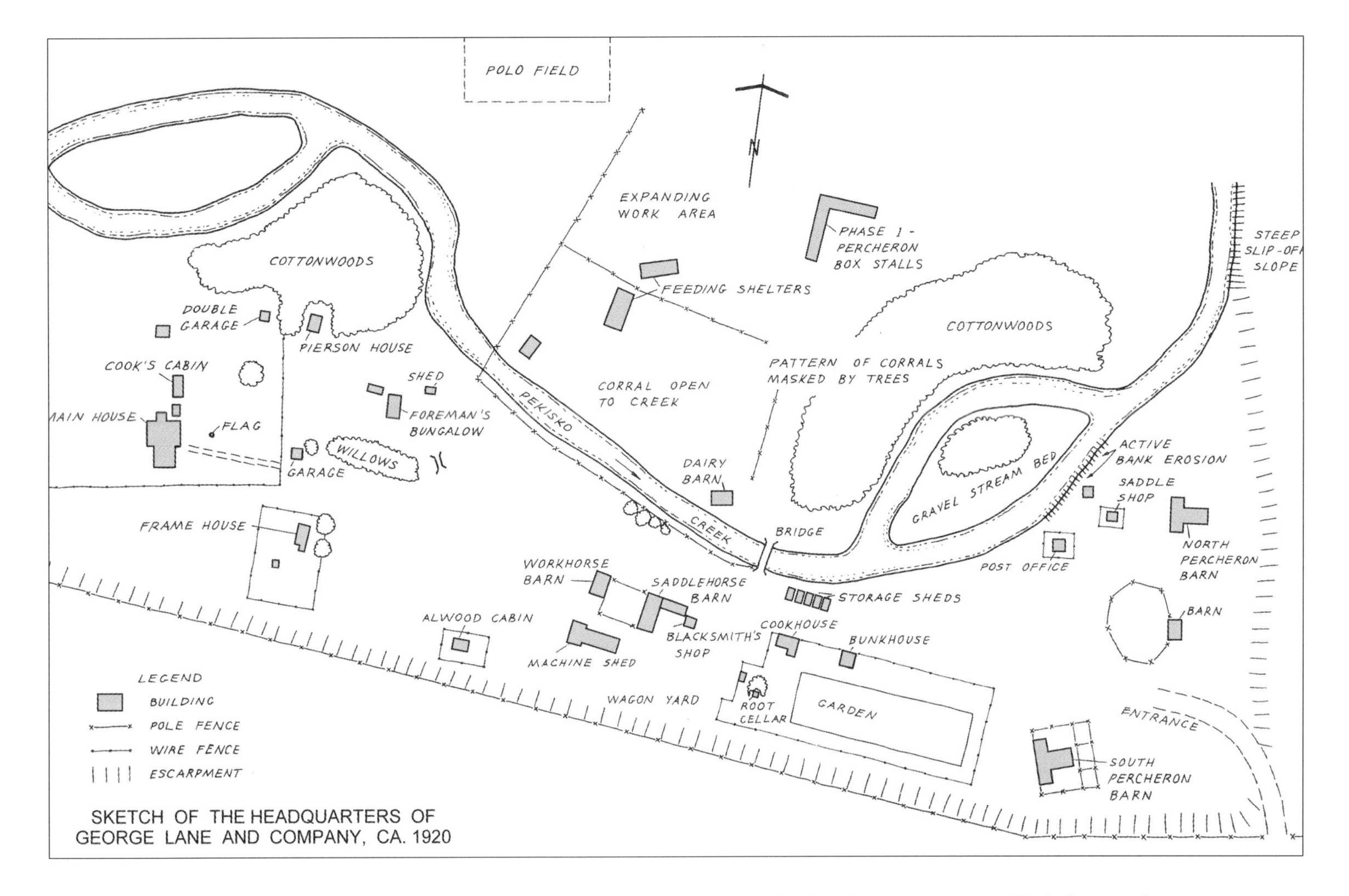

Map 5.2 The headquarters site in 1920, perhaps showing it at its most developed.

and horseback. As soon as robust motor vehicles became available, Lane purchased one, and soon a bridge was built over the creek to replace the ford, and garages were built on the ranch.[76] All these changes gradually reshaped both the 'streetscape' of the ranch headquarters and the landscape in the immediate vicinity of it [Map 5.2].[77]

On a Sunday morning at the end of January 1910, a fire broke out in the cookhouse at the Bar U.[78] Driven by a fierce wind, the flames soon engulfed the nearby bunkhouse. There were no serious injuries, and the ranch hands battled the fire until late afternoon. They managed to contain the grass fires which threatened to spread to other buildings. A report concludes by remarking: "No time was lost in replacing the buildings, for on Monday morning, building material was being taken from High River for a new cook and bunkhouse."[79]

5.7 Two cowboys lounge on Fred Andrews' bunk upstairs in the new bunkhouse; notice all the Percheron photographs and posters. Fred Andrws looked after the Bar U Percherons during the period after the First World War. Glenbow, NA 5652-70

5.8 Fred Andrews, a friend, and the cook's helper play with some of the many cats outside the back of the new frame-built cookhouse. Glenbow, NA 5652-61

5.9 Residential area of the ranch headquarters looking northwest, summer 1920. The log cabin in the foreground had just been completed for Fred Nash and his bride. Percheron mares and their foals receive some attention. Glenbow, NA 3752-46

The new cookhouse was a frame building. It was much larger than the one it replaced and fulfilled a number of functions. First, and most important, was the kitchen and the room where the men sat down to eat. The big stove was kept going, and warmth, coffee, and food were always available to returning riders. The cook had a room adjacent to the kitchen and easy access to the woodpile and the coal shed behind the house and to the ice house and other stores across the 'street.' The upper floor of the cookhouse acted as a dormitory or bunkroom for riders, haymakers, and chore boys alike. Straw mattresses on the floor and a few nails on which to hang their meagre wardrobes was the extent of the 'appointments.' This space, with its minimal level of comfort, remained in use, substantially unchanged, through the 1940s.[80] It was not until 1917 that a new bunkhouse was built about fifty metres to the east of the cookhouse. This signalled a considerable rise in the living standards of the permanent staff, for each had his own room in the new structure.[81]

A blacksmith's shop was built adjacent to the saddle horse barn, which was extended southward. Logs salvaged from the fire were used in both these projects.[82] Closer to the edge of the coulee bank, a long, low frame building was constructed to house the expanding inventory of agricultural machinery. Taken together, these new buildings filled in the area to the south of the two horse barns. It became the area where seasonal equipment like hayracks were stored and where wagons were worked on and maintained. On the bank of the creek, a short distance north and east of the cookhouse, a solid cabin was built of squared logs. This became a nerve centre of the ranch, for it housed both the post office and the ranch business office. Until this time both these functions had been performed in a room in the main ranch house. Just as new enterprises and new methods demanded additional space, so too the new regime "brought with it the demands of a larger and more complex business." Bill Yeo goes on to explain:

> The Bar U became for a decade or more, the business headquarters for a number of large properties, with livestock wearing more than a dozen brands, managed 'as one outfit.' The installation of the Percheron breeding and sales enterprise at the Bar U further enhanced Pekisko's brief career as a center of agribusiness in Alberta.[83]

F.R. Pike, described as "bookkeeper and general superintendent of purchases and sales,"[84] lived in this cabin, which provided some degree of security for the office, and at the same time some privacy and living space for an important officer of the company.

The main ranch house, or "Pekisko House" as Fred Stimson had styled it, was not substantially modified during Lane's period of ownership, probably because he and his family never lived at the ranch for an extended period. The house was used by Herb Millar in 1902, while his own bungalow was being built, and then it remained without a permanent resident for a number of years. Lane lived there on his frequent but short visits to the ranch. Mabel Lane, George's second daughter, had married the Reverend Mortimer Paulin in 1912, and they and their four children spent their summers at the Bar U from 1918 through the early 1920s. Whenever they were there, the rest of the family would come over from Willow Creek. The ranch house was also used to accommodate visitors and pro-spective buyers of Percheron horses. Alex Fleming and his wife Elsie and their young family lived in the house between 1908 and 1912, when they moved to High River. Thereafter, a permanent staff ensured that all was ready at a moment's notice.[85] Photographs taken during this period show that a low verandah had been completed along the eastern facade of the house and a glass-sided conservatory had been added on the south side. Gradually, over the years, the grounds adjacent to the ranch house had been cleared and smartened up. A white-posted page wire fence enclosed both a front garden and a large, flat, lawn-like area behind the house which reached westward to a line of ornamental conifers.[86] Carefully whitewashed rocks bounded the path to the front door. Pictures taken during the Prince of Wales' visit in 1919 show flowerbeds and a tall flag pole in front of the ranch house.

5.10 The "Big House" at the Bar U Ranch, eastern facade, c. 1919. Parks Canada, Norman Mackenzie Collection, 96.04.02.36

A frame bungalow with a hipped gable roof was built on a slight rise in the flood plain just east of the main ranch house. This was the home of Lane's trusted lieutenant Herb Millar. He and his wife Ella lived there from about 1903 until 1913, when Millar went

5.11 One of Fred Andrews' daughters looks a bit apprehensively at some really large hens. Glenbow, NA 5652-34

to manage the Namaka farms. His place as foreman of the Bar U was taken by Neils Olsen. Between 1913 and 1925 the bungalow became a centre of family life, and the four Olsen kids (and their governesses) were favourites with the cowboys. Another small frame cabin was built among the aspen trees close to the creek and north of the foreman's house. This was the home of the Pierson family; Bert, Mayme and their son Manville. Finally, a third frame house was built of the south side of the 'street.' This was the summer residence of the Paulin family. The house was also used as a schoolroom by the Olsen governesses.[87]

In the past, I have tended to think of the ranch community as being almost exclusively adult and male. This may have been true during the North West Cattle Company era,[88] but became less so after the turn of the century. From the arrival of Alex Fleming and his family in 1908, through the years when the young Olsens and Paulins explored every corner of the ranch, to the Alwoods, the Nashes, the Bakers, and the Nelsons, the ranch has nurtured families, and their presence in the background must have added something important to the bachelor life of the bunkhouse.[89]

During the first decade of the twentieth century, George Lane concentrated his Percheron breeding activities at the Bar U Ranch.[90] Mares and some purebred studs were brought in from the YT Ranch and were joined by horses imported from France or acquired elsewhere in North America.[91] The eastern end of the headquarters site, which had been empty until now, was adapted to accommodate the Percherons. As Barbara Holliday expressed it:

> What was once the exclusive domain of cowboys, their saddle horses and cattle, by 1909 included specialist horsemen, massive Percherons, and the barns, heavy duty corrals and buildings to support them.[92]

On the flood plain to the north, between the creek and the bank, a handsome T-shaped barn was constructed and fitted with box stalls for the Percheron stallions. A huge open hayloft above the stalls made feeding easier. Each stall opened onto an exercise yard constructed of massive posts and page wire.[93] This is where the great stallions of the Bar U lived for part of the year. However, after the breeding season was over, the mature stallions were turned out together into a large pasture field and allowed to run at large until the approach of the breeding season the following year.

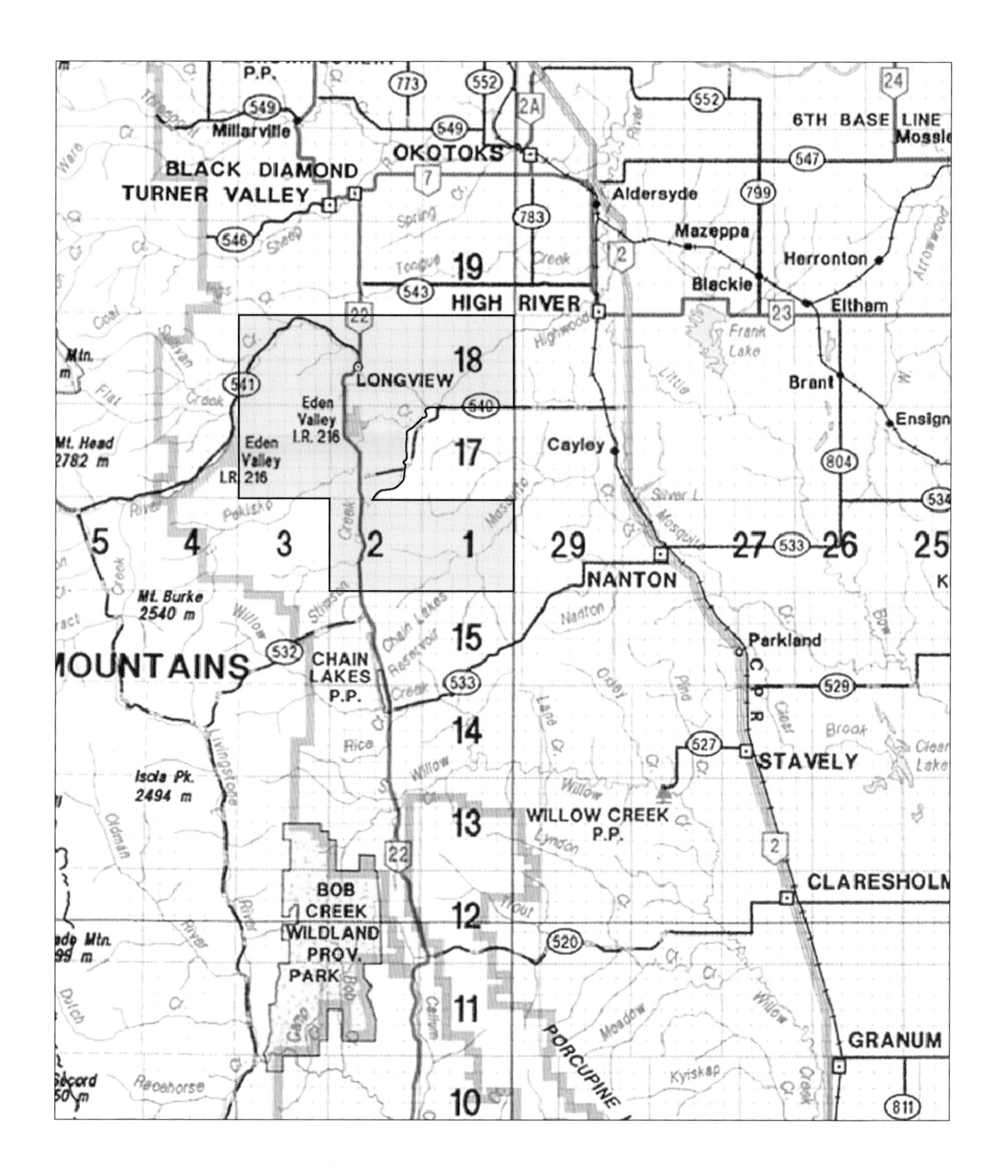

Map 2.5 Bar U leased land shown on a modern road map to show its extent. The leases covered almost seven townships, or 158,000 acres.

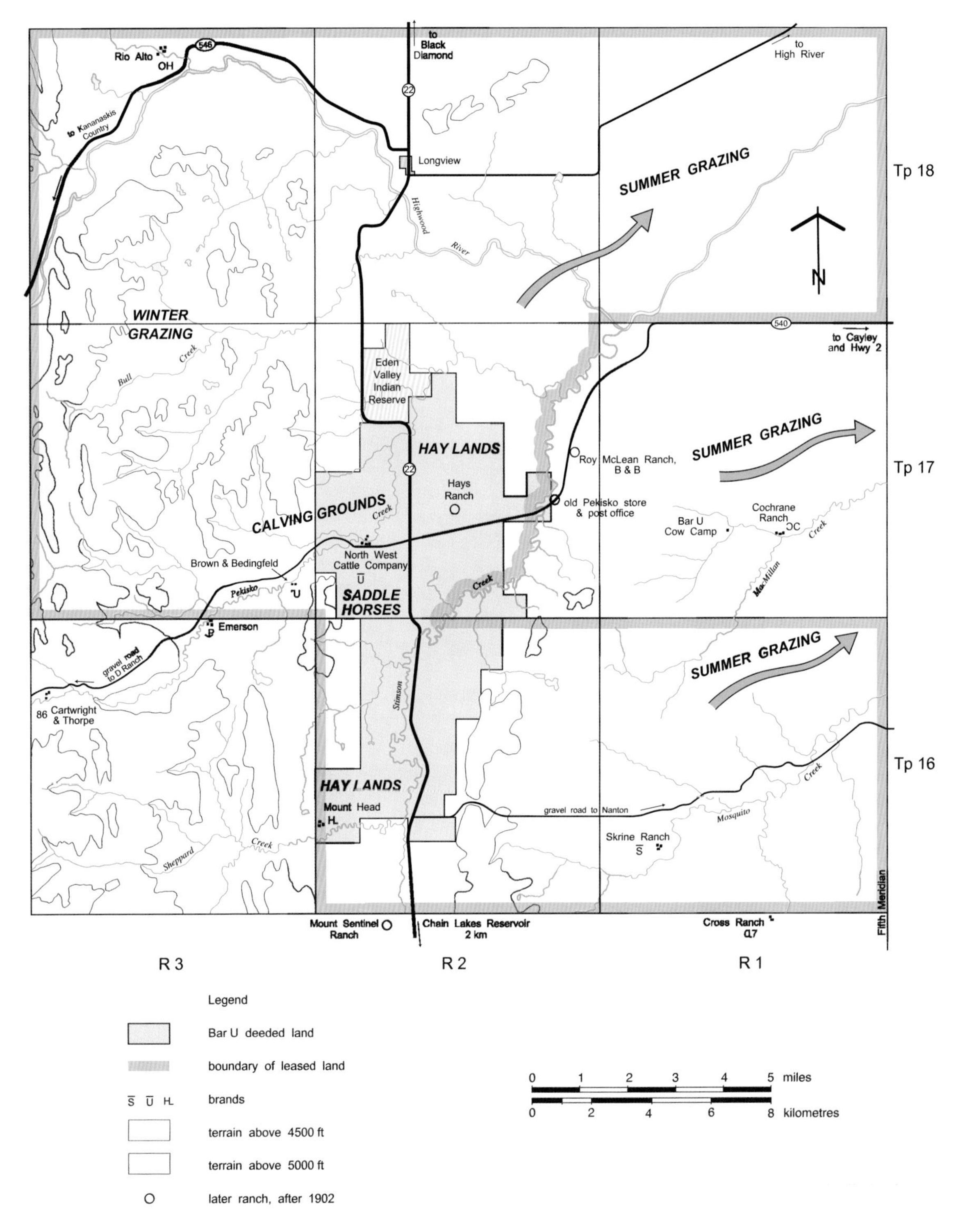

Map 2.6 Leased and deeded land in relation to topography, drainage and land use.

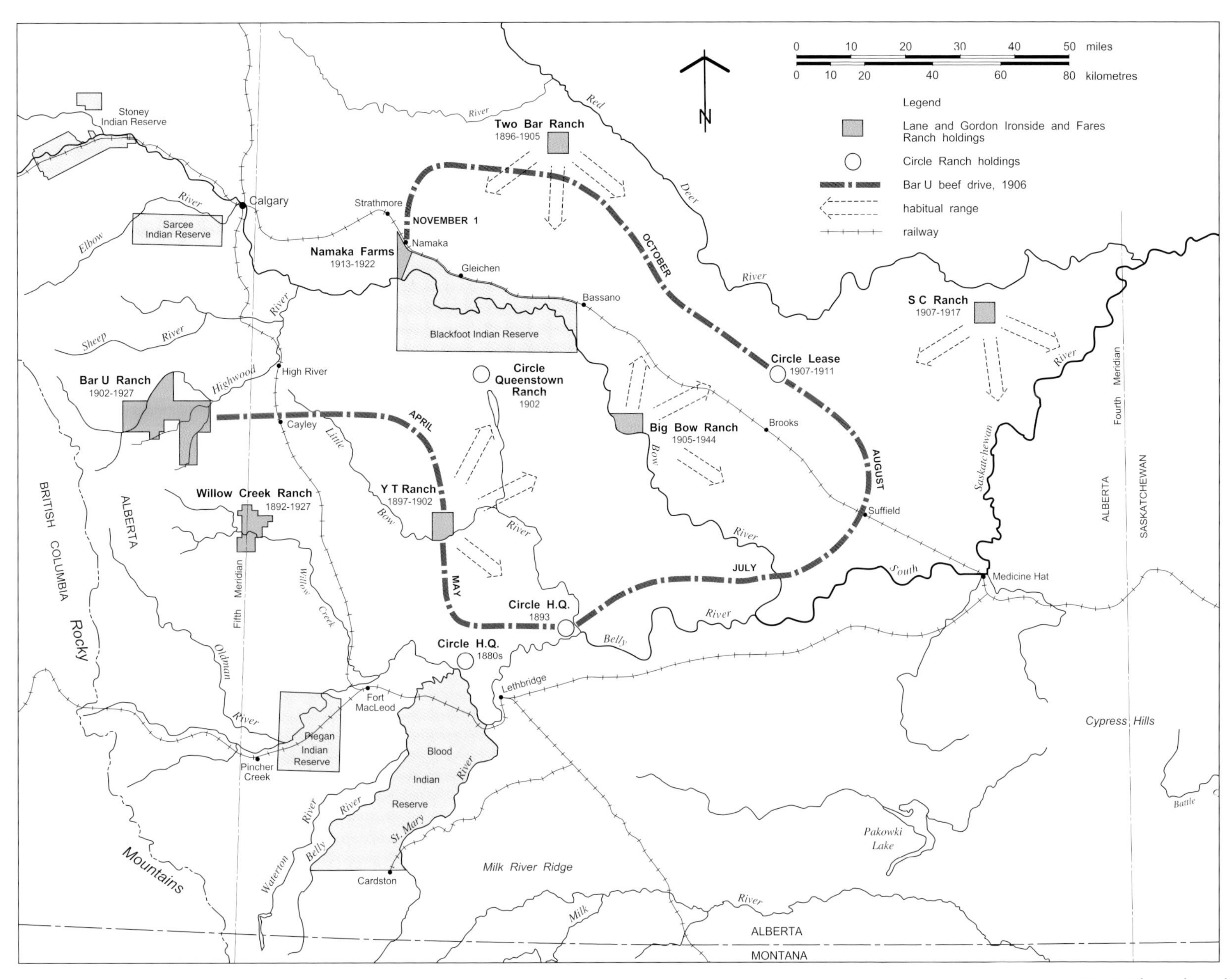

Map 4.1 The ranches and ranges of George Lane and his partners, Gordon, Ironside and Fares, c. 1906. 128

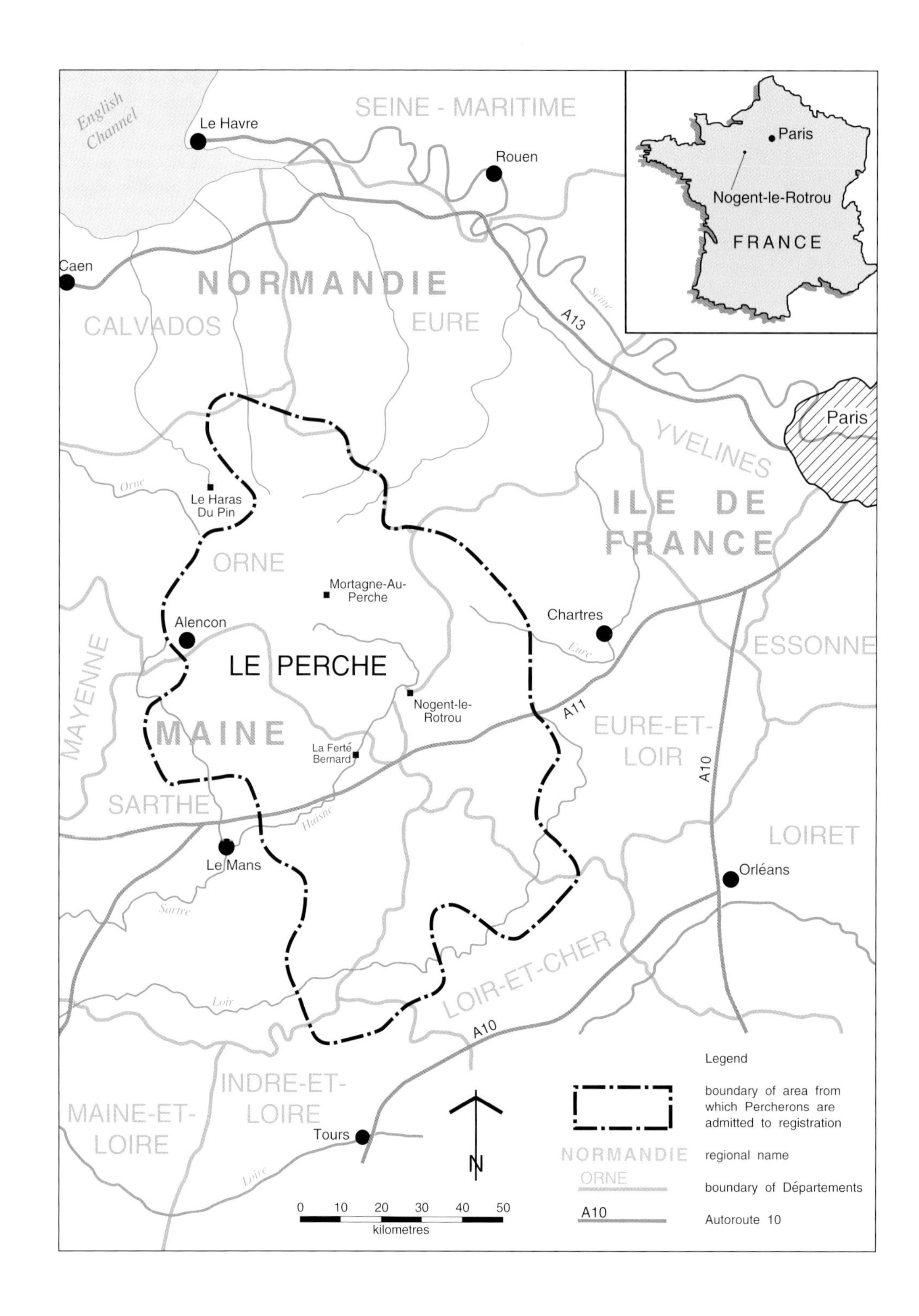

Map 5.1 The location of le Perche, a vernacular region which sprawls across the boundaries of several Departments.

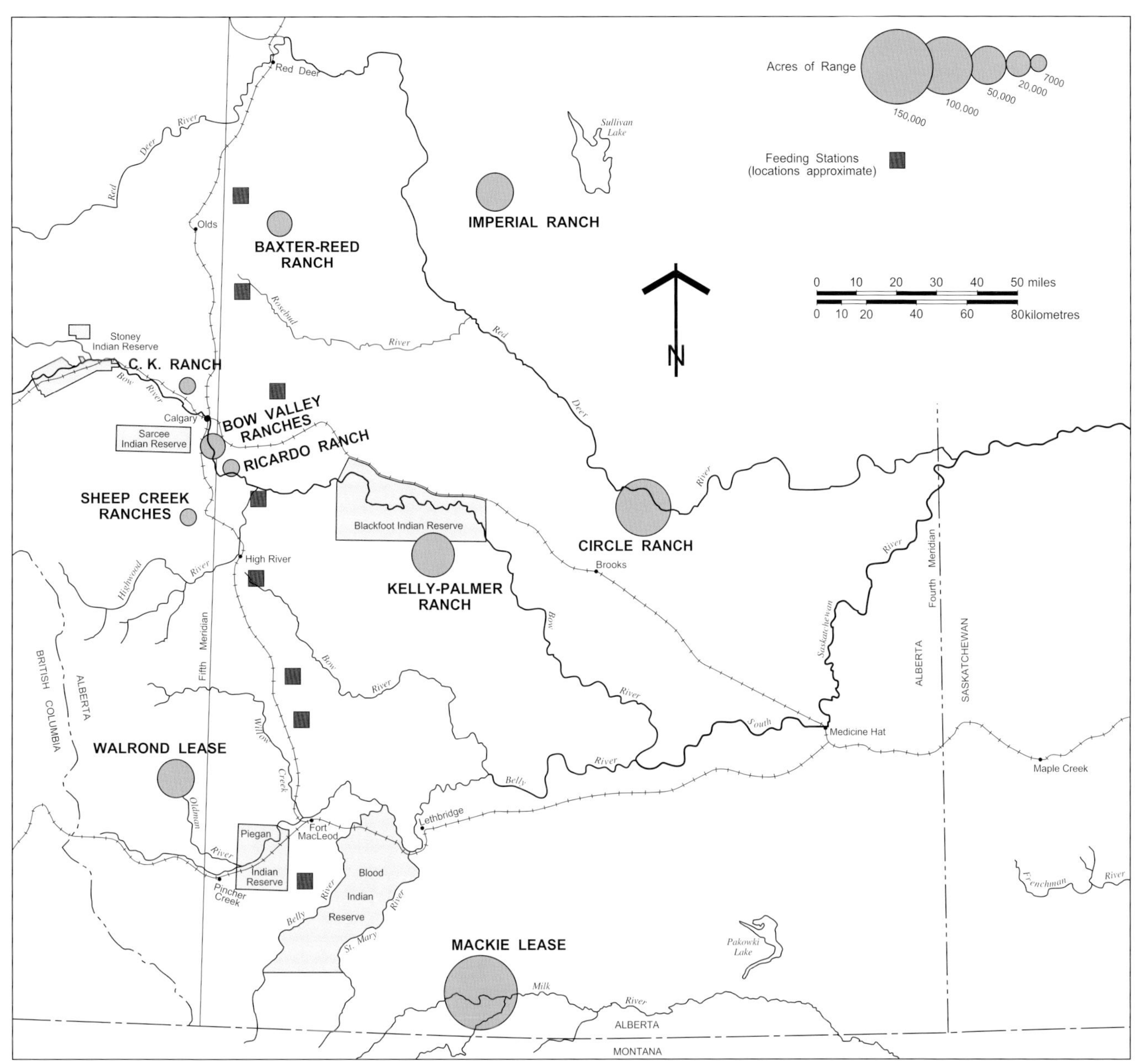

Map 7.1 Burns Ranches: early expansion, 1907–17.

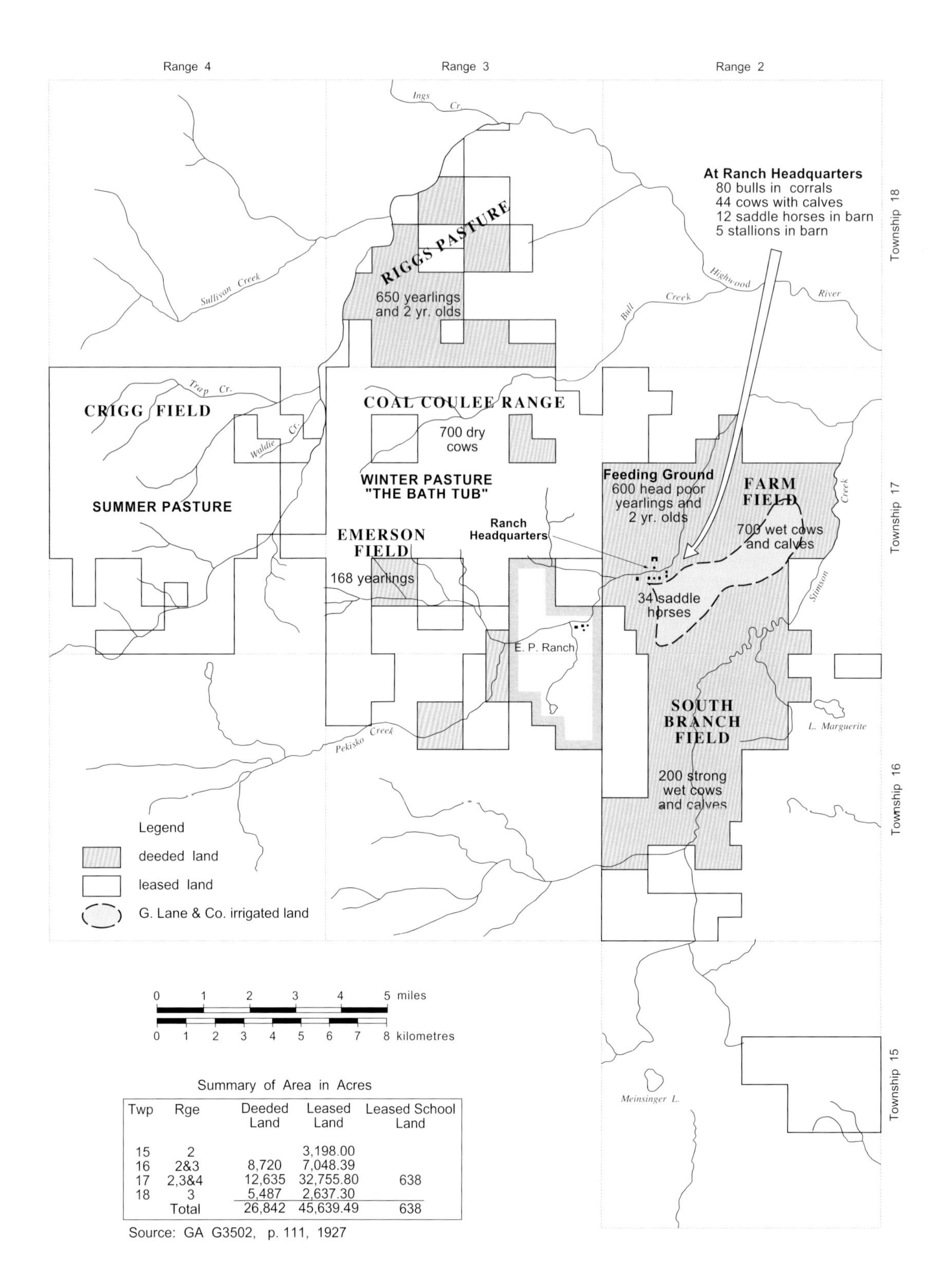

Summary of Area in Acres

Twp	Rge	Deeded Land	Leased Land	Leased School Land
15	2		3,198.00	
16	2&3	8,720	7,048.39	
17	2,3&4	12,635	32,755.80	638
18	3	5,487	2,637.30	
	Total	26,842	45,639.49	638

Source: GA G3502, p. 111, 1927

Map 7.2 Bar U ranch lands, 1927: showing disposition of herds.

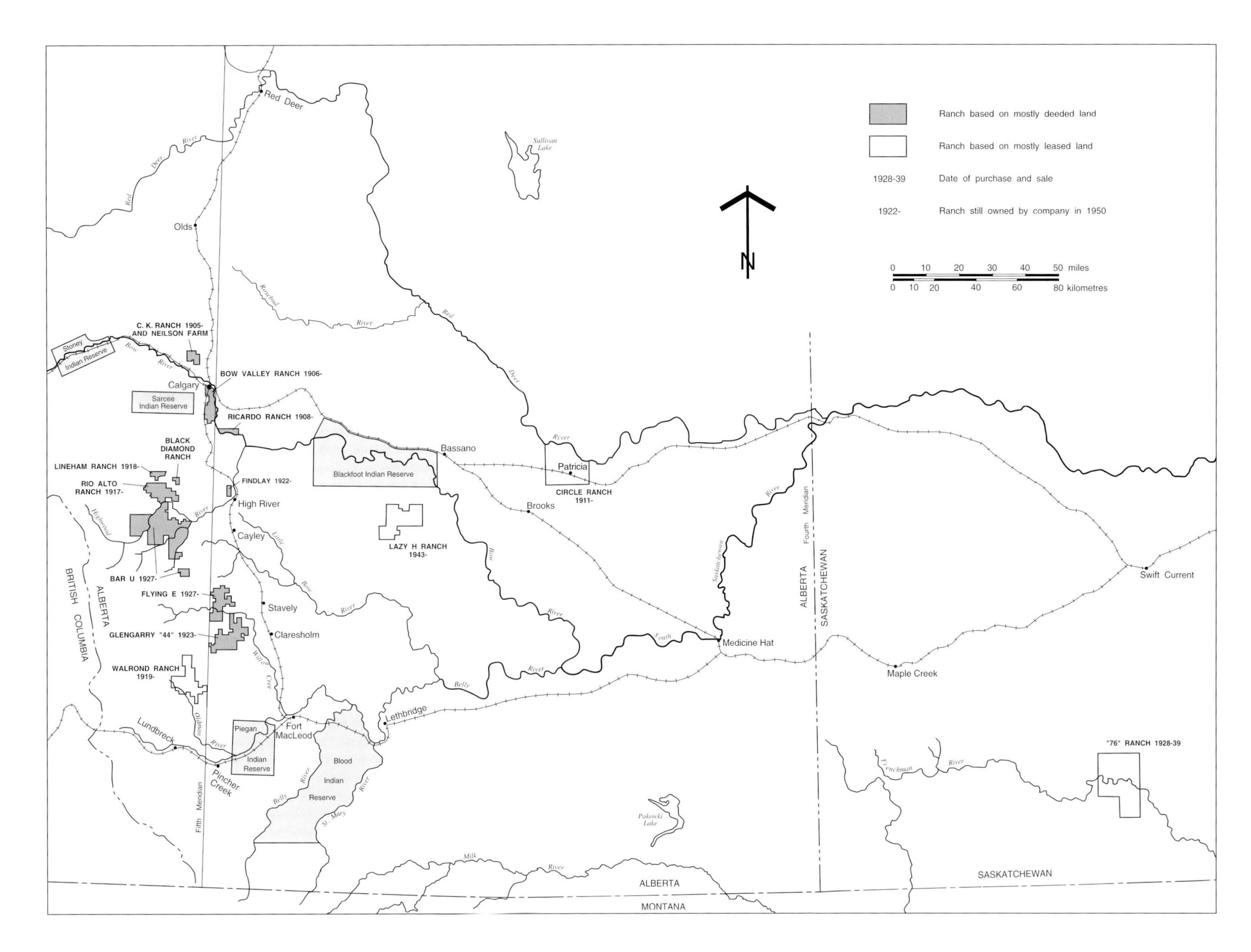

Map 7.3 Burns Ranches: later expansion, 1930s and 1940s.

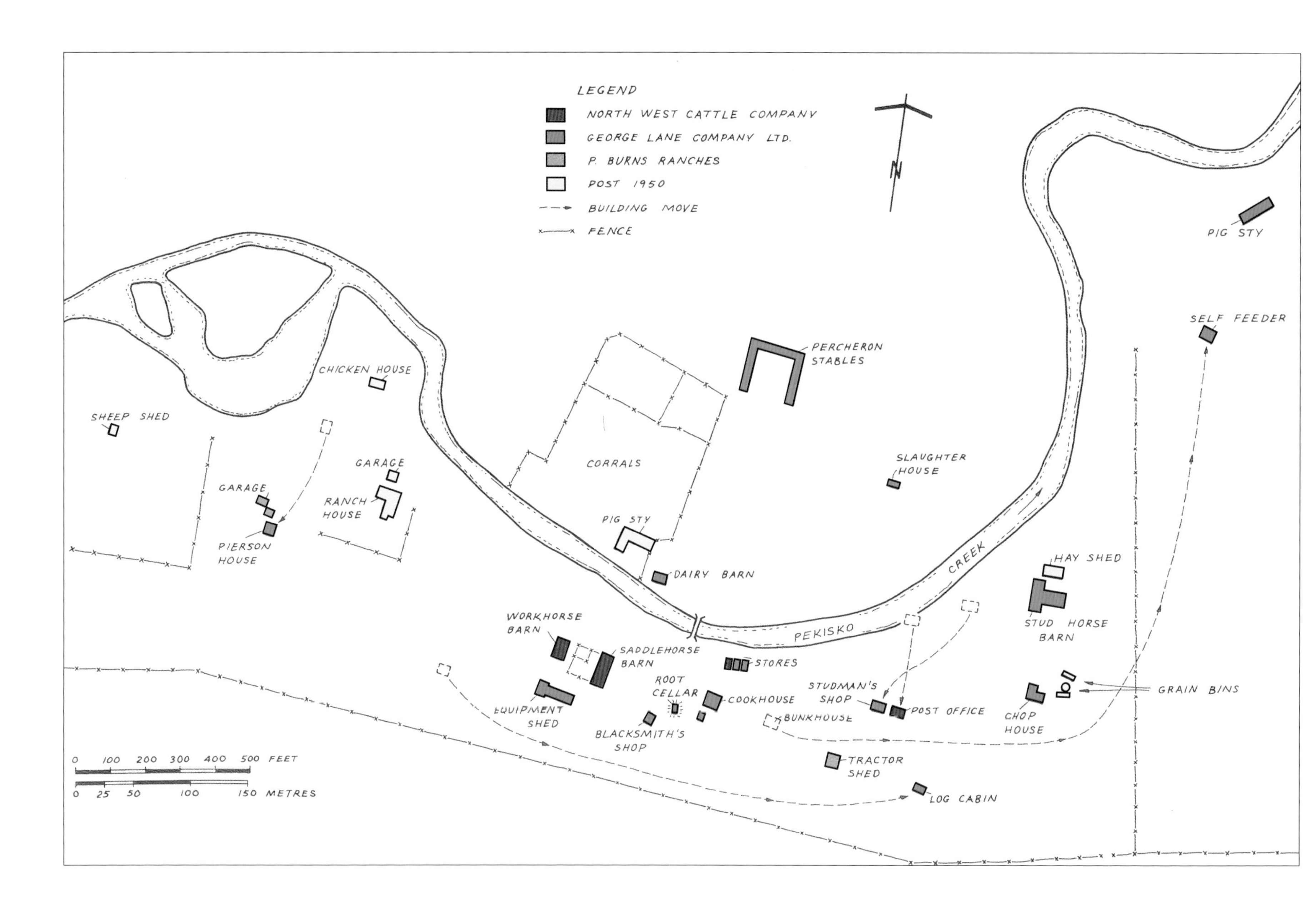

Map 8.1 Existing major buildings at the Bar U Rnach, by construction era. The headquarters site today. 222

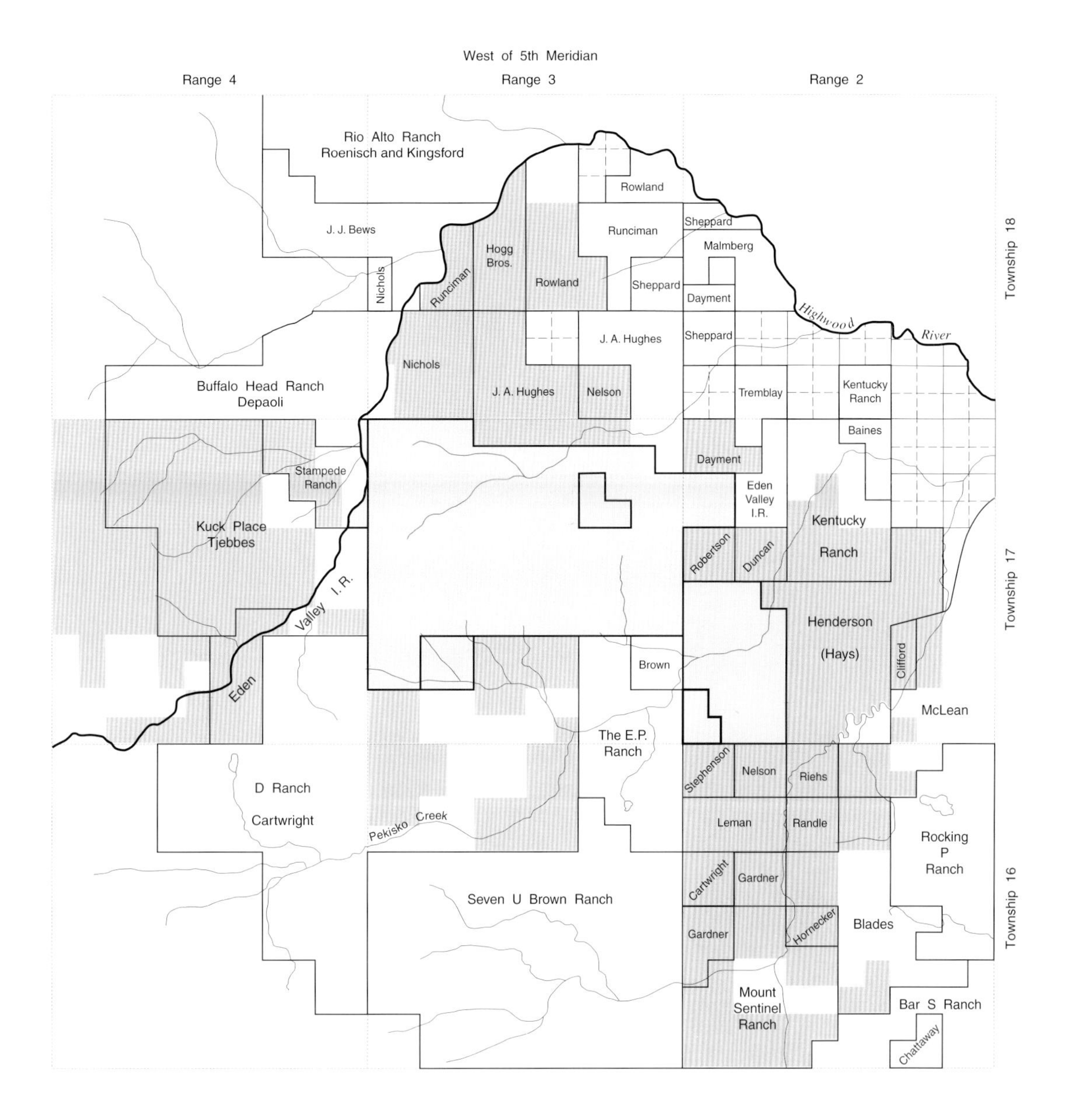

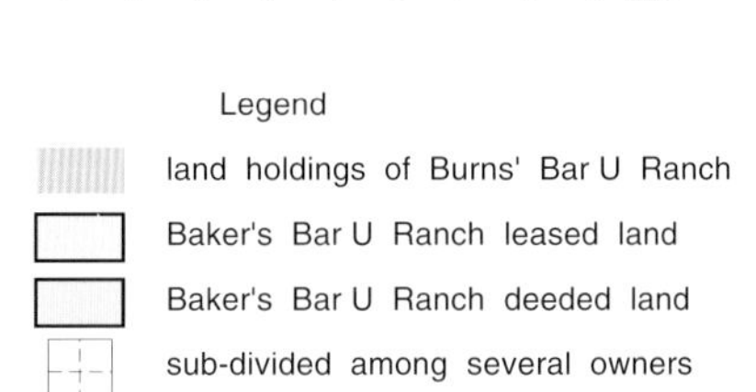

Map 9.1 The break up of the Bar U Ranch lands, 1949–50.

O.N. Grandmaison

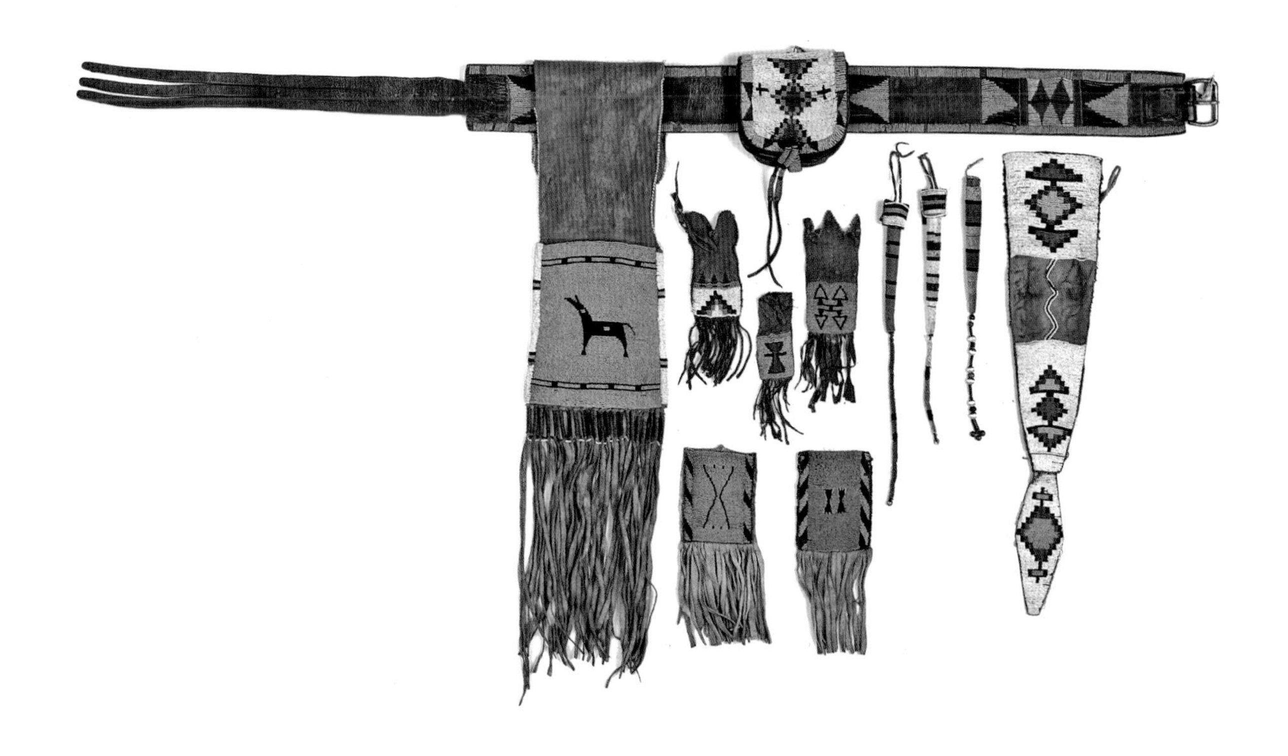

Plate 1 (facing page) O.N. Grandmaison, "July," George Lane, Bar U Ranch, Nanton, 1963. This picture is currently hung in The Ranchmen's Club, and is reproduced with their kind permission

Plate 2 Some examples of Plains Indian beadwork from the Stimson Collection of the Royal Ontario Museum. Pouches suspended from their belts were used to carry small items. A pipe bag with an animal motif and a cartridge case are shown on the belt; (upper row from left to right) three pouches for face paint, three awl cases, a knife sheath; (bottom row) two ration-ticket cases. Arni Brownstone, "Tradition Embroidered in Glass." With permission of the Royal Ontario Museum © ROM

Plate 3 (facing page) Charlie Russell, "George Lane attacked by wolves." The artist left out the carcass of the cow, on which the wolves were feeding, which allowed Lane to creep up close to them. Glenbow, Art Collection 65

Plate 4 Edward F. Hagell, "Bringing in the thin ones." One of the tasks of riders during the winter was to watch the condition of the herd and cut out weak cows and calves for feeding close to the ranch buildings. Glenbow, Art Collection 39, and permission of the Hagell family

Plate 5 Percheron Stallion at the Haras du Pin, the French National Stud Farm. Postcard, le Haras du Pin, France

Plate 6 (facing page) The courtyard at la Touche Percheron Stud, home of the Aveline family, showing the stable wing and loose boxes. Photograph by S.M. Evans

Plate 7 (following page left) Ted Schintz, "Stoney Indians Camping." Artists Ted and Janet Schintz lived in a cabin on the Highwood River close to the Eden Valley Reserve. Ted painted several portraits of Stoney people as well as landscapes and ranching scenes. Glenbow, Art Collection, 56.43.01 and permission of Michael Schintz

5.12 The Percheron show barn with the bunkhouse and the cookhouse in the background, fall 1917. Glenbow, ND 8-61

Professor W.L. Carlyle explained the details and the benefits of this regime:

> They are each given a box of oats night and morning to keep them in good thrifty condition but are turned out to graze each morning even in the coldest days of the year and they remain out all day.... Instead of vicious fighting, kicking and biting, as might be expected, our four breeding stallions appear to have real enjoyment in each other's society and are encouraged ... to take much more exercise than they would if each were in a small field by himself. Since this practice has been followed they are proving much more potent sires and the colts are possessed of much greater vitality and vigour at birth than was the case formerly when horses were stabled most of the time and exercised only on a halter.[94]

Close to the southern extremity of the flood plain, and to the south of the road leading down into the ranch, was the show barn. This was where the cream of the Percheron herd were prepared or "fitted" for the show ring. They underwent a rigorous exercise regime and received special food. It was in the corrals adjacent to this building that prospective buyers would be introduced to what the Bar U stud had to offer. When American Percheron enthusiast, Harley E. Salter, made the first of several purchases from Lane he found the process quite intimidating. He reminisced:

> I was taken down to the great circular corral into which around a hundred head of mares had been driven. Lane intimated that I could take my pick. I would locate a suitable beast, only to see one that looked better. Never in my career of buying Percherons did I feel so frustrated.[95]

Midway between the two main Percheron barns a third building was constructed. This structure was used to store feed and to prepare it for those horses which were receiving grain supplements. At some point during the Lane era an engine was added to provide power to chop grain, and corrals were built on the western side of the barn, which suggests that some feeding was done there. The final building in this area of the site was a smaller frame structure which was used as a residence for the studman and his assistant. The house included a workroom where harnesses could be kept in first class condition, while part of the building was used to store the show wagon.[96]

The working area of the ranch, across the creek, was also transformed by the arrival of the Percherons. A substantial U-shaped stable with twenty-six box stalls was built to provide shelter in which mares could deliver their foals.[97] An enclosed courtyard provided space for some exercise out of the worst of the winter wind. In the trees along the creek open shelters were thrown up and corrals were built where the yearling fillies and colts could be fed during the winter.[98]

The cattle corrals, which had been used since the early days of the ranch, were renewed and extended. A long, low shelter was constructed under the lee of the bank on the western side of the corral system. In addition, a substantial and rather elegant dairy barn was built on the bank of the creek just across the bridge. Norman Mackenzie was milking about twenty cows there by 1913.[99] Further afield, across the creek

5.13 Some hands enjoy a game of cards in the cookhouse. Herb Millar at the head of the table with his back to the camera. Glenbow, NA 5652-65

and to the north east of the headquarters site, a state-of-the-art hog barn was built using plans from Iowa; it was capable of accommodating several hundred pigs. It is probably true to say that the headquarters site of the ranch reached its fullest development during the years leading up to the First World War.

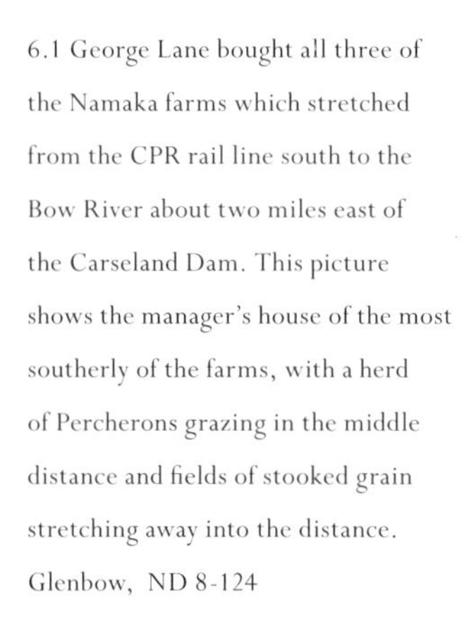

6.1 George Lane bought all three of the Namaka farms which stretched from the CPR rail line south to the Bow River about two miles east of the Carseland Dam. This picture shows the manager's house of the most southerly of the farms, with a herd of Percherons grazing in the middle distance and fields of stooked grain stretching away into the distance. Glenbow, ND 8-124

6
George Lane: At Full Gallop!

"Indeed, I am not aware of any evidence that any of the human experience we lump together under the heading of 'history' is determined by anything. A near life time of studying it has left me convinced that it happens at random, within limits allowed by a mixture of willpower and material exigency. Or else it happens chaotically, by way of untraceable causes and untrackable effects."[1]
– Felipe Fernandez-Armesto

We have already seen that George Lane was much more than "just a cowboy." He was an innovative, perhaps visionary, entrepreneur with enormous drive and energy. He thought big and was prepared to take risks. As settlers flooded into the 'last best west' during the first decade of the twentieth century and 'King Wheat' began to challenge stock rearing for the dominant position in the provincial economy, so Lane joined the stampede and became one of the largest grain farmers in Canada. Somehow, Lane also found time to play an important role in establishing and sustaining the organizations which represented the collective interests of cattlemen and horse breeders. His success in this sphere, and his growing personal prestige, made him an attractive political prospect and he was drawn, for a while, into the fringe of national politics. Finally, the scale of Lane's operations and his enthusiastic boosting of the west meant that he was much in demand as an expert witness for journalists, and he pontificated on reciprocity, mixed farming, and the progress of settlement.

Namaka Farms

In June 1913 George Lane purchased the Namaka Farm from its American owners for $250,000.[2] This property, on the CP main line east of Calgary, encompassed about ten thousand acres, of which four thousand were already in crops. Originally, the farm had been established in 1884 by Sir John Lister-Kaye as part of his Canadian Agricultural, Coal and Colonization Company. More recently, it had been developed as a large-scale grain farm by Vic Anderson of Calgary. He reaped two enormous wheat crops in 1908 and 1909 and sold the farm to foreign interests on the strength of this success. The next owner, John Adler, who represented Alabama investors, poured money into the property. He added farm buildings, including an elevator at Namaka, and further extended the cropped acreage using massive steam tractors.[3] The labour force during the growing season swelled to as many as fifty, and permanent employees included carpenters, engineers, and other tradesmen. Such an enterprise demanded careful and consistent on the spot management, which Adler as an absentee owner could not provide. Lane was able to purchase the farm as a going concern at a very reasonable price. The following year he rounded out his purchase by acquiring Strangmuir, the home place of the Military Colonization Company, which comprised a further two thousand acres. Thus his new holdings extended in a rectangular block from the CPR main line in the north to the Bow River in the south, with the Blackfoot Indian Reserve as the eastern boundary.[4]

Lane asked Herb Millar to move from the Bar U to manage the new enterprise. Between them they divided the twelve thousand acres into three farms.[5] Farm No. 1, with its headquarters just a mile and a half south of Namaka station, was a stock operation. Lane moved some of his Percheron mares there and focused on breeding superior work horses.[6] There was also a big dairy, and breeding sheds for more than a thousand pigs. Farm No. 2, some four miles further south, was a grain-farming venture with Alex Montgomery as manager. Oats, wheat, and barley were all raised, and as long as precipitation was adequate, the yields proved exceptionally high. Farm No. 3, the "Colonel Strange Place," was used for both grain and horses. The flat or gently undulating prairie was used for wheat and tame hay, while the breaks and bottomland along the river were grazed by stock.

6.2 Six Massey-Harris binders, each drawn by four Percherons, working on the Bar U farm east of the headquarters site. The field in the foreground is growing tame hay. Parks Canada, Hugh Paulin Collection, 96.11.01.33

With the outbreak of the First World War in 1914, Lane was poised to contribute to Canada's war effort by producing more and more grain and meat, while at the same time he benefited from a period of firm or rising prices. Lane seeded the largest crop in his experience in 1915, some seven thousand acres. He used 120 horses with fifty drills and harrows and sixty drivers and assistants. This was a prodigious investment; as the *High River Times* noted, "Mr. Lane will risk a small fortune in the soil this year."[7] For some years the gamble paid off. By 1917, Lane was cropping almost nine thousand acres on his various farms, and the value of his wheat crop was expected to be nearly $560,000.[8] The hay crop at Namaka was also exceptional, and a thousand tons of alfalfa were put up as well as huge amounts of prairie hay.[9] The Minister of Agriculture encouraged grain production under headlines like "Food Will Win the War." He urged farmers on: "Despite the difficulties we must produce more food than we have ever done before.... Plan to bring as much new land under cultivation for another crop as possible."[10] During this period, Lane vied with Charles Noble for the title of the largest grain grower in Canada.[11] Men like these large-scale grain growers could reap great rewards, but the short Canadian growing season meant that risks were equally great. Seeding had to be completed as soon as possible after the land became workable. Thus, in order to complete sowing in a timely fashion, a large number of teams and machinery had to be assembled along with a huge labour force.[12] The same was true at harvest time. Each crop year was a renewed gamble with heavy investments sunk in seed and labour in the expectation that nature would co-operate by providing 'rain in due season,' and a period of at least ninety days between the last frost of the spring and the first of the fall. Lane had amazing good fortune in that he enjoyed a sequence of four good years with high yields and rising prices between 1914 and 1917. Then the droughts began, and post-war deflation hit wheat prices, while labour costs rose precipitately. Lane's elaborate and diversified empire teetered on the brink of disaster.[13]

The Public Life of a Private Man

Lane never shirked his responsibilities to the communities to which he belonged. He saw the advantages of a strong cattlemen's association which could lobby Ottawa effectively as well as dealing with local matters relating to ranching. He was an officer of the Western Stock Growers' Association for eleven years and president in 1913 and 1914.[14] When that organization became somewhat moribund, Lane was the driving force behind the formation of the Cattlemen's Protective Association of Western Canada. This dynamic new association tackled the problem of mange with concerted vigour and soon had the disease under control. Thereafter, they became a highly effective lobby group tackling transport problems, land taxation, and the future of grazing leases.[15] Lane was equally energetic in promoting horse breeding in Canada. As well as establishing the Canadian Percheron Horse Breeders' Association,[16] he also played a role in the Alberta Horse Breeders' Association and was elected president of that body in 1912, a post he held for the next four years.[17]

Lane is best remembered as one of the "Big

6.3 Threshing oats at the Bar U, c. 1917. The threshing machine was driven by a steam engine, but horse power – provided by Percherons – is everywhere in evidence. Parks Canada, Olsen Family Collection, 96.03.01.54

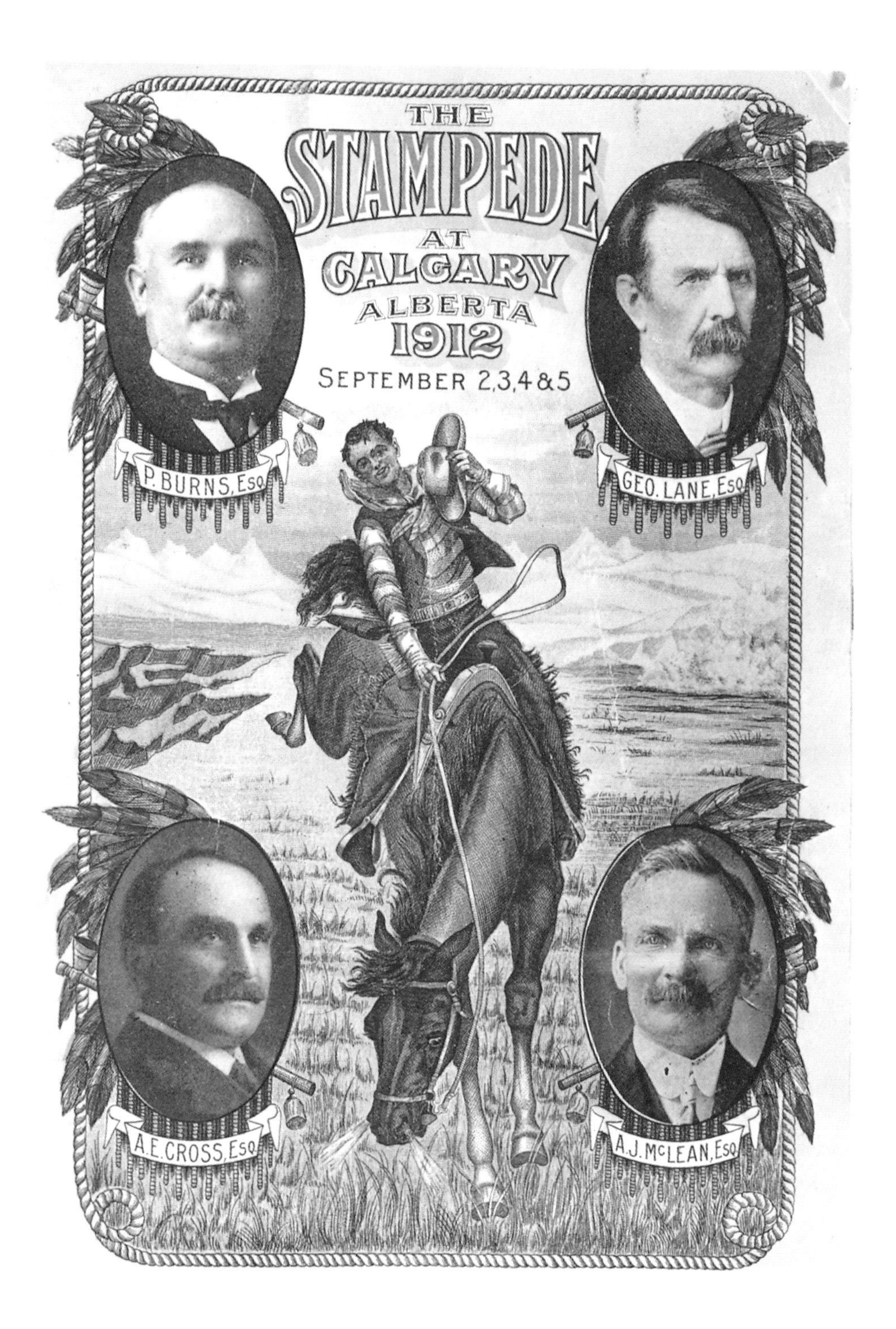

Four" who underwrote the first Calgary Stampede. He saw the Stampede as a perfect way to show off the achievements and the potential of the growing progressive west, while at the same time celebrating its frontier past.[18] It was Lane who approached Guy Weadick and asked him to explain his vision of a "Frontier Days" celebration, and it was Lane who sold the idea to Pat Burns and A.E. Cross. Donna Livingstone recreates the scene:

> The three ranchers delivered their instructions to Guy over a handshake: "Make it the best thing of its kind in the world – but everything must be on the square. We don't want to lose money if we can help it, but we'd rather lose money and have it right than make money and have it wrong." The celebration must recreate the atmosphere of the frontier west in an authentic manner, "devoid of circus tinsel and far fetched fiction."[19]

Each of the "Big Four" contributed $25,000, so Weadick had a 'war chest' of $100,000 to help him turn his dream into a reality.[20]

Much less obvious and public were Lane's consistent efforts to be a responsible citizen of the town of High River. He and George Emerson had built one of the first business blocks in the town, and he followed that by financing another block in 1909, which "would have

6.4 The "Big Four," who put up the money, grace the poster for the first Calgary Stampede in 1912. Glenbow, NA 604-1 graced any large city."[21] For many years Lane sponsored the local Agricultural Society and Exhibition and put up several prizes. He leased some land to the curling club for a nominal rent and made space available in one of his buildings for a young man's club.[22] In 1911, Lane accompanied two town officials to Ottawa to see if they could secure the service of a second railway. Six months later the Canadian Northern Railway added the town to its line.[23] Lane did well for the town, and it is fitting that his contribution is remembered in the beautiful and well-appointed park, developed on land he gave to the town and known as "George Lane Memorial Park."[24]

George Lane had become one of the most familiar figures in southern Alberta. Not only did his success as rancher, horse breeder, and large-scale farmer mean that his doings were constantly reported in the papers, but he was actively sought out by the press to comment on such issues as trade, transportation, mixed farming, and the future development of Alberta. The Liberal party in Alberta felt that he might be their best hope for breaking the long hold of the federal Conservatives over the southern region of the province. Lane was persuaded to run as the Liberal candidate for Bow Valley. The Conservatives underestimated Lane's popular support and began by ridiculing his lack of education and political experience. Soon, however, the increasingly strident articles in the *Calgary Herald* demonstrated their growing concern for the momentum that was gathering around Lane's campaign.[25] In due course Lane was elected and the paper was forced to report that the Bow Valley had gone "sweepingly" for Lane.[26] However, his tenure as a Member of Parliament was short-lived. Cabinet minister C.R. Mitchell had been defeated in the election, and Lane, having consulted his constituency, resigned his seat so that Mitchell could return to the house.[27]

The Lanes had moved from the Willow Creek Ranch to Calgary in 1897. Lane managed his increasingly diverse interests from a home in the city. By that time he had six children and Lillian and Mabel had been attending Sacred Heart Convent School as boarders for three years. Relocation to the city meant that the family could be together. Moreover, George Lane's business activities already transcended the management of a single ranch and demanded that he travel constantly all over southern Alberta. Calgary was the hub of the railway and telegraph systems, the two networks which Lane used to manage his growing agricultural empire.[28] On arrival in the city, he had rented a house on 7th Avenue, near the spot where the old Herald building stood. In 1899, he bought the former residence of Bishop Pinkham. It was a spacious seven-bedroom house, a little to the west of the centre of Calgary. Finally, in 1903, Lane purchased the Bangs home at 339 4th Avenue West. It had been built of sandstone for an English family, the Fews, and had a wide verandah on three sides. Elizabeth Lane described it:

> It faced north and had four front lots with a row of fir trees inside a picket fence, with another row of trees down the driveway to the back. The four lots reaching through to 5th Avenue were our lawn and garden. It was a large house with 7 bedrooms and one bathroom. It had a good sized reception hall in the centre, a large living room

6.5 The Pekisko district fielded a good polo team. They called themselves "The Geebung Polo Club", and practised on the level bluff to the west of the corrals. This photograph, of a pickup game with High River in 1909, shows left to right: Herb Millar, Bar U foreman; F.R. Pike, bookkeeper; A.H. Shakerly, a neighbour; and 7U Brown. Glenbow, NA 3627-8

on the west, and a dining room on the east; the south side of the house had a hall and stairs, a kitchen on the east and den on the west. The den and living room both had fireplaces. This was our last home … and it was my home for about 24 years.[29]

As the scope of Lane's enterprises broadened, so more and more clerical work was carried out from an office in Lane's home. While F.R. Pike continued to handle the books for the Bar U and Willow Creek ranches, Lane's correspondence was voluminous, and a separate set of records was kept in Calgary.[30] In 1916, Lane's various enterprises were incorporated under the title of George Lane and Company Limited. The company office was located in Lane's residence.[31]

Lane's "boys" and Happy Families

One cannot help but be impressed by the range of Lane's interests and by the role he played in the affairs of southern Alberta. These responsibilities demanded his time and attention, and he had little left for supervising the day to day management of his ranches. Instead, he delegated authority to the men who he had hand-picked for the job, and left them to run things. This policy clearly worked. Five senior men on the ranch: Millar, Fleming, Pike, Olsen, and Pierson, provided a hundred years of service among them! It was their expertise, their attention to detail, and their decision making that propelled the Bar U Ranch into its 'glory days' as one of the most famous ranches in Canada.

Herb Millar was the linchpin of Lane's management team at the Bar U.[32] They had been friends and business partners since Lane's arrival in Alberta.[33] Millar was more than an efficient foreman who knew the ways of cattle and the range; he was a confidant with a shrewd business sense to whom Lane could turn for advice. He could be relied upon to manage the big ranch with a minimum of supervision or fuss. Millar was forty years old when he returned from the YT Ranch to the Bar U in 1902. He was of slim build, stood about five foot ten and weighed around one hundred and fifty pounds. He had spent his whole life around cattle and horses and had mastered all the skills of the cowboy. Indeed, many acclaimed him as being the prettiest rider they had ever seen break an outlaw horse. Like his boss, Millar was also a legendary shot.[34] He could still show young horsebreakers a thing or two, and played polo regularly for the Pekisko Polo Club. Although he could still ride a bucking horse to a standstill, Millar recognized the importance of "gentling" horses for sale to the North West Mounted Police and the extra value that could be gained from training Percheron geldings thoroughly for use as draft animals.

6.6 Alex Fleming driving the Bar U six-horse show team at the Calgary stampede grounds, c. 1916. Glenbow, NA 2046-13

During his twenty years on ranches Millar had developed "cow-sense" to an extraordinary degree. Frank Dobie might have had him in mind when he wrote, " he savvied the cow – cow psychology, cow anatomy, cow dietetic – cow nature in general and cow nature in particular."[35] Herb was deeply imbued with open range attitudes in spite of his upbringing on a farm. He valued hardihood and the ability to rustle above all other bovine characteristics, and he was suspicious of innovations which he felt tended to pamper cattle and reduce their will and ability to survive.

Millar was a tough, taciturn man, short on tact and always prepared for confrontation.[36] His "people skills" would have turned a contemporary expert in labour management relations grey! He would fire a man in the morning on a whim, and rehire him in the afternoon. He would not tolerate insubordination, but the hands knew exactly where they were with him and respected his leadership. Millar could play other roles as well. He was a member of the High River Club, and was quite capable of entertaining prospective buyers there when Lane was absent.[37] Over the years Herb Millar exerted a tremendous influence on a generation of young cowboys and would-be ranchers. Of the many compliments he savoured on his eightieth birthday, probably none gave him more satisfaction than that offered by Harold Riley, who wrote:

> While engaged as foreman of the Bar U Ranch, many men who subsequently became ranch owners or expert riders, ropers and general all-round ranch hands were in a sense "pupils" at Herb Millar's school. Under his watchful eye they rode the range with him as their boss. From him they learned many valuable lessons in the art of handling livestock and range management.[38]

Alex Fleming brought style, charm, and humour to the Bar U community.[39] He was a handsome, vital man and, perhaps because of his early experience in the North West Mounted Police, he was always neatly turned out, his hair cut and his moustache trimmed. He conveyed a real presence sitting up on the box of a show wagon behind six matched Percherons. He enjoyed meeting people and was equally at home with princes and senators or with grooms and cowboys. Like Stimson before him, he was a gifted raconteur with a story to fit every occasion. He served the Bar U for sixteen years and played an important part in building the international reputation of the Percheron stud.

Alexander Fleming was born in Ontario into a well-to-do family. Unlike his close friend William Ives he chose to pursue adventure in the west rather than going to university. He joined the North West Mounted Police in 1881. A year later he was serving at Fort Walsh, and his remarkable gift for handling four- and six-horse teams brought him to the attention of his commanding officer, Major Walsh. Alex was selected to act as escort and driver to the Marquess of Lorne, the Governor General of Canada, on his tour of the west.[40] At twenty-one Alex was already hobnobbing with the highest in the land and holding his own in conversation around the campfire. Four years later he served at Duck Lake during the Riel Rebellion. In 1889, Fleming resigned from the force and went to work as a rider, first for the Oxley and then for the Cochrane Ranch along the Belly River. In 1904 he married Elsie Caldwell of Cardston. When the Cochrane Ranch was sold in 1906 the Flemings moved to High River, where Alex ran a livery service.[41] Here he deepened his acquaintance with George Lane and Herb Millar, and in 1908 Lane asked him to take charge of all the Bar U horses.

What an amazing opportunity it must have seemed! There were already more than two hundred purebred Percherons at the ranch, and Lane had the vision and the bankroll to support almost unlimited expansion. Fleming was responsible for every facet of the Percheron breeding program. He travelled to France on many occasions to seek out and purchase the very best stallions and mares. His special role was in training and driving the four- and six-horse hitches of Percheron geldings for which the Bar U became so famous. He drove with flair and gained the grudging respect of his fellow exhibitors. Lawrence Rye of Edmonton showed against Fleming for six consecutive years after 1913, and said of him:

> He always carried a long bullwhip and while holding all six lines in his left hand, he'd crack it over his horses' backs and then collect the reins in both hands. It was sort of rough in those days and he would make me mad when that whip would crack alongside my horses and just scare

> the daylights out of them. But say, Fleming was a driver. The best I can recall.[42]

Fleming's reputation among horsemen coupled with his expansive personality meant that he was a superb salesman. Whether he was dealing with an American buyer looking for a top flight stallion, or a local farmer interested in a pair of geldings, he did his best to ensure that they went home satisfied. He was much in demand as a show judge and travelled to New York City in 1916 for that purpose. He was an official with the Calgary Exhibition and Stampede from its beginning in 1912, and was the starter for calf roping and chuckwagon events right through until 1941.[43]

As Lane's empire grew, so Fleming became general manager of all the Lane ranches, and, when Lane incorporated in 1916, Alex was appointed Vice-President of George Lane and Company. Such responsibilities meant many hours in the office, but Fleming enjoyed nothing more than helping with the day-to-day operations of the ranch. He was a fixture at branding time, rode with the cowboys whenever he had time, and sometimes travelled to Chicago with cattle. Bert Sheppard described Alex Fleming during the roundup of 1922, and explained that he "spent a good deal of his time riding with us, probably because he enjoyed the work. He was a fine figure on a horse, riding a full flower stamped centre-fire Visalia saddle, and he had the upright seat of an old-time cowboy. He always packed a rawhide rope."[44]

Alex and Elsie Fleming raised three children, first at the Bar U Ranch and later at their home in High River. The boys shared their father's enthusiasm for horses and spent their summers working at the ranch and playing polo for High River.[45] The Flemings were noted for their involvement in the life of the community and for their warm hospitality and kindness. They made a home for Harry D'Aragon, a Bar U foreman, when he was incapacitated after a riding accident.[46] Alex Fleming's contribution to the Bar U as horseman, managing executive, ambassador, and exemplary human being, was enormous.

Frank Robert Pike served as bookkeeper at the Bar U for more than a decade. He was an unlikely "desk jockey." Born and raised in England, he had emigrated to the United States and – like so many other young men – drifted westward in search of adventure. After ten years working with cattle and horses in Wyoming, Pike embarked on the ultimate adventure and set out for the Yukon to make his fortune panning for gold. He passed through High River with a pack train in 1898 heading north. Somewhere north of the Peace River, Pike was able to help Frank Bedingfeld and Glen Campbell, and received an invitation to visit the Bedingfeld ranch on his return to civilization. In due course he took Bedingfeld up on his offer. During his stay he explored Calgary and the foothills and liked what he saw.[47] His experience as a rider earned him a job with the North West Cattle Company. When George Lane took over the ranch, there was an urgent need for a bookkeeper and Pike was promoted to fill this gap. It seems likely that he had spent some time getting a "commercial education," such as was offered in the late nineteenth century to prepare young men for careers in business offices. As Bill Yeo remarks, "If F.R. Pike did the typing, as well as the bookkeeping and the supervision of purchases and sales, it is plain that he was no ordinary cowboy."[48]

As the Bar U Ranch diversified and the labour force expanded, so Frank Pike's responsibilities grew. Just 'keeping tabs' on the labour force and insuring that the hands received their pay regularly must have been quite a task. The turnover was rapid. A man might be hired for a couple of weeks to help with a particular drive, while another might be "borrowed" from the Willow Creek Ranch to assist with dipping cattle. Moreover, much of the hay on the ranch was put up by contractors, and each of these needed a separate contract. In addition to managing the normal ranch purchases of foodstuffs and supplies for the ranch, Pike was at the Bar U during a period of construction. He arranged purchase and delivery of all the materials and building supplies required. Very little evidence of Pike's office work has survived except for a series of letters he wrote to the Commissioner of Irrigation. It is clear that Pike managed this correspondence himself and signed the letters on behalf of Lane and his partners. He was allowed a considerable degree of personal authority.

Lane relied heavily on Pike's skill as an organizer. When Lane and his friend W.B. Thorne established the Canadian Percheron Society in 1907, it was Pike who was appointed secretary, a post he retained until 1913. This meant that, in addition to maintaining the records for all the horses on the ranch, Pike was involved in correspondence with Percheron breeders throughout the country.[49] The publication of the first stud book of the Percheron breed in Canada in 1912 was an achievement not only for George Lane but also for Pike, who managed the large volume of registrations which the project entailed. Frank's role as secretary of the Gee-Bung Polo Club of Pekisko was more fun and less onerous. Pike played polo himself whenever he had an opportunity.[50] He also looked after the post office at the Bar U. Herb Millar's wife Ella was nominally the postmistress, but Pike lived and worked in the post office building and handled business with the local people who came to collect their mail.[51]

In 1913, Pike was hospitalized in Calgary with a serious illness. However, he returned to the ranch in June and took over the job of ranch manager on an interim basis after Herb Millar moved to the Namaka farm. His stay in hospital had pleasant and unforeseen effects on his life, for he married one of his nurses, and they moved to his homestead, three or four miles east of the ranch, close to Baines' "Pekisko Trading Post." From this location Pike continued to help with bookkeeping and other jobs at the Bar U on a part time basis.[52] The decade during which Pike had looked after the books at the ranch had been a pivotal one, and his contribution in a somewhat unglamorous role should not be underestimated.

Twenty-year-old Neils Olsen came to the foothills country in 1895 with a Swedish family from his home town near Malmo.[53] They joined the Nymans, who had been in Alberta for four years, and young Neils soon found work, first with H.N. Sheppard and then with 7U Brown.[54] With two years of experience around horses and cattle and a fair grounding in English, Olsen started his long association with the Bar U Ranch on the lowest rung of the ladder; he became nighthawk for the Bar U wagon on the roundup. This meant herding the saddle horses through the night, and then bringing them into a rope corral at the camp in the dark before breakfast, so that the riders could select their mounts for the day. Then Neils would help the cook break camp and would drive the bed wagon to the next campsite. It was hard work, and something of a "rite of passage" for a young man who wanted to be a rider.[55] Nevertheless, it must have been a thrill for the young Swede to be riding with the most prestigious outfit in western Canada. He had the opportunity to learn much from Charlie McKinnon, the wagon boss, about controlling and moving large herds of cattle. He was a reliable and conscientious employee, and in 1908 the thirty-two-year-old was made wagon boss on the Bow River range.

6.7 The Olsen family of the verandah of the foreman's house at the Bar U. Left to right: Emil, Albert, Elise, Irene, Ruth and Neils Olsen, 1921. Parks Canada, Olsen Family Collection, 96.03.01.25

Having made his mark in the new world, Olsen decided to pay a visit to his family in Sweden. On the boat he met and fell in love with Elise, who had been living with her family in Chicago, and was also going home for a holiday. They were married in Malmo and returned to western Canada in 1909.[56] Neils took over as 'range boss' when Charlie McKinnon left to start his own ranch. He was based at the Big Bow Ranch, and his job meant that he was away gathering cattle for weeks at a time. Elise went to stay with her Swedish friends in Red Deer, where she gave birth to her two sons, Albert in 1911, and Emil in 1914.

Late in 1914 Lane appointed Olsen as manager of the Bar U Ranch to replace Herb Millar. Neils and Elise and their two sons moved to the foreman's bungalow. It was to be their home for more than a decade. The thirty-eight-year-old Swede had already spent fifteen years with the Bar U. During that time he had experienced all facets of Lane's diverse operations. His particular expertise was with cattle rather than with horses.

The Olsen family flourished at the Bar U. Ruth was born there in 1917 and Irene in 1920. Both the memories of Ruth Olsen and the family photograph albums suggest that the children enjoyed an idyllic life at the ranch. There were three other families living on the ranch, and during the summers there were twelve children ranging from babes in arms to young

6.8 The Olsen children's governess outside their house, c. 1918. Later, she married Bob Armstrong, the Bar U bookkeeper. Parks Canada, Olsen Family Collection, 96.03.01.37

6.9 Fred Nash started working at the Bar U before the First World War. When he enlisted he was promised his job back when he returned. This photograph was posted in the cookhouse throughout his absence. He is shown in the uniform of the Canadian Mounted Rifles. Parks Canada, Reynolds-Leadbeater Collection, 96.08.01.05

teens.[57] In addition, the two Fleming boys were already working on the ranch, and there were frequent visits from the Gardner children of the Mount Sentinel Ranch and from near neighbour Josephine Bedingfeld. There was the creek to swim and fish in during the summer and to skate on in the winter. The Olsen children learned to ride on a buckskin pony named "Tiger," which was given them by the Bedingfelds when they sold their ranch to the Prince of Wales and returned to England in 1919.

Neils and Elise valued education highly and they were determined that their children should have every advantage in spite of the isolation of the ranch. They employed a succession of young women to act as governesses. This was much appreciated by the young men of the Bar U community. Bob Armstrong, the bookkeeper, married the first governess and they moved to Calgary, where he continued to work for Lane in the city office. Fred Nash arranged for his English fiancée to stay with the Olsens when he returned from the First World War in 1919. She taught the children for two years and the Nashes were married in 1921. Alice McIsaac, a young teacher from Prince Edward Island, was the next to arrive, and she was courted by Pete Hardy. When they married they settled on a farm near Okotoks.[58]

In March 1925, the Olsen family's life took a shocking and tragic turn. In Ruth Olsen McGregor's words:

> Father attended a horse show in High River and came home with a terrible cold. He took sick and was taken to High River Hospital and a few days later he died. Mother took ill while he was still at home and she was taken to hospital too. Mother wasn't told that father had died and Father didn't know that mother was in hospital. Two days later she died as well.[59]

Suddenly, pneumonia had made orphans of the four children. Albert, Emil, Ruth, and Irene were taken care of by Mrs. W. Buchanan of High River, who had been a close friend of Elise. It was a terrible shock, particularly for Emil, as his sister remarked, " his life had changed, taken away from home and life at the ranch, away from his horse, to live in town with strangers."[60] He never talked about this painful period. Ruth and Irene finished school in High River and then went on to nurses' training in Calgary.

Neils Olsen had managed the Bar U for the last decade of the Lane era. Under his leadership the Percheron stud had flourished, the cattle herd had been improved, irrigation had been extended and a significant acreage of the ranch had been broken and cropped for grain. Olsen had handled a growing labour force and diverse responsibilities with aplomb, and his young family had added a new and delightful element to the Bar U community.

6.10 Bar U ranch hands, c. 1917. Standing left to right: Arthur Mortimer, Carl Sonne, Jack Smith, Roran Bill, and Norman Mackenzie, the young man who was himself an avid photographer. Kneeling: Joe 'Jeff' Reynolds, and Mutt Berg. Parks Canada, Reynolds-Leadbeater Collection, 96.08.01.02

A Royal Visit

The apex of the fame and fortune of the Bar U Ranch was the visit to the ranch by the Prince of Wales during his Canadian tour in the fall of 1919.[61] The popular Prince moved across Canada on a tidal wave of patriotic euphoria. For a day or two the eyes of the world were focused on High River and the Bar U, and on the ranching way of life which Prince Edward said that he had enjoyed so much. The subsequent purchase of the neighbouring Bedingfeld Ranch by the Prince prolonged Pekisko's time in the public spotlight, and the visits of Alberta's royal rancher over the next few years all involved George Lane and the Bar U Ranch.

Later, the new owner of the Bar U, Pat Burns, was to supervise the management of the E.P. (Edward Prince) Ranch through the 1930s.

The canny British Prime Minister, Lloyd George, was the instigator of the idea of a post-war royal tour of the empire. He felt that the war-hero Prince of Wales might do much to cement imperial relationships before wartime camaraderie had dissipated. He was confident that "the appearance of the popular Prince of Wales might do more to calm the discord than half a dozen solemn imperial conferences."[62] Prince Edward accepted this charge with enthusiasm. It was just the kind of challenge that he was looking for. Before the war he had remonstrated with his father that as he had "neither the mind nor the will for books," the experience of going up to Oxford would be wasted on him. Would it not be better he argued, for him to be sent on a world tour? That way he could learn about people and places at first hand.[63] Now he grasped his opportunity with both hands, and the tour was to change his life.

From the time he first stepped ashore in the New World at St. John's, Newfoundland, Britain's oldest colony, the tour was a triumphant success. Everywhere, the crowds were enormous and responded to the Prince with warmth and enthusiasm. Indeed, it seemed that the mood and psychology of the people was different from that of pre-war crowds. There was more spontaneity and lack of inhibition apparent. Godfrey Thomas, secretary to His Royal Highness, reported to Queen Mary, "These were the most extraordinary days I have ever seen ... half the people seemed to have taken leave of their senses."[64] As Edward was to write later in his account of his adventures:

> The crowds proved so volatile and vigorous as to constitute at times an almost terrifying phenomenon. Uncontrolled, almost ferocious in their determination to satisfy their curiosity about me, they again and again broke through the swamped police lines. They snatched my handkerchief; they tried to tear the buttons off my coats.[65]

It is hard for us, bombarded as we are by sophisticated media, jaded by the machinations of political "spin doctors," and with a somewhat tarnished view of royalty, to appreciate the magnetism, the charisma, and the star quality of the Prince of Wales at this time. The moment could not be replicated. It was partly due to the unique qualities of the young man in the eye of the storm, and partly to the context of the times. The twenty-five-year-old Prince intuitively understood that relationships between monarch and people had been changed by the war and the social upheavals it had brought about. He was never a remote figure, a marionette mechanically bowing and waving, but rather a golden-haired young man who moved among the people and made you feel as you shook his hand that, for a moment at least, he was genuinely delighted to have the opportunity to meet you. Prince Edward was the first to try and give the aloof and mysterious institution of monarchy a new, popular, and "more democratic" image.

Among the crowds that came to greet him there were few adults who had not been deeply affected by four years of bloody conflict. Families had lost sons and fathers. Even in fortunate homes where men and women had returned from "over there," it was impossible to pick up where one had left off. There were deep scars which would never heal. The war had cast a dark pall of worry, deprivation, and tragedy over society.[66] Now, the lights had come on again, and here was a young man with an impish grin who had been there and had survived. He was a symbol of hope and of new beginnings.

At every stop across the country, the Prince spent as much time as was possible with veterans. He had been with the Canadian Corps in Flanders, and already knew thousands of Canadians. The correspondent of *The Times* of London caught the spirit of these occasions well in an account of a ceremony at which the Prince bestowed decorations on returning soldiers or on their nearest relations:

> For each old mother or father, all of whom seemed to look into his face as if he himself was their son, for each pathetic widow, for each wounded soldier, he had an especial word of sympathy and praise and understanding, and not for a short time either, so that before he gave them his left hand to shake after pinning on the medal won in the great struggle, he never stopped talking for a moment … for him and for the Empire this gift of human sympathy and kindliness is a great and valuable possession.[67]

The citizens of High River were up and about early on the morning of 15 September 1919. The station and nearby streets were liberally decked with flags, bunting and streamers, and a triumphal arch was built leading out of the station yard. During the afternoon local children were joined by others from surrounding schools. Two bands played and the children sang songs until the royal train finally appeared. After listening to some words of welcome from the mayor, the Prince devoted his attention to the veterans who provided a guard of honour:

> … shaking each by the hand, looking each in the eye and passing some complimentary remarks to each, and endeavouring to locate some personal recognition of possible slight acquaintance at the front, and in this latter it was stated that he was successful. The Prince by his thoroughly democratic appearance in both dress and manners, immediately enthroned himself in the hearts of the large crowd present, who were greatly impressed with his open manner and lack of strict formality.[68]

Then it was time to move on to the Bar U Ranch, where the Prince was to spend the next twenty-four hours. Prince Edward disappeared into George Lane's car and was driven off, followed by three other cars filled with royal staff, guns, and fishing tackle. They bumped and plunged over the country roads for twenty kilometres to the ranch. They arrived about 6 p.m. and had time to explore the Percheron stables and to pot a few ducks along the river before the evening drew in. Lane was proud of the fact that the ranch had its own generator which supplied electricity to all the main buildings. Every light was turned on to greet the Prince. Unfortunately, this overloaded the circuit and

6.11 His Royal Highness, Edward Prince of Wales, accompanied by George Lane and Neils Olsen, rides through the Bar U cow herd. The cattle are Shorthorns showing evidence of cross breeding with Herefords. The calves have achieved a good size in six months. Glenbow, NB 16-146

the ranch was plunged into darkness. Neils Olsen, the foreman, had to hurry out to try and rectify the fault, while lamps and candles were lit.[69] Fortunately this contretemps added to the spontaneity and gaiety of the evening. A home-cooked meal around a crowded ranch house table was in marked contrast to the stuffy formal dinners which had been so common on the tour. Conversation flowed as the Prince and his companions luxuriated in being out of the spotlight for a few hours. After dinner the table was cleared and some of the party enjoyed a hand or two of cards. For once, the Prince retired early to his room.

The detour to the Bar U Ranch was designed to provide the royal party with a brief respite from the rigours of the tour. Before he arrived in Canada the Prince had written somewhat diffidently to the Governor General:

> I hope you won't mind my saying that I should very much appreciate an occasional free day, particularly when I go west or when visiting some district where there would be a chance of some form of sport or recreation. I am by nature rather a crank on exercise and get very stale if I have to go a long time without getting any at all.[70]

Indeed, by the time he left Toronto the Prince was dangerously close to collapse, and desperately needed to spend a few days at a more relaxed pace. His exhaustion was added to by his inability to "switch off" and his tendency to stay up until all hours, talking and smoking incessantly.[71] The Prince was also determined that he should learn something of how Canadians made their living. He had visited factories, mines, pulp mills, and farms, and now he would have a chance to find out something about ranching. George Lane had visited England the previous year to arrange the sale of some Percheron horses, and he had met some highly placed men. Moreover, he was also well known in Ottawa. He was the natural choice to host the royal party.

In the morning the Prince was up early and slipped from the ranch unnoticed for a run. It seems likely that he went southward to climb the rounded hills from which he would have been able to look down on the wooded valley of Pekisko Creek, and the Bedingfeld Ranch, which he was soon to purchase, with the snow-capped Rocky Mountains illuminated in the morning sun. After breakfast the party rode out to the west of the ranch to watch the roundup and branding of some Bar U cattle. The Prince borrowed Mrs. Alex Fleming's horse with young Billy Ives' saddle. He is reported to have shown "horsemanship that was delightful to see."[72] He remarked that he was "darn glad he wasn't a calf," and declined to take a turn with the branding iron because, he said, "I don't understand the work and might inflict some unnecessary suffering on the beast."[73] The incident which impressed local people most occurred in the late morning when there was a severe shower. The Prince was offered a ride back to the ranch in a car, but declined, saying he would stay with the cowboys. He rode back soaking wet for lunch. In these circumstances it was fortunate that there were suitable refreshments (alcoholic) on hand to warm everybody up. A crisis had been averted, for, in the bustle of leaving the royal train, certain vital supplies had been left behind. Fortunately, Major Walker and Bishop Capers of Texas were to drive out

6.12 The Prince of Wales and George Lane pose on their horses during the Royal visit of 1919. Glenbow, NB 16-149

from High River in the morning. They brought with them the delinquent crates.

> … much humour was aroused when the Prince was told that the bishop had accompanied the refreshments and bestowed a benediction upon them by the imposition of his hands as the car jolted over the uneven trail. Seeing the humorous side of the incident, the Prince raised a glass to his lips and drank his lordship's very good health.[74]

In the afternoon, the Prince met with returning soldiers from the Bar U and neighbouring ranches under the Royal Standard which flew from the tall flagstaff outside the ranch house. The men shared their experiences overseas and were photographed with Prince Edward. Then he went out shooting with Dr. Carlyle and a couple of aides. They saw few birds, but the Prince was fascinated by the beaver dams along the creek, and his guide was hard-pressed to answer all his questions. The royal party left the ranch about 6 p.m., the Prince having written "Some Ranch!" in the guest book. As the Prince passed once again through the triumphal arch at High River to climb into his railway car "Killarney," he was heard to remark that he was going to return. The party arrived back in Calgary in time to attend a gala ball at the armoury in Calgary that night.[75] The Prince's stamina was truly amazing.

The brief visit to the Bar U Ranch had been an enormous success. The Prince remarked, "I spent 24 hours at the ranch, I wish it could have been 24 years," and he told reporters, "Ranch life is the life for me."[76] Colonel Henderson, who, as an aide to the Governor General of Canada, was responsible for insuring that the tour ran smoothly, was able to report: "The visit to the Bar U was an unqualified success. I am more than pleased with this in that some of the party were rather bored with the idea of going there."[77]

However, the most intimate and vivid evaluation of the visit comes in the Prince's own words, in two letters he wrote to Freda Dudley-Ward. The first was written at 10 p.m. in his bedroom at the Bar U, and the second on the train to Banff at 4 a.m., having just returned from the dance in the Calgary armoury.[78]

> … we had a one hour journey to High River where there was the usual mayor and his address to reply to, 20 veterans to inspect, school children to hear sing and crowds to wave to!!
>
> Otherwise there was nothing but to get in a car and drive to this ranch about 20 miles east [*sic*] of High River across the prairies with a dear old man called Lane who owns the Bar U Ranch which has its name from the branding stamp on their cattle (U)!!
>
> This is marvelous country sweetheart and how I wish you were here; we are staying the night with old Lane in his comfortable wooden house and I've got such a huge bed in my room angel that I shall feel quite lost when I turn in!! Guess this is the country and the life for me, sweetheart, if we could live together, though I don't think otherwise; I could never settle anywhere away from you, darling one, though the atmosphere of the west does appeal to me and attract me frightfully!![79]

6.13 H.R.H. Edward, Prince of Wales, greeting Bar U staff who had served in the war, September 1919. Notice that the Prince is using his left hand; his right hand had been badly bruised from overuse. On the left, Alex Fleming moves towards George Lane, while aide de camp "Fruity" Metcalf watches the Prince. Fred Nash, Bar U horseman, is shaking the Prince's hand, while Bob Armstrong, Bar U bookkeeper, waits second from the right. Parks Canada, Florence Fleming Collection, 95.07.01.06

He continued in the second letter:

> I've done such a lot since I stopped writing last night Fredie darling that I can hardly remember it all and needless to say the reason for the hour of writing is dancing!!
>
> But I must tell you about yesterday on the Bar U Ranch angel as I enjoyed it all so much and got a huge amount of exercise … I started with an 8 mile run at 8 am, 4 miles out to a dried up river and back, and at 10 am we rode off to "the roundup" of cattle which was an amazing stunt; I rode a nice locally bred horse in a "stock" saddle … we were in time to help the cowboys and Indians round up the last odd hundreds of cattle and they collected close to 2000 head I guess; it was quite good fun and I got lots of hard riding doing my best imitation of a cowboy chasing refractory calves which wouldn't be driven into the kraal to be branded etc. … it gave one a small insight into ranching life on the prairies!!
>
> But of course I should have stayed at least a week on the ranch to get a proper idea of it all; I was walking fields for prairie chicken for 2 hours after lunch though neither Legh (who was with me) or self shot anything!! Still it was all fine exercise and we were sorry to have to motor to High River at 6 pm to catch our train back to Calgary; its a real good life that ranching, darling, though a vewy hard one and one's got to be real tough to take it on as a living though it pays if one can make good.

> We dined on the train and have been dancing since 10 pm, a huge party given by returning officers and the best we've struck during this trip.[80]

If the royal visit to the Bar U Ranch had been an isolated incident, it would still have merited attention as a high point in the long history of the ranch. In fact, the visit started a forty-year link between the Bar U and Alberta's royal rancher. Prince Edward was so entranced by his time at the Bar U that he determined to purchase a ranch to use as a "get away" in the midst of the western country he had come to love.[81] While the Prince travelled westward to Vancouver and Victoria, George Lane acted as his agent to arrange the purchase of the neighbouring Bedingfeld Ranch. Edward was able to announce his new venture in his farewell speech in Winnipeg:

> I shall not say goodbye to western Canada, but only au revoir. I think this western spirit is very catching, at least I know that I have been caught very badly. I feel so much at home here by this time that I want to have a permanent home among the people of the west, a place where I can come sometimes and live for a while. To this end I recently purchased a small ranch in southern Alberta and I shall look forward to developing and making it my own.[82]

He also wrote gleefully to Freda Dudley-Ward from the train traveling eastward across Canada:

> … I've heard that the fact and the news that I've bought a ranch in Alberta has gone down vewy well and has been the success that I hoped it would!! … I've got it for cheaper than I ever expected to and I'm going to send stock over from my home farm in Cornwall. Gud it must make you laugh darling to hear your little boy doing the heavy rancher and agricultural stunt knowing how gloriously ignorant about it all he is.[83]

The implications of this purchase and the subsequent development of the E.P. Ranch for the Bar U were considerable. Edward relied heavily on Lane for advice and expected him to "keep an eye on it as it's next door to his!!"[84] When the Prince visited his ranch in 1921 and 1923, the Lane family played a central role in the celebrations and the Bar U Ranch was drawn into the media frenzy attracted by its neighbour. Lane's successor, Pat Burns, took on his mantle and headed the advisory council which managed the E.P. Ranch for Edward during the 1930s.[85]

The Allans had established a large corporate ranch which had gained a reputation for good management and consistent returns under the management of Fred Stimson. Lane had extended the scope of ranching operations enormously. His Percheron stud had acquired an international reputation on its own account. However, it was the royal connection – the implicit cachet of "By Appointment to His Majesty" – that first raised and then sustained the profile of the Bar U Ranch on a world-wide stage and made the ranch familiar to Canadians from coast to coast.

There is something of the solemn inevitability of a Greek tragedy about the events which engulfed George Lane and his enterprises after the First World War. Elements of the physical, economic, and political environments conspired to strip him of almost everything he had built up. Over the years Lane had demonstrated unusual prescience: he foresaw trends and manipulated change to his advantage. Now, suddenly, the forces he had risked everything to challenge overwhelmed him.

The first blow was the death of James Gordon, of Gordon, Ironside and Fares, in December 1919. The firm was in financial difficulty even before the end of the war.[86] Management had failed to adapt to the virtual cessation of the live cattle trade with the United Kingdom, and their strategy of investing heavily in meat-packing went disastrously wrong. The settlement of Gordon's estate revealed the extent of the company's collapse. Lane's shares in Gordon, Ironside and Fares, with a nominal value of $560,700, were worthless. The company had bank debts of about $3.5 million, while its packing subsidiary owed a further $3.3 million. The firm's only assets were their land holdings, including a half share in the Bar U Ranch. As part of the settlement of the estate, Standard Trust negotiated the sale of this holding to George Lane for $650,000 in cash, and the assumption of $100,000 in debts. Lane had to arrange a mortgage with Osler and Hammond to meet this huge new demand for capital.

The second blow struck at the profitability of Lane's cattle enterprises, which had been the stable foundation for his other investments. Cattle prices fell from $15 per cwt. in June 1920 to $6.84 in July 1921. An animal worth $70 in 1918 was valued at $30 in 1924.[87] This meant that Lane, who marketed about five thousand head of cattle a year, was deprived of at least $1 million in potential returns over the next few years. Cattle inventories in Canada had risen from about six million head in the years before the war to ten million in 1918; a slump was inevitable.[88] Even more serious were rumblings of protectionist sentiment in both Europe and the United States. At the height of the wartime boom, Lane had warned cattlemen that things could change with the stroke of a pen. He pointed out that "the most dangerous thing that stockmen have to fear is a change in administration on the other side of the line that would slap a duty on everything."[89]

In May 1921, the newly elected Republican administration in the United States passed the Young Emergency Tariff.[90] Live Canadian cattle were subjected to a 30 per cent tariff at the US border. The door was closed to the Chicago market. For the past decade it had absorbed both the massive four-year-old steers shipped by ranchers, and the younger, lighter stockers marketed by Canadian farmers. Overall, the value of shipments out of the country dropped from $21 million to $3 million. Falling prices, rising costs, and uncertain markets squeezed many stockmen out of business. The cattle industry quickly became "a hollow eyed, weakened, ghost of its former self."[91] With his ranches fully stocked, Lane was forced to continue to ship to the United States, although the tariff cut deeply into his profits.[92] His wife remembered the situation well some twenty years later. She explained:

> For a long time Chicago had been the market for George Lane and Company. In 1910 the price of their range cattle broke the record in the Chicago stock yards for grass fed cattle, but in the 1920s the duty paid to the US government for selling in their market was enough to have paid the interest on the bank loan ... [George] felt politics were shaping economic changes and working to the detriment of men of his type.[93]

Worse was to follow. The dry summer of 1918 meant that cattle approached the winter in poor condition and feed was desperately scarce even before the winter weather began.[94] Snow fell early in November and accumulated. Although the temperatures during the winter were not far below average, the mean daily temperatures did not rise above freezing until well into May. On 15 April 1920, Lane told the *High River Times*: "There is more snow in Alberta now than I have seen at this time of the year in 35 years. The cattle cannot get out on the range to feed, and our supply of feed is getting very low."[95] Hay had to be shipped in from Ontario, Manitoba and northern Alberta, creating enormous price increases, from $30 a ton in November to more than $60 in March. Lane bought $30,000 worth of cottonseed cake in Galveston, Texas. He had hoped to ship it by sea to Seattle and thence through the mountains to Alberta. He ended up having to move the feed all the way by rail, and paid $630 to $650 per car for freight.[96] Lane moved much of his beef herd from the Bow River lease to Namaka and fed them there throughout the winter. Cattlemen who managed to nurture their stock through the depths of winter despaired as feed ran out in the spring. One veteran rancher ruefully noted:

> In the spring, if we could be said to have had a spring that year, the cattle lay down and died – hundreds of cattle had been kept alive to this point with feed worth almost its weight in gold. It would have paid us to have gone out and shot what cattle we had that fall.... No one in the business has quite recovered from that year.[97]

A.E. Palmer, of the Dominion Experimental Station at Lethbridge, echoed this evaluation. "The most disastrous winter of them all was 1919/20.... The actual losses of livestock were not heavy, but the financial loss was terrible."[98] The winter had a number of long-lasting effects. The calf crop was small; many stockmen had shipped stock into north Alberta or across the line; others had sold out: taken together this reduced the capacity of southern Alberta to produce young stock. In a more narrow sense, the cattle-rearing operations which had so often supported Lane's expansive plans with windfall profits had become a liability.

Matters were even more desperate at Namaka farms, and with Lane's heavy investment in large-scale wheat farming. As we have seen, the first years after his purchase of the farms were blessed with above-average precipitation; however, the next six years were drought-ridden.[99] The flood of grain to the elevators dwindled. Yields in 1918 were less than half what was anticipated and fell further in 1919. The 1920 crop was something of a respite, but prices had started to decline. No. 1 Northern Wheat, which had received $2.04 in Winnipeg in June 1921, was worth only $1.13 one year later, and this was not the nadir.[100] Drought, grasshoppers, hail, blowing soil, and falling prices gave farmers in the early 1920s a foretaste of the "Dirty Thirties."[101] The only thing going up was wages. Monthly wages for farm labourers rose from $40 in 1914 to $95 in 1919.[102] By 1923, workers in threshing outfits were asking $5.50 to $6 a day, and much of the available help was inexperienced and careless. Lane reported offering a couple of engineers in Calgary $12 per day to help run steam engines at Namaka. His offer was turned down.[103] The family farmer could retrench, become as self-sufficient as possible, and even seek seasonal work off the farm, while awaiting better times. Large-scale farmers were caught between escalating costs and declining returns, and the banks were inexorable in their pursuit of interest payments. Lane's affairs spiralled downward into a sea of red ink.

The fragmentary evidence available suggests that Lane kept fighting to put off the inevitable collapse. In October 1923, he and Bob Armstrong, his bookkeeper, had a meeting with the Osler and Hammond Trust Company. This was the company which had provided the money to pay for Gordon, Ironside and Fares' share of the Bar U Ranch. The original mortgage was for $650,000 and was issued in September 1920. A year later debentures were issued, probably in lieu of interest, and now Lane sought to avoid foreclosure by adding ten further parcels of land (7,840 acres) to the mortgage.[104] The move bought a few more months of independence. But, in 1924, Bruce E. Elmore, manager of the Dominion Bank in Calgary, became an officer of George Lane and Company, and from then on it was the bank which owned the assets and dictated policy. It was not until after Lane's death that the extent of the financial debacle became apparent. In January 1926, Elmore, the secretary and treasurer of the company, made the following declaration to the Department of Municipal Affairs:

> ... the said George Lane and Company Limited, is insolvent and is indebted to the Dominion Bank in a sum in the neighbourhood of not less than one and a half million dollars.
>
> That its assets, including the payments to accrue due under agreement for sale of the Bar U and Willow Creek Ranches to P. Burns and Company Limited and under agreement for sale to a Mennonite Colony of the Namaka Farm, do not exceed the value of one million dollars.
>
> That it is operating merely for the purpose of realizing its unsold property and when these properties are sold, the company will be wound up.[105]

When Lane's will was probated, the worth of his property was under $3,000, and his 9,980 shares in George Lane and Company, with a par value of $100 each, had no value whatsoever.[106] Although the "Elizabeth Lane Trust Fund" which George had planned to look after his wife, did not come into being because the shares were valueless, it seems that the family home in Calgary was in Mrs. Lane's name, and she was adequately provided for.[107]

In 1922, Lane told his wife that he was tired, and that "if he dropped off it would be all right."[108] During the journey to Pasadena California to spend the winter, Lane suffered a mild stroke. He recovered sufficiently to spend the following summer in Calgary, but in September 1923, the *High River Times* reported:

> Mr. George Lane left hurriedly for the coast last week in compliance with his physician who advises a complete change of climate for the benefit of his health.[109]

One wonders whether it was the Alberta climate or the looming clouds threatening all aspects of his business dealings which was affecting Lane. He rallied in the spring of 1924 and was able to travel to Chicago with his wife to see his eldest daughter and his grandchildren. Then the Lanes returned to the Bar U and spent a quiet summer surrounded by family. There was a glorious burst of Indian summer during September, and "on a beautiful autumn morning about 6 o'clock he died quite suddenly while having his pillows arranged."[110]

A few days later the country was hit by a violent fall snowstorm. Calgary received fifteen inches of heavy wet snow, and many roads were rendered impassable.[111] Nevertheless hundreds of friends and business associates made the effort to get to High River for George Lane's funeral, which was held in the town hall. A special train from Calgary, provided free by the Canadian Pacific Railway, carried many mourners, including the members of the Bow River Masonic Lodge. The pallbearers were almost all old ranching friends: J.H. (7U) Brown, Pat Burns, A.E. Cross, Archie MacLean, Dan Riley, and the cabinet minister Charles Mitchell. The ceremony, conducted by the Reverend G.A. Dickson of Knox United Church in Calgary and assisted by the Reverend W. McNicholl of High River, was simple but impressive.

The newspaper reports of Lane's death and funeral tended to headline two aspects of his life: his links with the pioneer days of the province and his friendship with the Prince of Wales.[112] Among all the glowing obituary notices, it was Pat Burns' that pinpointed Lane's most important attribute and stressed his contribution not to Alberta's past but to the province's present and future:

> The dominant characteristic of George Lane was his clear vision of the development of Alberta and his marked foresight in taking advantage of that development. At a time when few … saw any future for the country beyond its then development as a ranching country, his mind was constantly working out plans for the future, and many a forecast … was fully attained when the farmer arrived and the cities began to grow. George Lane was one of those men to whom the west owes much. For forty years he lived in Alberta and gave to its progress and development all his energies and foresight, and an inspiration for progress and better things which have gone far to build up prosperity at home and advertise our province abroad.[113]

Lane had taken over a large corporate ranch in 1902 and made it the most famous outfit in Canada. Over a period of more than twenty years he had integrated the cow calf operations of the Bar U with his ranches in the mixed-grass prairie zone, which specialized in

6.14 George Lane with a good bag of prairie chickens at the Bar U ranch house, c. 1920. He was sixty-four years old. Parks Canada, Catherine Gunn Collection, 96.09.02.01

fattening cattle on natural grass which would have otherwise gone unused. He had built up a world-renowned Percheron stud, which had sold purebred horses back to Europe to help re-establish the breed there after the devastation of the First World War. During his tenure, the headquarters site had been transformed, the Bar U farm had been established to produce grain and tame hay, and an irrigation project had been initiated. Lane had left an indelible impression on the big ranch.

7.1 Cattle from Pat Burns' ranches approach his feedlot and packing plant on the Bow River in east Calgary. Burns controlled each link in the production chain from the ranch to the butcher's shop. Ann Clifford, *Ann's Story*, p. 92

7.2 Pat Burns' feedlot and packing plant, c.1910. Glenbow, ND 8-317

7
Pat Burns Takes Over, 1927–37[1]

"By re-envisioning how people have been a part of place, then we can acquire a new way of seeing history, and in that vision, a route toward a better understanding of ourselves."[2] – James E. Sherow

The Bar U Ranch was purchased by Patrick Burns in 1927 and became part of an integrated and complex food-processing system, solidly based in Alberta, but extending westward through the Kootenays to the coast of British Columbia and eastward across the prairies. The term "agribusiness" has been coined during the past two or three decades to embrace the variety of ways in which the "mass production" methods, adopted with such success in the manufacturing sector, have been applied more gradually in agriculture.[3] Mechanization and the application of science to production techniques were part of this development, but perhaps even more characteristic has been the establishment of vertical and horizontal linkages. A food corporation today controls every facet of production: land and farms; warehouses and cold storage; transportation equipment; processing facilities; and wholesaling and retailing outlets. At the same time, risk is spread by diversification into other sectors of the economy; perhaps into forest products, packaging and paper, or mining and refining. It is surprising to find that Pat Burns' operations displayed all these characteristics in the decade before the First World War. Vertical integration involved the operation of some ranches which concentrated on breeding calves, while others focused their attention on fattening yearlings and steers. In turn, these ranches supplied the Burns stockyards, where cattle were sorted and dispersed to a variety of markets. The most important of these was the Burns' abattoir and meat-packing plant. From there, meat and a variety of meat products was distributed through Burns' chain of retail outlets. All the links in this long "value added chain" from the ranch to the butcher's shop were owned by Burns.

In 1928, only a year after he purchased the Bar U Ranch, Burns divested himself of his meat-packing business when he sold it to Dominion Securities Corporation of Toronto. If the split between Pat Burns the rancher and the new company had been complete, the chain of connections linking one stage of production with another would have been broken and the Bar U Ranch would have been only peripherally linked to agribusiness. Such was not the case. Pat Burns remained chairman of the board of directors of the new company and his nephew, Michael John Burns, became president and general manager.[4]

Thus the close linkages between the Bar U and the dominant market in western Canada – Burns and Company – were maintained.[5]

Pat Burns took an active interest in every facet of the management of his ranches. He spent a lot of his time at the Bow Valley Ranch and was reluctant to miss major events, like roundups, taking place at his other properties. Sometimes, he would come out from Calgary to the Bar U Ranch several times a week.[6] Moreover, the overall philosophy of ranch management at any of the Burns places was based on the down-to-earth attitudes of the "Old Man." Other things being equal, Burns favoured natural grass over tame hay, and both over feed grains.[7] He tended to be suspicious of innovations which involved tying capital up in machines or buildings, or, indeed, in any innovation which increased the costs of cattle production.

After Pat Burns died in 1937, Burns Ranches Limited was run by a board of directors chaired by John Burns. As the economic momentum of the country began to pick up after the long Depression, so more land was broken to grow grain and new scientific methods were applied to animal husbandry. The Burns era should, therefore, be divided into two sub-periods: the first, from 1927 to 1937, was characterized by the "hands on" management of Burns himself as he struggled to maintain his ranch enterprises through the trials of the Depression. The second period, from 1937 to 1950, saw an experienced management team responding to developments in both technology and consumer demand to maximize returns during an exciting period of rapidly rising food prices.

The straightforward chronology of the period from 1927 to 1950, during which the Bar U Ranch was managed by Burns and his successors, was riven by two occurrences of immense significance at every scale: personal, regional, national, and global. I refer, of course, to the Great Depression and the Second World War. The impact of the Depression on the ranching country in general and the Bar U in particular is examined in Chapter 9, while the influences of wartime on prices and labour costs are discussed in Chapter 10. In this chapter Burns' early years and his involvement in the cattle business are described to provide the context for his purchase of the Bar U Ranch in 1927.

The Irish Drover[8]

If Fred Stimson, the first manager of the Bar U, was best described as a stockman, and George Lane, the second owner, as a cowboy, then Pat Burns, the third owner of the great ranch, deserves the ancient and honourable title of "drover."[9] He spent most of the first half of his long and active life buying and collecting small lots of cattle and moving them to market. He was concerned primarily with the delivery and sales of livestock rather than with the production of animals. After his hardscrabble upbringing on a poor homestead in Ontario, twenty-three-year-old Pat set out for Manitoba in 1878. He took up a homestead near Minnedosa, but worked at all manner of jobs to build up his stake. The enterprise which seemed to bring him the best returns involved purchasing a few head of cattle in the countryside and moving them to Winnipeg for sale. By 1885, Burns had finally "proved up" his homestead, and turned his full time attention to trading in livestock. The Riel Rebellion had raised the price of beef from 4 cents per lb. to 14 cents per lb., and the completion of the Canadian Pacific Railway brought with it the possibility of selling western stock in eastern Canadian cities. The prospects looked good.

7.3 Interior of one of Pat Burns' shops, decorated for Christmas. Glenbow, NA 1149-4

Burns traversed the country throughout Manitoba and westward into the North-West Territories, making contacts with farmers and buying their surplus animals. Like the drovers of Highland Britain, he lived a nomadic existence, "a tent, a boarding house, a hotel room or a railway coach were his homes."[10] Like them, too, he had a collie dog to help him manage the cattle. Those who knew him during this period of his life all mentioned the zest and enthusiasm with which he tackled each new day. He often trailed cattle during the day and then did his own butchering at night. His biographer, Grant MacEwan, wondered when he found time to sleep and said that he "kept the hours of an alley cat!"[11] Most of his cattle purchases were made on the understanding that the farmer would be paid when Burns had sold the stock. The trust which people placed in him was built up gradually over years of dealings, and his growing reputation for honesty, reliability and responsibility was to become one of his greatest business assets.

It was as a contractor who supplied beef to the labour crews which built Canada's western railways that Burns made the transition from being an obscure local cattle buyer to an entrepreneur who rubbed shoulders with some of the most powerful 'movers and shakers' in Canada. In short order he began to measure deals in thousands of dollars instead of tens and hundreds. In 1887, William Mackenzie and his partners Donald Mann, James Ross, and Herbert Holt secured a railway construction contract to drive a line from Quebec through Maine to the ports of the Atlantic seaboard. Mackenzie had grown up in Kirkfield, Ontario, with Pat Burns, and although he

7.4 Pat Burns at the height of his powers in his early forties. Glenbow, NA 3965-65

was some five years older than Pat, he remembered him from their shared school days and from the hours that they had spent in the fields picking potatoes. Aware that Burns had already acquired considerable experience in the livestock business, Mackenzie gave him the opportunity to provision the labourers who were to construct the line. It was during the year he spent on this contract that Burns worked out the techniques he was to employ again and again. He learned to establish a mobile slaughtering facility which could move easily as the railhead was extended. He employed a reliable butcher to prepare the meat, and handled the buying and droving himself.[12] His operations were flexible, capable of duplication should several gangs of labourers be employed at different locations, and required very little outlay of capital. Burns spent his scarce money on livestock and supplies.

The success of the contract in Maine led to whole succession of other contracts with Mackenzie and Mann. The first was the Qu'Appelle, Long Lake and Saskatchewan Railway, and that was followed by the Calgary and Edmonton Railway, and finally, in 1897, the CPR's Crowsnest line. Each contract enhanced Burns' reputation for reliability and ingenuity. Somehow or other he always managed to get the job done, and the camp cooks were never short of beef.

MacEwan described the thirty-five-year-old Burns as having "the face of a cherub and the arms of a stone mason."[13] He was only five feet seven inches in height, but he cut a fine figure on a horse, as his torso was long in relation to his short legs. Burns was vital and quick in all his movements, and radiated nervous energy. He found it hard to sit still and would pace up and down thinking through a problem and muttering to himself. When excited he was apt to stutter and repeat expressions like, "look it, look it now ..." or "to be sure, to be sure ..." He was physically very strong, and his hard life in the saddle kept him in the peak of condition. He did not smoke or drink, and had no patience for prolonged social chitchat. On the other hand, he could always find time for his employees and was seldom brusque or impatient.[14] Throughout his long life Burns never paid any attention to the clock and to normal business hours. He kept after things until they were settled, and made up for his long days and short nights by snatching naps whenever and wherever he could. Burns more than compensated for his lack of formal education by his ability to listen carefully, question keenly, and store information. He had a prodigious memory and a great facility for figures. Where other businessmen might have relied on written reports, Burns was always on the move, tracking down his lieutenants at the job site and finding out exactly what was going on. During his youth and middle age, Burns showed himself capable of numerous acts of endurance and he was proud of his robust constitution and fleetness of foot. On one occasion he was thrown from his galloping horse when it stepped in a gopher hole. Both his wrists were broken and he was quite alone. Despite the pain, he caught his pony and rode eighteen miles to a doctor.[15] Burns maintained a daunting pace as he moved from place to place and enterprise to enterprise, encouraging, inspecting, and correcting. It was typical of his impatient energy that when automobiles became available he liked to be driven as fast as possible, although he never learned to drive himself.

Twelve-year-old Harold Riley met Burns for the first time around 1890 when his family was supplying vegetables to the cooks along the Calgary and Edmonton Railway line. He explained:

> The work crews had all eaten their mid-day meal in the big marquee tent. When we pulled up to the tent, the cook called to us to unhook and feed our horses and come in for dinner, which we promptly proceeded to do. We had just got nicely started at our meal when there was the sound of another team approaching. This proved to be the team and democrat driven by Pat Burns. In the democrat he had the carcass of dressed beef neatly covered with a canvas. Once more the cook poked his head out and exclaimed: "Hallo Pat, unhook and come in for something to eat."
>
> In a few minutes, a well built, muscular man entered the marquee. He was dressed in a pair of somewhat soiled overalls, supported by a

MANUFACTURERS, WHOLESALERS, ETC.

P. BURNS & CO.

LIMITED

Head Office and Plant - - - East Calgary

Packers, Exporters and Provisioners

Manufacturers of

"Imperator" Hams and Bacon

"Shamrock Brand" Leaf Lard

Hams and Bacon

PACKING HOUSE PRODUCTS AND PROVISIONS

Dressed Meats all bear the "BLUE STAMP" of Government Supervision and Inspection.

Cured Meats and Lard---Our reputation stands behind the farmers, **"Imperator"** Hams and Bacon and **"Shamrock" Brand"** Leaf Lard, Hams and Bacon.

Poultry---Large, Assorted Stock of Eastern and Alberta Poultry.

Fish---Fresh, Frozen and Cured.

Butter and Eggs---"Shamrock" Butter and Fresh Eggs.

Oysters, Sauerkraut, Mince Meat.

Sausage Makers, Sausage Casings.

Chicken Feed, Fertilizer Manufacturers.

Consignments Solicited---Of Choice Butter, Eggs and Poultry. **Highest Prices Paid.**

7.5 An advertisement for Pat Burns and Company illustrating the number of products which he handled. Glenbow, NA 2047-1

> leather belt, and wore a greyish-blue shirt, also somewhat soiled, without a collar, and an old battered felt hat. He was introduced in typical western fashion as "Pat Burns."[16]

This is not the place to recount in detail the spectacular growth and diversification of Burns' interests from the turn of the century until the First World War. The story of his entrepreneurial gambles and his flourishing business empire has been told by his biographers. Among other things, he carried out a successful government contract to provide beef to the Blood Indian reserve; he developed a chain of butchers' shops to supply the isolated Kootenay mining communities; and he sent beef on the hoof to the Yukon during the gold rush. However, the linchpin of his business operations was the development of slaughterhouses and meat-packing facilities, at first in Calgary, and later in all the major urban centres of western Canada. It was perhaps inevitable that Burns would tackle the production of stock as well as handling distribution and marketing.

New Ways of Rearing Stock

Burns was drawn into ranching by the need for a steady flow of stock to meet his railway contracts and later to supply his retail stores and packing plants. A reliable year-round flow of market ready cattle was essential. His strategies in ranch management were a reflection of his knowledge of the market and turned conventional wisdom on its ear. The prevailing system was to fatten cattle during the spring and summer, using the seasonal flush of prairie grass, and sell them in the fall in the pink of condition. This produced a cycle of scarcity and glut which was clearly reflected in the prices paid to ranchers.[17]

Burns and his friend and partner Cornelius J. Duggan developed an innovative system which involved buying cattle cheap in the fall, feeding them through the winter, and selling the steers in prime condition in the spring when the prices were highest. This system depended on obtaining thousands of tons of hay. The problem was, how to do this without employing a costly army of workers. Burns and Duggan resolved this dilemma by careful organization rather than by capital investment. They quartered the country and made arrangements with innumerable incoming homesteaders to put up wild hay. Many pioneer families earned their first cash from putting up hay under contract to Burns. If a settler did not have a mower or a rake Burns would loan him enough to buy the implements and accept repayment in hay. Typical was Louis Hammer, who came from Missouri in 1901 and cut and stacked hay for Burns each year from 1901 to 1910. Payment was $1.75 a ton. "Everybody for miles around made hay for Burns," and Olds became known as "Hay City."[18]

As the scale of operations increased, Duggan arranged with more isolated farmers to look after small bunches of cattle during the winter instead of hauling hay long distances. Using a tiny deeded land base of only a couple of sections as feeding stations, the partners were able to "winter over" several thousand head of cattle. For a decade or so, this arrangement benefited the settlers, who could begin to break and crop a portion of their homesteads and still have time to cut and stack wild hay wherever it grew abundantly. During the summer Burns' cattle ranged far and wide over the area of unsettled land eastward from the Calgary to Edmonton railway line. The right to graze this land was not conferred by any formal arrangement. Burns did not take out grazing leases at this time. Capital was tied up in beef cattle rather than in land, buildings or fences, or in a breeding herd. When supplies of stocker cattle from the local area proved inadequate, they were acquired from Manitoba and Ontario.[19] Steers might be held over for a number of years. One herd of more than a thousand head was described in the following terms: "five-year-old and six-year-old steers ... undoubtedly the finest bunch of cattle wintered in Alberta."[20]

The methods worked out in the Olds district could be adapted for use throughout the foothills, wherever there were incoming homesteaders willing to put up hay and abundant unclaimed grass for summer grazing. The size and number of Burns' itinerant steer herds continued to grow through the 1890s. Fred Stimson reported in 1901 that, "Mr. Burns is feeding about 10,000 beef cattle."[21] By 1904, his company had arranged to have forty-five thousand tons of hay put up which was to be fed to between twenty thousand and thirty thousand head of steers at ten different feeding stations.[22] What was innovative about the methods used by Burns and Duggan was that arrangements were made to feed all animals, with the objective of ensuring that stock emerged from the winter ready for sale. On other foothills ranches hay was only fed to high-value imported bulls, and to cows and calves which needed particular care to bring them through the winter. Neither the labour force nor the infrastructure at a typical ranch site was sufficient to enable an outfit to feed a whole range herd even in an emergency. Burns' and Duggan's strategy of "wintering over" was an intermediate step toward the use of feed lots and the use of grain to fatten cattle. Burns and Company were to make a contribution to these developments in the years ahead.

Burns viewed the influx of farm settlers, which began to swell around the turn of the century, as both inevitable and as something to be encouraged. As we have seen, he used settlers to put up hay for his cattle. Unlike the majority of ranchers, who bemoaned both the end of the open range and the insecure tenure provided by the remaining grazing leases, Burns saw each incoming farm family as potential customers for his food products. Moreover, he never relied on leases, but rather depended on the unused grass between and around growing settlements. Nevertheless, as more and more land was broken for crops and clusters of farms round stopping points on the railway expanded and merged, so farmers increasingly resented the presence and depredations of what they called "pirate herds," which trampled their crops and ate the grass which might have sustained their teams and dairy cows.[23] For a variety of reasons the pendulum was swinging against ranching and in favour of "King Wheat." Homestead entries soared.

7.6 Unlike many in the cattle business, Pat Burns welcomed the influx of farm settlers after the turn of the century. He recognized that every new family represented more customers for his products. Glenbow, NA 3818-2, Front cover Canada West, 1921

Cheap land, rising grain prices, extensive railroad construction, and a cycle of favourable climatic conditions, led to the rapid extension of the cropped acreage. In Alberta, 187,618 acres were in field crops in 1901; by 1910 the figure was 2.1 million acres, and the following year it reached 3.37 million acres.[24] The returns per acre from growing wheat were so much greater than they were from raising stock that small landholders got out of cattle raising. The number of cattle in the west began to drop in the early years of the new century. The killing winter of 1906–7 accelerated this trend. The gradual transition, which Burns had foreseen and supported, from open range methods to more intensive mixed stock farms, was disrupted. The smaller farmer-ranchers, to whom Burns had confidently looked for continuing supplies of cattle, rejected stock raising and threw all their energy and capital into wheat farming.[25] Shipments of cattle from Medicine Hat out of the west jumped from 11,815 in 1906 to 38,438 in 1907.[26] These heavy shipments, coupled with the losses of the winter and a very low calf crop, led to a precipitate decline in the number of stock on the range. Burns reacted to this challenge with vigour and determination. He did what he could to stem the tide and to ensure a steady supply of cattle for his meat-packing plants and shops. Over the next few years he purchased a number of ranches and became, for the first time, involved in breeding stock rather than merely fattening cattle acquired from other sources.

Burns' first, and one of his most long-lasting attachments, was to the Bow Valley Ranch. He bought this ranch, which had originated as "the government farm" in the 1880s, from William Roper Hull in 1902. It was included as part of a deal which also involved a chain of retail meat markets and some city property.[27] Hull had built a fine house on the property where he entertained distinguished visitors to Calgary. Burns used the ranch as a base where bunches of cattle bought throughout southern Alberta could be held until they were needed at the packing plant. This arrangement worked so well that Burns acquired six adjacent ranch properties over the next few years until his contiguous holdings amounted to some 12,500 acres, flanking the Bow River for eighteen miles.[28] Cattle were delivered on the hoof or by means of the railway and the productive river bottomlands furnished abundant hay for winter feeding.[29] These first ranch purchases could be viewed as necessary support facilities for Burns' meat-packing plant, they did nothing to meet the shortfall of stocker cattle.

The huge Mackie lease did address just this problem. It seems likely that Burns was running some cattle in the Milk River area even before the bad winter of 1906–7, but in 1910 he extended his operations there by taking over a lease of three and a half townships from A.T. Mackie of Pembroke, Ontario.[30] The outfit ran both a beef herd and a breeding herd for a total of about seven thousand head. This was an entirely different kind of enterprise from one of Burns' "cow camps." Firstly, it was much bigger, and secondly, it involved a breeding herd and all the attendant risks of calving. The methods used on this property, which encompassed some 150,000 acres between the Milk River and the international boundary, were reminiscent of the open range. The stock was wintered on the good bunch grass of the Milk River Ridge and little hay was put up. The cow herd calved on the range and 5 to 10 per cent losses were regarded as acceptable. The scale of the operation is indicated by the fact that P. Burns and Company shipped twenty carloads of cattle from the ranch one fall.[31] It was also a profitable enterprise, earning profits of $80,216 in 1916.[32]

In 1910, Burns moved to formalize his presence on the unoccupied range between the Little Bow and the Big Bow Rivers. He leased four properties with the help of Douglas Hardwick. The total acreage was about seventy-one thousand acres centred on the present location of Lake MacGregor and reaching northward to the Blackfoot Indian Reserve and eastward to the Bow River. The new properties were thrown together and referred to as the Kelly-Palmer Ranch. Herd size fluctuated between twenty-five hundred head and four thousand head according to the carrying capacity of the drier mixed-grass prairie. Informal arrangements with

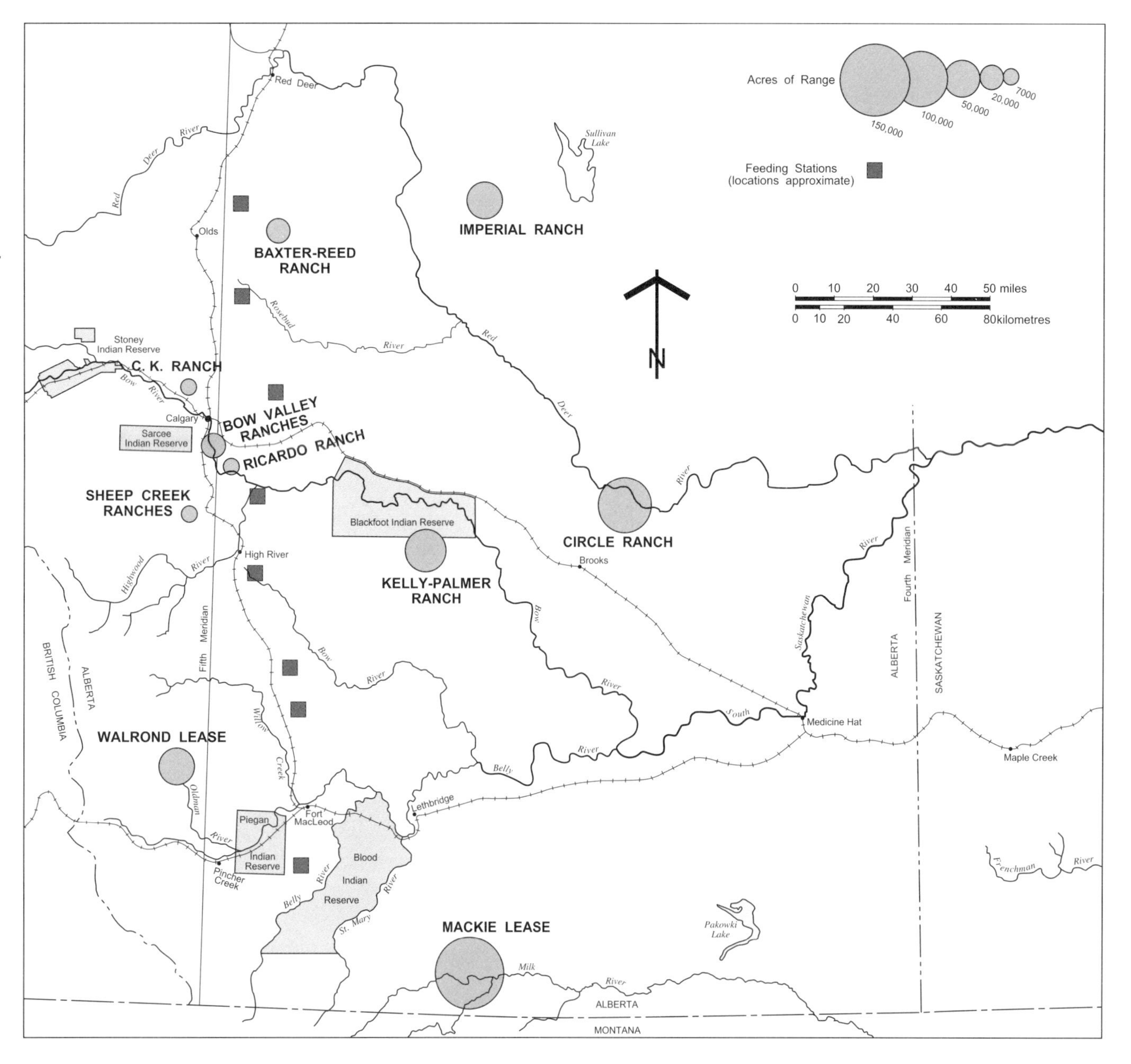

Map 7.1 Burns Ranches: early expansion, 1907–1917. The location of feeding stations estimated from Newspaper Accounts. Sources: Drawn by the author from a variety of sources, see also colour section. GA M7771, f.8-59, "Memorandum re Company History," 1945, contains a useful hand coloured map of ranch locations.

the Blackfoot permitted grazing along the southern margins of the reserve. In addition, to the east of his old base near Olds, Burns took over the Imperial Ranch lease comprising forty-eight thousand acres. Foreman Ted Gardner built the herd up to seven thousand head by wintering cattle with neighbours.

It was during this period of expansion in ranch ownership that Burns established long-lasting connections with two very famous spreads, the Circle outfit and the Walrond Ranch. "The Circle" was the colloquial name given to the famous ranch established by I.G. Baker and the Conrad brothers of Montana.[33] They had profited from the first lucrative contracts to supply the government of Canada with beef to feed the Indians and the police. From these great Montana-based companies many of the smaller ranchers had obtained their foundation herds. As settlement encroached on their initial range between the Little Bow and the Bow, so the Circle outfit moved to the mixed-grass prairie along the Red Deer River north and east of Brooks. As a major producer of beef cattle, the Circle had long dealt with both Burns, and Gordon, Ironside and Fares of Winnipeg. When the Conrads finally decided to sell their Canadian herd to Christian Bartch in 1911, they passed their leases and headquarters infrastructure on to Burns and Company, who held the property until 1950, first leasing from the Canadian Pacific Railway and then from the Eastern Irrigation District.[34]

The Walrond Ranch was, like the Bar U, one of the four or five great corporate ranches which dominated the first two decades of the cattle industry in the foothills.[35] It brought together Canadian expertise in the person of Duncan McEachran, the general manager, and allied it with British capital represented by Sir John Walrond. In spite of the bitter enmity which grew up between the Walrond and neighbouring small ranchers, the company flourished until the first decade of the twentieth century. In 1908 the cattle herd was sold to Burns for $250,000, but the company continued to operate as a landholding company.[36] It held some thirty-seven thousand acres of deeded land. Leasing the property to Burns provided enough income to pay the land taxes while land values rose. Burns ran both steers and a cow herd at the Walrond for a total of about twelve hundred head of cattle. The balance sheet for 1914 makes it clear that almost all his capital was tied up in cattle not in "improvements." This was the way Burns liked to operate and profits from the ranch rose from $15,391 in 1914 to $21,335 in 1916 and $29,785 in 1917.[37]

By the outbreak of the First World War, Burns controlled about 450,000 acres and ran more than 30,000 head of cattle. His holdings were greater than those of any of the great corporate ranches of the so called "golden period" of the open range.[38] Map 7.1 shows the location and extent of Burns' ranches and feeding stations at this time. The region around Olds, the site of his first small land purchases and experiments with "wintering over," remained important, but most of his big leaseholds were in the drier mixed-grass country where the challenge from wheat farming was less intense. These extensive herds did something to compensate for the declining production of young cattle on farms during the years after the bad winter.

Conditions had changed radically by 1917.[39] High wartime prices had enticed farmers back into stock rearing. Ranchers too had restocked their places, and cattle numbers were growing again. It was no longer so necessary for Burns to produce cattle for his packing plants, the market would do it for him. He moved quickly before the cycle peaked, and sold several of his ranches and most of his stock while the prices were still high. F.M. Black, the treasurer of P. Burns and Company, is reported to have explained this move as a response to rising land prices and a desire on Burns' part to raise capital for reinvestment in other ventures.[40] It is perhaps impertinent, more than seventy-five years after the event, to challenge the logic of an observer who was on the spot. Certainly, it is clear that the subsequent expansion of Burn's packinghouse operations into Saskatchewan, and the establishment of his wholesale fruit and vegetable distribution system, required a lot of investment. However, because the ranches sold were all based on leased land, it is not clear how their sale would have realized a large sum of capital. In my view, it was the disposal of the cattle on the leaseholds not the land which would have realized about half a million dollars for reinvestment. After the sales, Burns was left with his various smaller ranches close to Calgary, along with the productive Walrond leases and the huge but semi-arid Circle outfit.

Burns had been producing stock, as well as buying them and distributing them, for more than twenty years. He and Duggan had demonstrated that steers could be wintered over on a large scale and return handsome profits. He had used unoccupied grassland in a manner reminiscent of the open range. As agricultural settlement spread and branch lines proliferated, so Burns acquired a number of ranches in regions which he regarded as unlikely to be attractive for some time to wheat farmers. His large herds did something to maintain a pool of breeding stock in the west and his staunch championship of ranching helped to sustain the industry through a difficult period.[41] It was to be some years before he expanded his ranch holdings again, but when he did, he reached for some of the brightest jewels in the foothills.

Buying the Bar U Ranch

On 1 November 1927, P. Burns and Company made an offer to purchase the Bar U and Willow Creek Ranches from George Lane and Company. It was a huge deal involving 37,000 acres of deeded land, 58,000 acres of leased land and 6,500 head of stock.[42] The value of the offer was $723,000. Matters were complicated by the fact that Lane had been dead for two years and his heavily mortgaged properties had been run by the Honourable Archie McLean on behalf of the Dominion Bank. It is important for our purposes to separate as far as possible the assets of the Bar U from those of the Willow Creek Ranch, because this sale allows us a rare opportunity to make an inventory of the Bar U, which can be compared with its former extent in 1902 and 1882, and with developments which were yet to transpire. The sale in 1927 provides a benchmark, and is therefore worthy of careful scrutiny. The offer to purchase was worded as follows:

> ... to purchase from you all farm and ranch lands owned by you and your rights to all lands leased by you in the province of Alberta, contained in

7.7 A formal portrait of Patrick Burns taken in 1927, the year he purchased the Bar U Ranch from the estate of George Lane. Glenbow, NA 1149-1

> what are known as the Bar U and Willow Creek Ranches ... together with all cattle, horses, and other livestock thereon, [here Burns made an additional handwritten note, "including hay, green feed and equipment,"] inclusive of this year's calves, at the price of $408,001.00 for the said lands, $50.00 per head for all cattle [calves to be thrown in free of charge] and $40.00 per head for all horses [foals to be thrown in free of charge]....[43]

This letter was accompanied by a cheque for $10,000 as a deposit, and the offer was accepted on 10 November 1927. The land inventory shows that there were 26,865.60 acres of deeded land at the Bar U. The map of the property, prepared to facilitate the sale, reveals that the extent of the freehold lands differed little from those purchased from the Calgary and Edmonton Railway Company around the turn of the century. The lands spread north and eastward from Pekisko Creek and the ranch headquarters, and followed Stimson Creek southward as far as Sheppard Creek. The other big block of deeded land took in the southern banks of the Highwood River as it turned south in Township 18, Range 3, West of the Fifth Meridian. At $11 an acre, this asset was valued at $295,515. [Map 7.2].

The leased land, which Lane had originally obtained from the federal government in 1905, was also a key asset. It consisted of 45,000 acres and linked the two major areas of deeded land. All the rolling foothills country to the north and west of the ranch headquarters was held under lease and was ideally suited for wintering cattle. The south-facing windward slopes of these hills were quickly cleared of snow by chinook winds, while the valleys contained stands of willow and aspen which provided shelter. The whole area was watered by springs and creeks. Further west, across the Highwood River, leased land reached up toward the high country of the Forest Reserve. This was good summer grazing for cows and calves as the lush summer grass produced plenty of milk. All this grazing land was available for a very modest annual rental and the existence of this pool of natural grassland meant that the flat deeded land could be cropped and used to grow hay. The overall carrying capacity of the ranch was greatly enhanced.

The key importance of the leased lands is hinted at in a letter written by W.E. Corlet, secretary to P. Burns and Company, to Messrs., Munson, Allan, Laird, Davis and Company, the Winnipeg Solicitors who were handling the transaction. He wrote:

> … there has been one point that has been overlooked, which is the most important of all, and that is, that no word is said as to the assignment of the leases, and *they are the most important part of the ranch.*[44]

The following March, as matters moved toward their conclusion, Corlet again wrote

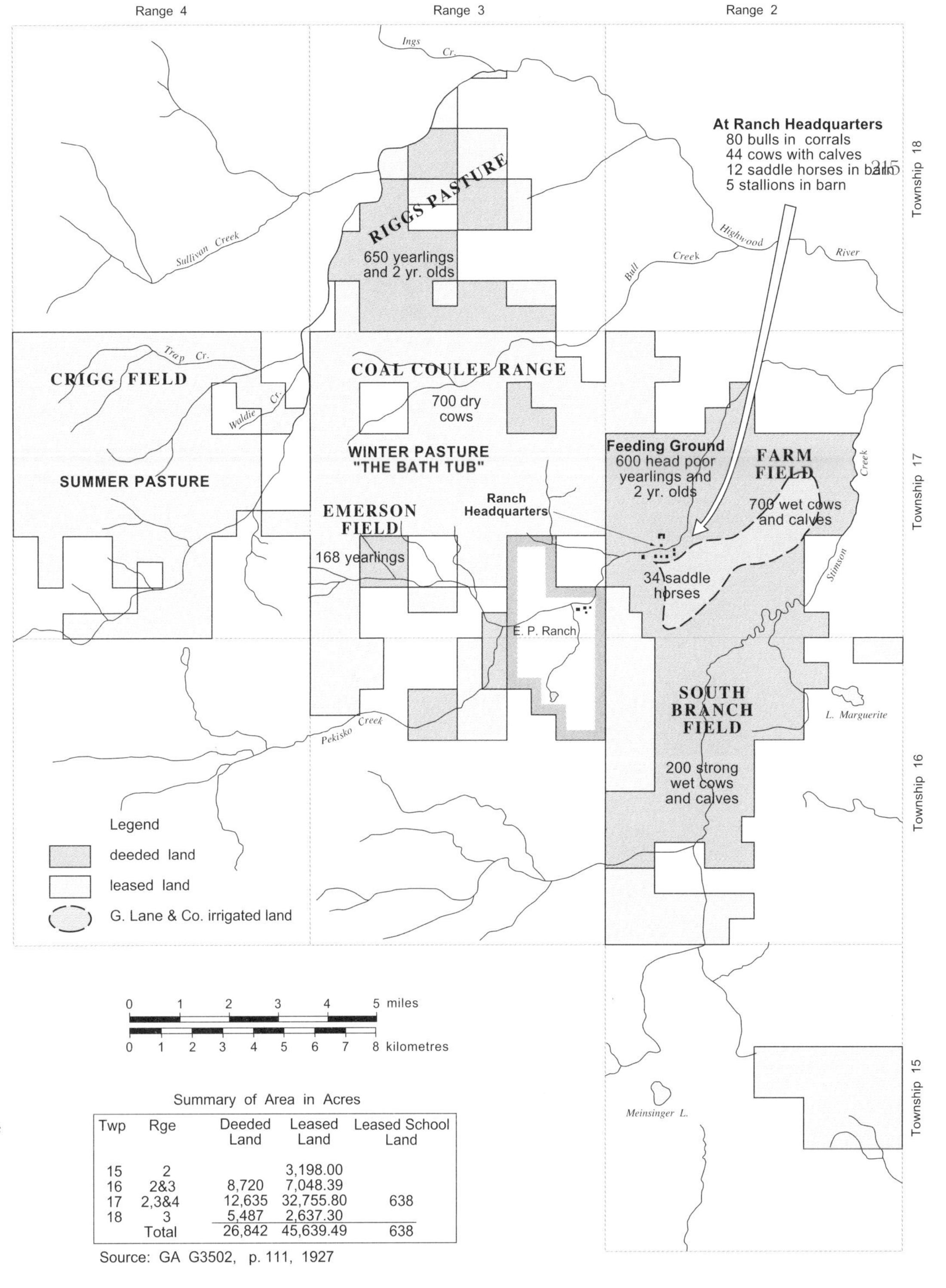

Summary of Area in Acres

Twp	Rge	Deeded Land	Leased Land	Leased School Land
15	2		3,198.00	
16	2&3	8,720	7,048.39	
17	2,3&4	12,635	32,755.80	638
18	3	5,487	2,637.30	
	Total	26,842	45,639.49	638

Source: GA G3502, p. 111, 1927

Map 7.2 Bar U Ranch Lands, 1927: Showing Disposition of herds. See also colour section.

concerning the leases. Munson had suggested that Burns and Company hold the leases from the bank to avoid the hassle of a series of assignments. But Corlet disagreed:

> We do not agree with your suggestion as to the assignment of leases from the George Lane Company to ourselves, to be placed with the Bank until the balance is paid up, as the regulations at the end of the stipulated time for the payment of deferred payments might have been changed. We would much prefer what you describe as the more cumbersome method of having the leases assigned to us.... There is the further difficulty that at the end of seven years the George Lane Company might not be in existence.[45]

It is clear that Burns and his advisors set great store by the leases, and if the deeded land was worth almost $300,000, the right to use 45,000 acres of leased land was probably valued at $60,000.

The ranch lands were subject to two bond mortgages, one held by the Security Trust of Rochester, New York, and the other by a local firm, the Osler and Nanton Trust Company. Some $88,000 was owing on the Security Trust Mortgage, but the agreement to sell made it easy to redeem these bonds. Similarly, all bonds under the Osler and Nanton Mortgage were held by the Dominion Bank and could be discharged at any time.[46] Thus the solicitors were able to arrange the transfer of titles in a timely manner. A cheque for $32,000 was sent by Corlick to the seller on 27 December, so that the deal could be finalized by the end of 1927.[47]

As soon as his offer was accepted, Pat Burns dispatched J.H. Johnson, one of his most trusted buyers of cattle, to make a careful inventory of the stock at both ranches. Johnson spent two weeks counting cattle and assessing their condition. He submitted a written report on 9 December 1927. It is clear from this letter that the seventy-four-year-old Burns accompanied Johnson, as he carried out at least part of his task, for he writes, "the 518 bunch we looked at together west of the ranch ..."[48] According to Johnson there were 4,025 cattle at the Bar U and about a thousand calves. There were also three hundred head of horses. He expressed concern about the condition of 1,220 yearlings and two-year-olds which had been pastured over the river during the summer. He reported that this herd had been turned out too early in the spring, and had never really "picked up," he thought that lack of salt might also have contributed to their poor condition. In his opinion the bulk of this herd would need feeding to bring them through the winter. On the other hand, the 750 cattle grazing the Coal Coulee range were in excellent fettle and constituted "the bank of the herd." During the hiatus between the death of Lane and the purchase of the ranch by Burns, one might have expected the bank to sell off valuable assets, particularly stock. This does not seem to have been the case with regard to cattle, although some valuable Percheron horses may well have been sold. In 1927, the Bar U was grazing 4,300 mature animals on 71,000 acres, for an average of 16.5 acres for each animal unit. This stocking ratio had remained remarkably constant throughout the history of the ranch, and was to remain so for the next twenty years.

With some $300,000 worth of stock changing hands, it was essential that buyers, sellers, and their managers on the spot, should be absolutely clear as to which brands should be transferred and how the cattle should be marked. When Lane had purchased the Willow Creek spread, the famous "Flying E" brand was not available. Hence, during his period of ownership, the ranch was always referred to as the "Willow Creek Ranch" or even "the Willow Creek Bar U." Now Burns was able to write to the Recorder of Brands and revive the old brand which had been cancelled.[49] Lem Sexsmith, who had just been installed as manager, was told: "... for the future we are all going to call the Willow Creek Ranch the Flying E Ranch."[50] All of Lane's cattle on either ranch bearing the Bar H brand would be vented and branded Bar U in the spring, while cattle bearing the Bar U brand at Namaka, or on the Bassano place, would be vented and rebranded Bar H.[51] This would distinguish Burns' newly acquired herds at the Bar U and Flying E from those of Lane's estate or the herds of Lane's sons which were being run along the Bow River.

The land and stock on it made up most of the value of the two ranches. Buildings, fencing, and the added value of land which had been broken to the plough were also taken into account at the settlement. The six stables at the Bar U were valued rather more highly than were the bunk house and the dwelling houses! There were also thirty miles of fencing valued at $100 a mile. Together these improvements were worth $28,000.

The Burns organization took over the operation of the two ranches on 1 December 1927. The managers received circular letters from head office telling them exactly what was expected: "... we are writing you this letter, in order that you may know how we usually handle the accounts etc. on our ranches."[52] The letter makes two things clear. Firstly, there was an established modus operandi for Burns' ranches which had been worked out over a thirty-year period. Secondly, the organization was hierarchical, with clearly defined responsibilities at each level. For example, Frank McLaughlin was responsible for the day-to-day running of the Bar U. He was empowered to hire and fire and to arrange for supplies from local merchants. However, pay cheques were sent out to the manager from head office for every full-time employee, and receipts and accounts had to be submitted each month.

Wondering Why

Why did Burns purchase the Bar U? He had reached an age when many men would have been quite content to retire and watch others manage the business and run the range. His motives have been interpreted in a variety of ways. Grant MacEwan writes as follows:

> ... the Bar U captivated him and he got the idea that George Lane would have wanted him to own it. If more reason were needed, he was worried that the famous ranch would fall into hands which would not bring adequate safeguards against abuse and exploitation of soil and grass.... Although it was not always recognized, this man of thrift and industry was at heart a conservationist.... He wanted to insure that the

> Bar U would be protected the way George Lane would have done it, and so he bought it.[53]

This explanation suggests that Burns made the purchase because of a sentimental attachment to Lane. But neither Lane nor Burns were given to close personal relationships. Burns' glowing eulogy for Lane stresses his determination and foresight as a pioneer, but makes no mention at all of personal links.[54] If the purchase was tied to the death of a close friend, why did Burns wait two years? Moreover, it is hard to accept the idea that Burns bought the two ranches to prevent them from being overgrazed. Admittedly, his whole life had been bound up with grass and his ability to assess its carrying capacity from year to year and season to season. He was a conservationist in the sense that overgrazing was clearly bad business, but Burns bought and sold all kinds of properties, including ranches, like counters or poker chips in his business plans. It is hard indeed to imagine him acting from such poorly defined, although admittedly altruistic, motives.

Albert Frederick Sproule suggests a more personal reason for the purchase. He points out that Burns was brought face to face with his own mortality in 1926–27. The seventy-four-year-old entrepreneur was very sick with pneumonia during the winter. He was also struggling with the intensifying symptoms of diabetes. Sproule goes on to suggest that to divest himself of the day-to-day concerns of the meat-packing business, and to take up instead the management of ranches, "may have seemed like a pleasant way to spend his remaining years."[55] Moreover, Burns had lost his wife a few years before and was having a difficult time with his twenty-one-year-old son Michael. The young man had made it abundantly clear that he wanted nothing to do with the factory or the food outlets, while intimating that he was prepared to spend some time at the ranches. These insights grew out of Sproule's in-depth interviews with many of those closest to Burns, and are not to be set aside lightly.

However, I contend that the radical reorganization of his business made by Burns during late 1927 and 1928 are wholly in keeping with his personality and the way that he had lived his life up to that point. Sickness, age, and his family concerns may have triggered action, but the underlying motive for the new adventure seems to me to have been more deep-rooted. All of Burns' formidable energy was concentrated on his business interests. He did not marry until he was forty-seven years old, and the failure of his marriage must be attributed, in part at least, to his single-minded dedication to his work. Nor was he a "clubman," like Fred Stimson, who derived pleasure and satisfaction from socializing with his fellow leaders in agriculture, industry, and politics. He was sometimes a patron of the arts, but neither music, art or theatre ever became an important part of his life. On a daily basis his social and business lives were merged together. The people he saw most of and enjoyed most were his trusted lieutenants like Duggan, Kelly, and Farrell, and the "boys" who worked for him in his garden, at the ranches and stockyards, and with whom he had established an enviable relationship based on mutual respect. He often remarked that although he was rich in monetary terms, his brother was rich in terms of his family and children.[56] To some extent Burns made his employees a substitute for a large and close biological family. He made room in his organization for nephews and cousins, and helped favoured employees with the education of their children, who were later given jobs. Thus Burns' whole life was narrowly focused on his business interests and it was from them that he gained fulfillment.

A key to understanding Burns' business dealings throughout his long career was his continuous search for new challenges. He played "the game" not solely for monetary rewards, but for the satisfaction it gave him. Once a new acquisition was established and working well, he looked for the next area to expand into. A few examples will illustrate this. When supplies of cattle were threatened during the first decade of the twentieth century, he acquired ranches to ensure a continuing flow of fat stock; when he saw an opportunity to expand his meat-packing enterprise geographically, he sold the ranches and redeployed his capital. During the 1920s, returns from meat packing languished, so he expanded into butter and cheese. By 1924, he controlled 80 per cent of the cheese production in Alberta and 40 per cent of the butter output. The next challenge was in the supply of fluid milk. Burns experimented in Ontario, and when he was ready, he introduced Palm Dairies throughout the west in 1926. The purchase of several fruit wholesaling companies led to the establishment of the Consolidated Fruit Company, and, by 1928, Burns owned fifteen fruit houses. This "track record" does not provide much evidence of a slowdown in Burns' enthusiasm for business dealings as he passed from his sixties into his seventies.

Perhaps then, the purchase of the Bar U and Willow Creek Ranches was part of the next exciting new vision: a string of Burns' ranches from the Oldman River to the Bow. He had already leased the famous Walrond Ranch, and had purchased the Glengarry "44" Ranch in 1923. Its lands almost bordered the Willow Creek spread. With the vast Bar U lands as the corner stone, Burns' range would dominate the remaining island of prime ranching country on the Porcupine Hills and the valleys of the foothills. Moreover, the Bar U was still, saving only the E.P. Ranch, the most famous ranch in Canada, and to maintain or even to enhance its prestige would be a real challenge. The fact that the ranch had a royal neighbour, whom Burns had met several times, especially during the 1927 Royal Tour, probably made the ranch more attractive to him. In fact, Burns did establish close contacts with Prince Edward and became chairman of a committee that advised the Prince of Wales on the management of his ranch from 1931 until his death.[57]

This interpretation links the purchase of the Bar U and the Willow Creek Ranches with the sale of Burns' meat-packing, wholesaling, and retail empire seven months later. Indeed, it is hard to avoid this conclusion, since the financial ramifications of the two deals were inextricably bound together. Burns needed to free up some capital for the ranch purchases and for their subsequent development. Although he had immense assets, his money was all "gainfully employed." Expansion in a new area meant that capital had to be obtained by selling something off. For example, when he purchased the "44" Ranch it was in exchange for the Burns Block on 8th Avenue in Calgary. Thus Burns embarked on a major restructuring of his affairs. He sold all his packing houses and his wholesaling and retail outlets to Dominion Securities of Toronto for

$15 million. The only assets he retained were his ranches, which were consolidated in "P. Burns Ranches Limited" and "P. Burns Holdings Limited."[58] The transaction specified ten ranches: the Bow Valley and C.K. along with adjoining properties close to Calgary; the Ricardo; the Circle; the Walrond; the Bradfield; the Lineham; the Glengarry "44"; the Bar U; and the Flying E. The vision of a string of ranches from the Oldman to the Bow was almost realized. The only gap between Burns' string of ranches was between Sheep Creek and the southernmost of his properties along the Bow River, and this gap was only about twenty miles in extent [Map 7.3, colour section]. Because of their geographical proximity, the operations of these ranches could be integrated in a number of ways. The Circle Ranch was further to the east, but its location on the CP main line meant that young stock from the foothills ranches could be fattened on the protein-rich mixed grasses along the Red Deer River.

As one considers the "portfolio" of ranches Burns now owned, one cannot escape the conclusion that the shrewd but romantic old entrepreneur had a profound sense of history. Was it an accident that he controlled the Bar U, the Walrond and the Circle Ranches, three of the most significant names from the Golden Era of ranching?[59] In addition he owned two other foothills ranches with long histories, the Flying E and the Glengarry "44." This conjecture, that Burns was drawn to acquire ranches which had played an important role in the evolution of ranching, is strengthened by his purchase of the "76" Ranch on the Frenchman River. This acquisition took place in 1928, even before the final details with Dominion Securities had been worked out. There were few more famous and historic brands in the Canadian west than the "76." It was originally introduced into Alberta by Moreton Frewen's Powder River Cattle Company, which had moved from Wyoming to the Alberta Foothills in 1886.[60] The brand was passed to Sir John Lister Kaye's grandiose Canadian Agricultural Coal and Colonization Company when it purchased Frewen's herd.[61] More recently the brand had been acquired by Gordon, Ironside and Fares in the aftermath of the bad winter of 1906–7.[62] The new ranch was situated in the country immortalized by Wallace Stegner in his classic reminiscence *Wolf Willow* , and had a history comparable in its own way to that of the Bar U.[63] The property consisted of some eighty-five thousand acres of leased land and a few hundred acres of deeded land on which the ranch house and corrals were located. Situated as it was in the arid heart of Palliser's Triangle, Burns was unlucky to purchase the ranch when its carrying capacity, always low, was to be further reduced by the droughts of the 1930s. It was his only ranch property which consistently lost money.[64]

In the late 1920s Pat Burns decided to concentrate on purchasing and managing a series of ranches. While most of these were in the broken foothills country along the eastern slopes of the Rocky Mountains, two were located in the mixed-grass prairie to the east. It seems likely that in choosing his ranches, Burns was drawn by the history and reputation of an outfit as well as by its economic viability.

8
The Bar U during the 1930s

"The serious study of humankind's story will always need the particular and the general, the contingent and the profound."[1] – Paul Kennedy

The basic layout of the Bar U headquarters remained very much the same throughout the Burns era. Nevertheless, there were important changes in detail and function. Burns acquired thirty-eight buildings from the Lane estate; twenty years later there were twenty-six buildings still extant.[2] During Burns' lifetime neither the mindset of the "Old Man" nor the state of the economy encouraged new building. Burns demonstrated almost fanatical zeal to cut out waste and to avoid extravagance, while at the same time showing himself quite prepared to spend large amounts of money to achieve important long-term goals. After his death, grain farming became increasingly important at the ranch in response to wartime prices and government incentives. Several buildings were adapted for grain storage, while others were used to house and maintain agricultural machinery [Map 8.1].

The most dramatic change at the site was the loss of Pekisko House, the main ranch house. In the spring of 1928, Herb Millar was making his way from Namaka toward the Bar U to take over the management of the ranch on behalf of the new owner. He stopped off at the Pekisko store to visit with his old friends the Baines. He was told that he had better stay the night, as the main house where he had been planning to sleep had burned down the previous day.[3] Hugh McLaughlin was fourteen years old at the time, and recalled the fire quite vividly:

> The house burned down maybe a month or so before we left ... on a Sunday morning about 11 o'clock. I remember the time the boys used to get back from hauling hay. They had all kinds of help and they had a real bucket brigade going pretty quick, but they never seemed to be able to get to the source of it. But we got 95% to 98% of the stuff out of the house. Didn't seem able to stop the fire. It had been built onto a lot, just by the general look of it. We chopped holes in the roof and put on scads of water with the men that were around there. But we couldn't seem able to get it stopped.[4]

Although the efforts of the amateur fire crew were unsuccessful in saving the house, they played a role in containing the fire, which could easily have spread through the long grass between the main house and adjoining buildings.[5]

The house had played its major role as the home of the first ranch manager, Fred Stimson. It was by no means a pretentious building, but its twin gables, framed by the barns and storehouses of the ranch's "village street" and backed by the Rocky Mountains, had been a focal point of the site. It had not been a permanent home for the Lane family; they had lived in Calgary and spent most of their summers at the Willow Creek Ranch. Nevertheless they had spent some happy family times there when the Paulin family

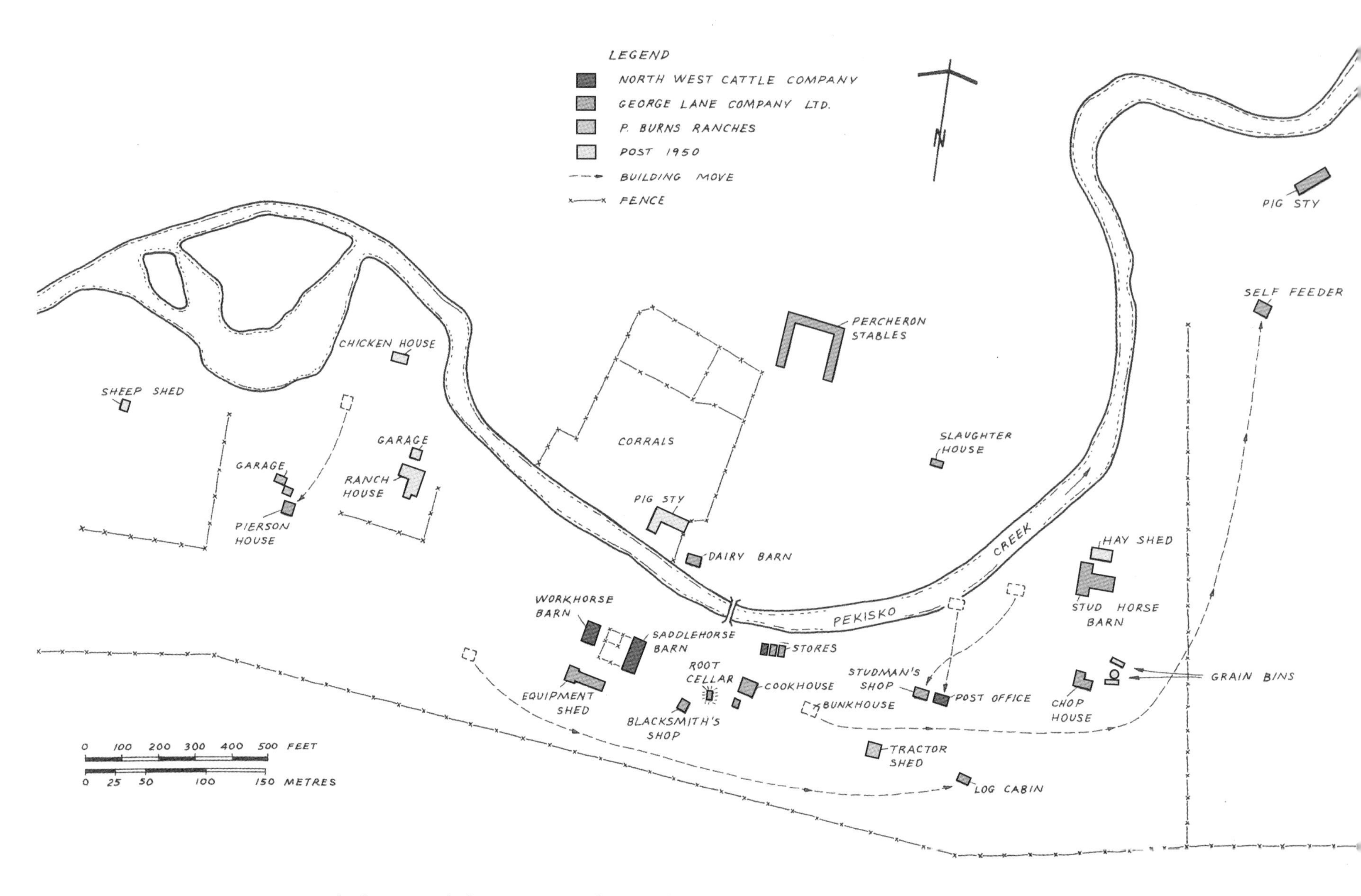

Map 8.1 The Headquarters site today. See also colour section

had occupied their summer place at the ranch, and it was there that George had died in 1925. The small dining room had been used to host a variety of distinguished guests over the years, including the Prince of Wales, the Crown Prince of Japan, and the Duke of Devonshire, Governor General of Canada. The loss of the old house, which dated back to the foundation of the ranch in the 1880s, leaves an unavoidable gap in the interpretative story told at the Bar U Ranch National Historic Site.

Some oblique air photographs, taken during the summer of 1931, provide an overview of the site at this time.[6] The scar left by the fire at the main ranch house is still visible. The frame dwelling which had

8.1 This oblique air photograph of the Bar U Ranch headquarters site, taken in 1931, still shows the scar where the main ranch house burned down in 1927. The foreman's house is in the middle of the picture just south of Pekisko Creek. The layout of the corrals and the U-shaped barn to the north of the river is shown clearly. University of Calgary Collection, 82J 1931 C, A3507-54

served as a summer home for the Paulin family and as a schoolhouse for ranch children had disappeared too. Ann and Raymond Clifford joined the staff at the ranch in 1931 and moved into the old Pierson house, set back in the woods close to the creek. No one had lived there for some time, so it required a thorough paint job and some carpentry to make it livable. Ann Clifford recalled:

> A door and ceiling in the kitchen had to be repaired where Michael Burns and a friend were target shooting from the living room. Quite

8.2 Pekisko Creek in flood, spring 1923. Serious bank erosion meant that the storage shed on the right had to be moved immediately, while both the post office and the studman's house were relocated later. Parks Canada, Hugh Paulin Collection, 96.11.01.08

> a hole in both the door and ceiling. At first I thought it was mice....[7]

Their new home was nearly washed off its foundations by the spring floods a few months after their arrival, so the house was dragged out of the woods into its present location in 1932. Tom McMaster and his wife Maud were living in the log cabin under the bank to the west of the work horse barn. McMaster had worked as horsebreaker and rider at the ranch for some years and was a great help to Raymond Clifford as he strove to get to know the huge spread.[8]

Rapid stream erosion threatened buildings located on the outside of the meander where the river turned northward. First the ranch office, and then the "stockman's residence," were moved southwestward until they were roughly aligned with the store houses on the north side of the Bar U Trail. This added to the street-like layout of the central area of the site. Locational changes were accompanied by functional changes too. In July 1927, the Pekisko post office was moved to the South Fork Trading Post when Mrs. Mary Baines became the postmistress.[9] During the 1930s, the old post office building on the ranch was still used occasionally to accommodate visitors and as a depository for office papers. However, as Ann Clifford assumed more responsibility for keeping the books on the ranch, the real "office" was her residence and the old log building became a general storehouse. Once Burns had bought the Bar U he made it clear that the ranch would revert to its former character as a cattle ranch. Horses would remain a vital part of working life on the ranch, but would not be a major economic product. Expert horsemen like Bert Pierson became a luxury, and there was no longer a need to keep a 'stud horse man.' On the other hand, the many work horses had a great deal of harness which had to be maintained if the horses were to do their work efficiently. The ranch employed a harness repairman for three or four months of the year.[10] The stud horse men's house became a workshop and a place to keep a light wagon. At the eastern end of the site, the show barn and its corrals had disappeared.[11] The chophouse, however, was still in place and its function had been enhanced by the installation of a large stationary engine.[12] This was used to grind feed and also to power a saw to cut firewood. The page wire corrals were still in good repair around the stud horse barn and along the top of the rise where the visitor centre now stands. Across Pekisko Creek to the north, the air photo shows the layout of the working area: the cow corrals and the long, low cowshed nestled under the bank to the west, the U-shaped horse barn with its sheltered courtyard and attendant corrals, and the colt barn close to the stream. The spring floods had removed vegetation and deposited gravel on the flood plain on the inside of the meander and may well have damaged or destroyed the sheep barn and other sheds which had been sited under the cottonwood trees.

8.3 The eastern end of the headquarters site showing the new location of the post office and the studman's house. Parks Canada, Alwood Rousseau Collection, Rosettis 6397-14

Riders, Cooks, and Choreboys[13]

The working community at the Bar U included a core of year-round employees, many of whom had been with the ranch for a long time. This group numbered twelve to fifteen and included riders, cooks, chore boys, the manager and foreman and their families. During the growing season extra hands were signed up to put in the crops and put up hay. From May to October twenty to twenty-five men lived in the cookhouse and the bunkhouse, while the Bar U farm housed another eight to ten. In addition, two or more contract haying outfits might be working in various parts of the huge ranch, and an encampment of Stoney Indians on the Emerson Flats to the west of the ranch provided a pool of skilled labour for occasional needs and for fencing contracts. Thus, during the growing season, the ranch was home to eighty or so people.

The key figure guiding the fortunes of the Bar U during the 1930s and 1940s was Raymond Clifford. He directed the day-to-day operations of the ranch for more than twenty years, and it is therefore important to know something of the man and his background. Raymond's father, Antony Clifford, married one of the Gibson girls, a sister of Pat Burns' mother.[14] "Dad" Clifford worked at the Gordon Ranch and at the Burns' feedlot. Raymond too started working for Burns when

he turned fourteen years old.[15] He was a rider on the Ricardo Ranch to the east of the Bow River. Here, cattle were held and finished until they were needed at the packing plant and Raymond often helped drive bunches of cattle over into the stockyards.[16] Later he worked at the meat-packing plant, and as he grew up and filled out he was chosen to deliver beef to markets around Calgary. A showy team of horses was used to draw the wagon, and the sides of beef were neatly wrapped in muslin. Young Clifford lived at the Shamrock Hotel at this time and thought nothing of keeping the day's returns, which often amounted to several thousand dollars, under his pillow for the night.[17]

By 1927, Clifford was driving a truck between the company warehouse at the Calgary stockyards and the various Burns ranches. It was his job to see that each outfit got the barbed wire, salt, seed grain, and hardware for which they had indented to head office. After a year or two, Raymond knew all the roads between the foothills ranches and their seasonal peculiarities. It was not surprising that Pat Burns chose him to be his chauffeur. This job involved maintaining a fleet of three or four cars, being on call at all times, and driving Burns – as fast as humanly possible – where he wanted to go. But it meant a lot more besides. Clifford acted as a sounding board for the "Old Man's" ideas, and it was he who had to follow up to make sure Burns' instructions were implemented. Burns insisted on Raymond accompanying him on his tours of inspection and listening to his discussions with ranch managers. "Come on Raymond, how are you going to learn anything sitting in that car?" Burns would say.[18]

8.4 Raymond Clifford with a favourite horse. For some years he was a cowboy before becoming Pat Burns' chauffeur, and in due course foreman at the Bar U. c. 1928. Ann Clifford, *Ann's Story*, cover.

Raymond married Ann Smith in July 1929, and the couple settled on the MacInnes place in Midnapore, not far from where he grew up. He had a responsible job and the complete trust of his employer. His prospects were excellent. However, this cozy career path was soon to be disrupted. Burns took him aside one day and said: "You know Raymond, you have to go back to the ranches!"[19]

Burns had watched progress at the Bar U since his purchase of the ranch, and had become concerned that stock losses there remained rather high and that the herd was not showing the overall improvement he had expected. He was afraid that there was something wrong at the top. Burns had persuaded Herb Millar, who was sixty-six years old, to return from Namaka Farms to the Bar U, where he had served George Lane for so long.[20] Now Millar's energy and zest seemed dulled and the big ranch lacked leadership and direction. Thirty-year-old Clifford was given the difficult job of initiating change while Millar remained nominally in charge. He was appointed "cow boss," while Millar remained overall ranch manager.

8.5 Raymond Clifford sitting tall on one of his favourite horses "Napoleon" in the midst of the Hereford herd which he had done so much to transform and "breed up". In the background are numerous hay stacks on the land east of the headquarters site. Parks Canada, Alwood Rousseau Collection, Rosettis 1342-12

Raymond was not at all sure that he wanted to leave his comfortable, clean, "suburban" job as chauffeur to take on an awkward assignment at an isolated location. He questioned whether he had enough knowledge and experience with cattle, and wondered how his young wife would adapt to the change. Still, Burns was persuasive, and Raymond agreed to go to the Bar U for a couple of years to try and turn things around.[21] He was to stay for twenty years and to effect as much change in the day-to-day management of the ranch as had Fred Stimson or George Lane before him.

Pat Burns had proven once again his gift for picking the right man for the job. Raymond Clifford was a big, strong man, and, as is often the case, he was gentle, quiet, and patient with both men and animals. Many young men came to work at the Bar U straight from school, and Raymond would show sixteen-year-olds the way to put on a horse harness or the way to doctor an inflamed hoof. Those who worked for him remember him with respect and affection. He was firm but fair, a good man to work for. One employee who had knocked around foothills ranches said, "You couldn't beat him" as a boss.[22] Clifford used to have breakfast in the cookhouse with the men so that he could give orders for the day's work. His last chore at night was to check the bunkhouse to make sure that all the men were accounted for. He was proud that during his time they never lost a man at the Bar U.

Ann Clifford seems to have adapted to life on the Bar U quite smoothly. She had been brought up on a homestead in the Burdett area west of the Cypress Hills, and knew something of isolation and the endless round of household chores.[23] When she arrived at the ranch there were only two other women living there: Ella Millar, who was about sixty years old, and Maud McMaster, who proved to be a good friend to Ann. For several years, after these two wives had left the ranch, Ann was the only woman living there. It was not until her sister arrived in 1937 that Ann enjoyed continuous female companionship. Her social isolation must have been increased by her role as "the cow boss's wife." She was expected to wear a dress and had to keep her comfortable jeans for occasional picnics with Raymond.[24] Her contacts with the men were limited to a daily visit from the chore boy who brought her milk, and conversations with Hop Sing, the Chinese cook, who helped her get supplies and meat from the cold house. In those early years she was accompanied everywhere by her German Shepherd dog, Captain Fleetwood. As soon as she had learned to cope with the new McClary kitchen range, and to do the weekly wash, Ann began to enlarge her role bit by bit. She raised day-old chicks and turkeys; she became foster mother to orphaned colts, and later raised many lambs on the bottle. She protected her charges with her trusty ".22," and enjoyed recounting how she dispatched a weasel which had been after the chickens.[25] From the first, Ann helped Raymond record information about the cattle herd: where each bunch was, and how many head had been put on straw stacks with which farmers. After the Millars retired in 1934, she took over the payroll and the books. This was a complex task. She had to keep track of thirty or more employees and to record the jobs which each performed on monthly time sheets.[26] The wage cheques for regular employees were made up at the company's head office in Calgary. Very often Raymond and Ann would go to the city to pick up the cheques and to transact ranch business. The stock books were equally time consuming. It was not merely a case of translating the "tally books" from branding and recording the number and gender of the new calf crop. Each movement of stock onto and off the huge ranch had to be noted. In this way the Bar U was credited with yearlings sent to other ranches like the Circle or the Walrond.[27] Saddle and work horses also had their own books. On the domestic front too, Ann soon gained skill and confidence. She got used to preparing lunch for Pat Burns and his party, and took pride in recounting how he would set aside the carefully balanced picnic lunch suitable for a diabetic, prepared for him by his housekeeper, in favour of her beef tenderloin with gravy and all the trimmings![28] When asked in later years whether she was lonely during her years as the only woman on the ranch, she smiled and shook her head and explained that she was far too busy to be lonely.[29]

Herb Millar and his wife Ella lived in the manager's house, which was located on a slight rise in the flood plain to the east of the site of Pekisko House. Herb was back on the ranch where he had spent the first decade of the twentieth century. It was hard for him to grasp how much had changed. Fences and farms had filled in the open prairie westward toward Cayley and High River. Millar found it difficult to understand why it took the men so long to bring in cattle, forgetting that stock had to be worked slowly along road allowances instead of being driven straight across the open prairie.[30] He was still imbued with the philosophy of the open range era. At a gut level he felt that it was wrong to pander to stock. He felt that harsh winters acted as a natural cull for the herd, as only the hardiest animals survived. In his view a huge ranch employing extensive methods should be able to accept 5 to 10 per cent losses and still be profitable. It was hard for him to appreciate how slim the margin of returns over expenses had become. Although he could still put in a day in the saddle when he had to, Millar was getting more and more deaf. This cut him off from contact with people round him and added to his taciturn, even somewhat morose, behaviour. Gordon Davis remembered when he and Millar spent a long day taking salt to stock over the Highwood and did not exchange a word all day![31] Norman MacKenzie described Millar as "a cranky man, very set in his ways."[32] Once, the blacksmith made a remark to Millar. Herb considered it for a while, and the blacksmith bawled out again and added, "Do you hear me?" Millar replied, "I heard you the first time, and you'll hear from me after dinner," which he did, collecting his time.[33] On another occasion Herb's hearing problem had even more unfortunate results. Burns had bought a flock of sheep from the Minto Ranch. Carlyle at the E.P. Ranch was to take the young ewes, while Burns got the rams, which he wanted for another of his ranches. He phoned Millar and asked him to cut out the rams from the flock and to have them ready to ship when he arrived. When Burns turned up, the rams were no longer rams ... they had all been "cut" or castrated.[34] Millar made himself useful around the home place. He looked after the Delco plant in the basement of the bunkhouse, and it was here that he cured ham and bacon for the ranch. He also did his best to keep the books and handed out paycheques to the men. Herb was an inveterate card player and enjoyed a game of poker in the cookhouse, especially on Sunday nights after the men had been paid. He would play until midnight and still be up at 5 a.m.[35] Millar's wife Ella was quite a character. She had a parrot and whenever the telephone rang the bird would shriek, "Herb!" "Herb!" She was "death on magpies. She had a shotgun, and every time a magpie came close to that house she'd blow them out."[36]

Another key personality at the ranch was Hop Sing, the Chinese cook. He had gone to work at the ranch for George Lane and had stayed through the period when the ranch was run by the bank. He was a tireless worker and an excellent cook. Ann Clifford remembered him with affection.[37] His first job had been at the R.L. MacMillan Ranch, and Mrs. MacMillan had taught him many of his skills. Hop Sing was a representative of a group of Chinese men who served the ranches of the foothills faithfully and well. For example, Pat Burns became dependent on "Charlie" at his Calgary residence, not only to cook but also to milk the cow and help in the garden.[38] The Mount Sentinel Ranch, a close neighbour of the Bar U, employed Quon Yock, while the Browns at the 7U Ranch arranged for Yock's brother to come over and cook for them.[39] Many of these Chinese ranch cooks knew each other and got together occasionally at the Dart Coon Club in Calgary, a fraternal organization of Chinese men.[40]

8.6 Bar U hands in front of the drygoods store, c. 1920. Wong Hop the cook and his helper stand on the right. Parks Canada, Norman Mackenzie Collection, 96.04.02.04

8.7 A meal in the Bar U cookhouse. Table in the foreground left to right: Jack Crowe, Oscar Brandham, Paul Moller, Jimmy Duff and the top of Billy Hughes' head. In the background left to right: Tom Feist, Ben Jarvis and Cliff Young with his back to the camera. Ann Clifford, *Ann's Story*, p. 41.

It is hard to imagine the hours which Hop put in looking after the Bar U work force. Through the winter he might be cooking for twelve or fifteen men, but visitors were frequent and they were always made welcome. During the summer and fall, there would be twice as many mouths to feed, and the long days and the huge extent of the ranch meant that riders often got back to the cookhouse long after the regular evening meal had been cleared away. Nobody ever went to bed hungry, but these occurrences did cause some grumbling! "Cowboys, cowboys, sleeping out there??" Hop was heard to remark.[41] He was so good at his job that he earned the respect of the men. One said: "Hop, the Chinese cook, was quite a worker … I couldn't speak more highly of him."[42] Breakfast consisted of eggs, bacon, hot cakes, porridge, toast, and coffee. "Dinner," eaten at midday, was roast beef, fresh vegetables, potatoes, and gravy, followed by rice pudding, tapioca or pie. Supper was at 6 p.m., and was another hot meal of meat and potatoes. Ham and chicken provided some variety, while beef which had been put up in jars provided a fast fallback if unexpected visitors showed up.

Young Jack Peach, who visited the ranch with his contractor father, remembered the warm atmosphere of dinner time in the cookhouse:

> Riders who came in to eat wore sweaty shirts, grubby hats and spurs ... manners were in short supply ... but they were always very cheerful, it was always very friendly, you were welcome, so come on in and eat and when you're full, on you go on your way. In the center of the table was a turntable made of a reconstructed wagon wheel ... on the wheel would be the meal ingredients: grey dappled enamel coffee pots, condiments and other needs: salt, pepper, catchup, vinegar, bowls of butter, jugs of milk, jugs of syrup, loaves of sliced bread, bowls of jam, china tureens of mashed potatoes, cooked cabbage, mashed turnips, platters of sliced meats, bowls of gravy etc. Among these main course items were big bowls of rice pudding, plates of sliced pies and cake, because each man took his place at the table wherever there was an empty chair, since there were no sittings or specific times. The cook kept a constant eye of the situation on the wheel, whipping items off and hastening to the kitchen for refills, replenishing cold ones with hot, bringing in more gravy and whole loaves of sliced bread, butter, and replacement pies ...[43]

Apart from preparing these huge meals, Hop found time to plan and do much of the work on the large garden which was laid out to the east of the cookhouse and the bunkhouse. When the cabbages, carrots, turnips, and potatoes were harvested they had to be cleaned and stored in sand in the root cellar behind the cookhouse. Usually, enough vegetables were put up to last all the year. Hop also looked after the chickens in the storehouse closest to the bridge. He was a busy man and had no patience when an old biddy hen would not settle on a nest full of eggs – she had her neck wrung in short order.[44] Hop also churned cream to make butter and rendered beef fat whenever there was a butchering to make tubs of suet. Nor was Hop's kitchen one which minimized labour; running water was not installed until the year he left the ranch, and water for drinking and cooking was drawn from an outside pump.

As one tries to picture Hop's busy days, the degree to which the Bar U provisioned itself becomes more and more apparent. The primary staple was, of course, beef, and the cowboys were responsible for slaughtering a three-year-old steer every two weeks or so. Similarly, pigs were slaughtered in batches of eight to ten and were cured into ham and bacon. Chickens were 'harvested' as necessary. They also provided plenty of eggs, for there was often a surplus which was passed on to the Stoneys.[45] Meat was stored in the ice house, which was filled with ice from the lake on the E.P. Ranch during the winter. Pekisko Creek was full of trout at this time, but unlike visitors to the E.P. Ranch, nobody had the time or the inclination to provide fish for the community on a regular basis.[46] There was very little fresh fruit in the diet at the Bar U. Very occasionally, saskatoon berries would be obtained from the Stoneys. Finally, the ranch provided itself with fuel. Massive piles of firewood were cut from willow, poplar, and old corral rails. Loads of local coal were hauled from Coal Coulee on the bend of the Highwood River. The value of this self-sufficiency was recognized by the company. In a brief to the government, their lawyer wrote:

> At the time the ranches were purchased they were not self-supporting. There were no hogs, very little poultry, practically no vegetable gardens, and at most of the ranches no cows were milked. Now, hogs are kept at all points, chickens are raised, lard, bacon and hams are produced, vegetables required are grown, and the ranches have been made as self-supporting as possible.[47]

Of course, there were some staples that had to be purchased from outside. They were obtained in wholesale quantities from Calgary, or later from Jenkins Groceteria in High River and Gallups in Longview. The tin-lined dry goods store house would hold commodities such as a ton of flour, half a ton of sugar, 100-lb. bags of coffee beans, 50–100 lbs. of tea in large boxes lined with lead foil, cases of dried prunes, apricots and apples, raisins, macaroni, tapioca, and rice. Whole rounds of cheese were kept in the cool store.[48]

The men at the Bar U, whether they were riders or part of the hay crew, worked hard and put in long hours. They were sustained by large amounts of wholesome food. It is a tribute to the cooks at the ranch that there were few negative memories of the food among those interviewed. The relative self-sufficiency of the ranch and the scale on which it purchased necessary staples did much to isolate the Bar U community from the harsh economic times suffered by the people of western Canada. During the Great Depression, wages were slashed to less than they had been during the 1880s, but the food was still plentiful and the accommodation sparse but adequate. As for wartime shortages, Lee Alwood remembered:

> We were never short. We had more stuff during the war than we had before. We used to have that "meatless Tuesday," and we'd have to get chicken or fish or something. The guys liked that 'cause they had a change from three year old beef.[49]

There were usually six or seven "riders" employed at the Bar U during the 1930s. They were at the top of the "pecking order" among the work force and had their own rooms in the bunkhouse. These were competent men who had served on many ranches and could turn their hands to a variety of activities. Some, like Oscar Brandham, stayed with the Bar U for many years, while others moved on after a year or two to try something new or to explore a new region.

Gordon Davis was typical. He was a farm boy from Manitoba. In his teens he caught a freight train westward and spent some time 'riding the rails.'[50] He then worked on ranches in the Nanton and Willow Creek districts, but he had an independent streak and worked for the Forest Service during the summers at a fire lookout high in the front ranges of the Rockies. He would pack in all the stores he needed, including water, and proved to be one of the few men who could cope with the total isolation. Raymond Clifford employed Davis as a rider in the early 1930s. Gordon said his "main love in life was horses," and this was one thing that drew him to the big ranch. He had a string of Bar U saddle horses for his exclusive use, and worked to bring them on and train them as good cattle horses. As he said: "When you work cows every day, a horse that doesn't do anything else, he gets pretty good and he gets to know something."[51] He remembered one winter when he and two other men "rode on the lope" keeping an eye on fourteen hundred head of cattle. During the bitter winter of 1936, they fed thirty-eight tons of hay each day and five hundred bushels of oats. It snowed from Christmas Day until 14 February and was very cold. The riders put aside their hats in favour of woollen caps which covered their ears. One early morning when Gordon rode out of the ranch it was 51 below. He was with an old-time cowboy named Shorty Merino, who had worked for the Douglas Lake Cattle Company in British Columbia and had drifted to Alberta and found work at the MacCleays; now he was at the Bar U.[52] Davis recounts:

> He had an old pair of Angora chaps on that were just about wore out, an old coonskin coat with a rope around it, and he was riding down the HL, eight miles down there to open water holes, every day. Somebody talked about a Chinook. Shorty replied, "Well if one of those chinooks come up, those old time chinooks, these rags would beat me to death before I got them off."[53]

Some winters Davis ran a trapline up into the mountains and sold the furs to the Hudson's Bay Company in Calgary. On another occasion he spent the winter cutting corral fence posts for the ranch and skidding them out with a borrowed Percheron work horse. Listening to Gordon talk of his time at the Bar U during the 1930s, one gets the picture of an independent-minded man who had the experience to work with a minimum of supervision. Over a period of years he would work for Clifford and then do something else, only to return to the Bar U when it suited him.

Other riders during the period included Dunc Fraser, Tom McMaster, Roy Boulton, Jack Carol, Dick Andrews, and Jimmy Lyons. The work these riders did differed little from that of cowboys generations before. During calving, in April and May, they rode through the herd and gave help where necessary. Sometimes a calf had to be pulled. George Read explained that he had trained his horse to help pull calves:

> You had the leg started and then you put the rope on it. You had a small rope, not quite as big as your little finger and you put it round the feet. Then you hooked that to your lariat and that hooked on the saddle horn. Then you just tapped on the rope and told him to back up and pull the calf out. When the calf came out and the rope slackened, then he'd step ahead and slacken the rope some more. I trained the horse to do that, not many you can train to do that.[54]

During the 1930s, Clifford contracted out bunches of cattle to farmers in the neighbourhood to be fed on straw or hay during the winter. In most cases the farmer would feed the stock and keep the water supply open, but in cases where there was an absentee landlord a Bar U rider was sent to do the job. Billy

Hughes spent a lonely winter in 1934 "batching it" in the mobile cook car on some land outside Nanton looking after a bunch of heifers. His weeks there were not all that different from the time spent by an earlier generation of cowboys in "line cabins." Again, Ann Clifford recalled a winter night when she received a telephone call with the news that five hundred head of Bar U cattle were drifting south in a blizzard from a farm east of Cayley. She managed to contact Raymond, who turned out all the riders to try and halt the drift and trail the herd back to the straw stacks.[55] This emergency was reminiscent of some of the grim battles fought with the elements during the winter of 1906–7.

The Bar U Ranch employed a number of men who performed specialized functions at the ranch. Earl (Slim) Clarke would turn up every spring and turn his attention to a new generation of cow ponies. He worked with patience and 'gentled' the young horses rather than 'breaking' them. When he had ridden each foal a few times, the young horses would be attached to a rider's string and would begin the long process of learning to work cattle. Slim also trained the work horses. First he would break them to the halter, and then he would hitch them up to a wagon with a huge black gelding named Dan. Slim and Dan would then work together to educate the novice. When Slim pulled on the reins, Dan would lean gently on the colt's shoulder and he had to go where they wanted him to.[56]

There was a lot of work for a blacksmith at the ranch. Not only did the saddle horses and work horses need frequent attention, but farm machinery and haying equipment also required maintenance. In the early 1930s there was a full-time blacksmith from the Ukraine on the ranch, later Burns employed a man to visit all his ranches in turn. Several of the riders could cope with emergencies.[57] As the years went by, so the skills of a mechanic became as important as those of the blacksmith.

The life and work of the Bar U community was made smoother by the hard work of older men who had worked for Lane and who had been promised continuing employment by Burns. Ironically, they were referred to as chore boys. Jimmy Duff was well into his sixties. He had been a good rider in his time and had even shown horses for Pat Burns. However, he had been badly banged up in an accident. Nevertheless, Jimmy made himself indispensable around the ranch. One rider remembered:

> Jimmy milked the cows, looked after the chickens, cleaned the barns and bathrooms at the bunkhouse. He rose every day at 4 a.m. to milk the cows, and there was always a pitcher of milk and one of cream on the round table in the cookhouse to drink or to use in your coffee. Jimmy smoked a pipe and had a Scottish accent, although he had a weak voice … he knew every horse on the ranch and which horse went best. He enjoyed looking after the stallions, and would lead them down to water at the creek.[58]

Paul Moller was another chore boy. As a young man he had been a member of the Danish Royal Guard.[59] Now he was nearing seventy. He was a handyman, good at carpentry or painting; he peeled corral rails and hauled manure from the barns. Whenever an

8.8 Bar U cattle and horses feeding on hay in pasture east of the headquarters site. Parks Canada, Norman Mackenzie Collection, 96.04.02.67

extra team was needed to haul hay in winter or to help in the farm fields, Paul was a willing standby. He was known as the "Eaton's agent" by the men because of his expertise in making up orders from the Eaton's mail order catalogue, a service he would perform for anybody who asked him.[60]

Instilling New Attitudes

Burns ran the Bar U primarily as a cow-calf operation. A herd of about three thousand cows produced between sixteen hundred and eighteen hundred calves annually. About five hundred cows were culled from the breeding herd each year and replaced with two-year-old heifers. The steer calves and surplus heifers yearlings were shipped to other Burns ranches in the spring to make room for the new calf crop. Raymond Clifford transformed the Bar U cattle herd during his tenure as "cow boss." He achieved this not so much by revolutionary or innovative ideas, but rather by changing attitudes and insisting that every operation be done with care and attention to detail. A 'laissez-faire' philosophy was replaced with 'hands-on' management directed to achieve clearly defined goals. The "calf-crop" of a ranch represents the interest or return on investment of the whole outfit.[61] Clifford worked patiently to ensure that there were enough potent bulls to service the herd, and he culled infertile cows ruthlessly. Gradually the calving rate began to rise. Under old-fashioned "open range" conditions it had hovered between 50 and 60 per cent, now the company secretary was able to report:

> During the last ten years the percentage of calf crop has steadily increased to 85% as a result of weeding out cows that were unsatisfactory breeders and of modernizing the ranching methods of the company.[62]

What this achievement really entailed was a cow boss who got to know all the cows in his herd and kept systematic records. Closely allied to improving the calf crop were changes to ensure that fewer calves were lost at birth and during the first month of their lives. Again, the changes were incremental: cows due to calve were drawn into a field adjacent to the ranch. They were fed as much as they could eat. Bulls were carefully chosen to make sure that calves were not oversized, for smaller calves presented fewer problems at birth. Riders moved through the herd as frequently as possible and assisted cows in trouble, but the number of riders compared to the size of the herd did not increase, so there were simply not enough men to keep an eye on all the calving animals.[63] However, a change of attitude was achieved and all the riders did all they could to minimize losses.

As he looked back over his time at the Bar U, Clifford said that at the big ranch a cattleman "wanted for nothing." Hay, water, and winter grazing were all available in abundance.[64] And yet it is clear that he understood, albeit instinctively rather than scientifically, the ecological limits beyond which grazing was not sustainable. His "cattle books" show the way in which he moved the various elements of his herd around. Cows wintered in the south field were moved onto the farm field and fed hay; in March there were 157 "hospital cows" being fed hay; and from 10 April , three hundred head were being fed three loads of rye bundles per day.[65] The aim was always to maintain the health and vigour of the cattle while at the same time saving huge acreages of natural range to be cut for hay.[66] The lease west of the Highwood River was particularly useful in this respect, and about a third of the herd spent their summer there. The cattle would be pushed high up the creeks into the forest reserve and a rider or two would pack in salt and keep an eye out for predators.[67] During the winter, Clifford took a leaf out of Pat Burns' book and sent bunches of cattle to winter over on straw with neighbouring farmers. Typically there would be a bunch at Meadowbank and others at the Findlay Place, the Alwood place and the Larson place.[68] For the most part the labour was provided by the farmer, but in some cases a rider was sent with the cattle and boarded with the farm family. During the drought years Clifford expressed the opinion that the farmers "saved the cattle."[69]

Over the years Clifford radically altered the genetic mix of the herd. When he came to the ranch the herd was basically Shorthorn, although the effects of cross-breeding with Hereford bulls was obvious. The cattle were all large, big-boned and displayed a variety of colours. In contrast, photographs of the herd taken in the 1940s show a remarkable degree of homogeneity. The blocky polled Herefords all show the same conformation, with straight backs and well-developed flanks. The mix of bulls was changed until in 1941 there were eighty-one Hereford bulls, eighteen Shorthorn and six Aberdeen Angus. Good use was made of the pedigree Hereford bulls at the Bow Valley Ranch and of the herd of purebred cows kept there.[70] In making this shift toward a Hereford herd, Clifford was following the trend set by his fellow cattlemen. There was an overwhelming feeling that the Hereford could provide superior meat for a more discerning market, while at the same time demonstrating exemplary characteristics on the range. Herefords rustled well and were good mothers. Perhaps too the challenges of life on the range had been somewhat reduced and allowed a shift from tough mobility toward a rather less mobile animal which gave superior beef. The breed had been part of the cattle industry in the foothills since the 1880s when Senator Cochrane imported Hereford bulls for his new spread along the Bow River.[71] Close to Pekisko, S.M. Mace was a steady contributor of purebred Hereford bulls to the Calgary bull sale from 1906 onward, while the Glengarry "44" Ranch had been home to one of the famous "Gay Lad" Hereford bulls during the 1920s.[72] The changes in the percentage of each breed of bulls sold at Calgary demonstrate the growing dominance of Herefords and the demise of the Shorthorn. In the decade 1911–20, the figures were 56 per cent Shorthorn and 31 per cent Hereford; during the next decade, 1921–30, Shorthorns were still slightly more popular at 49 per cent than Herefords at 39 per cent; but from 1931 to 1940, Herefords grew to 56 per cent of sales while Shorthorns shrank to only 32 per cent.[73]

When the Bar U Ranch was purchased by Burns in 1927 it was in a somewhat rundown condition. Lane had died two years before, and his last years had been dogged by poor health and financial confusion. The great ranch had continued to run on the momentum of its established traditions, but both the herd and the infrastructure had deteriorated. After Raymond Clifford joined the ranch in 1931 steady progress was made toward rebuilding. Clifford's patient eye for detail and his ability to attract, direct, and motivate an experienced labour force had a profound effect on every facet of ranch operations. Stock losses were reduced, calving ratios increased, and Herefords gradually replaced the Shorthorns favoured by Lane. That every step of progress and each gradual innovation was achieved in spite of the extraordinary stress caused by the Great Depression makes Clifford's achievement all the more remarkable.

Pierre Berton called it "perhaps the most significant ten years in our history, a watershed era that scarred and transformed a nation."[74] That decade from 1929 to 1939 when Canada wrestled with the combined effects of devastating drought, a slump in commodity prices, and the complete disruption of world trade. Berton's words are echoed by those of Francis and Ganzevoort, who called the Depression decade:

> ... possibly the most significant and memorable decade in prairie history. The very phrase conjures up awesome pictures of defeat, despair, long lines of unemployed, boxcars covered by men 'riding the rails'; the boredom of the relief camps; windfall apples and surplus cod; whirling plagues of dust and grasshoppers; the interminable heat of prairie sun.[75]

In retrospect the impact of the Depression in the prairies proved to be "as important in the development of a regional society as the era of the pioneer."[76] It reshaped patterns of population, the institutions of government, and the relationships between the various levels of government and the governed. But above all at a personal and family level the experience of the "winter years" traumatized a generation of prairie people.[77] The battle for survival was fought on many fronts, in soup kitchens in urban areas; in "hobo jungles" down by the railway tracks; and on the struggling farms of newly arrived eastern European immigrants who were forced into a subsistence-like existence in order to get by. However, individuals could only grasp their local corner of the general disaster, and we, like them, tend to have but a single image of the Depression, a kind of Canadian *Grapes of Wrath*, in which the tragic fate of wheat farmers in Palliser's Triangle occupies centre stage. Ranchers and cattlemen suffered too, although their story is less well known. In this chapter we will discuss the environmental and economic context of the Depression in the prairie provinces before focusing on the experiences of Burns ranches, and the Bar U in particular.

A generation of scientific work has enabled us to set the droughts of the 1930s into the cyclic pattern of climatic change reaching back millennia.[78] Instrumental data for western Canada extends back only a century, but the use of "proxy" climatic data derived from pollen analysis and tree-ring records enables us to reconstruct with a fair degree of precision changes in temperature and precipitation over the last four hundred years. Alwynne B. Beaudoin concludes that drought has been part of the environment on the prairies for most of the post-glacial period. Wallace Stegner echoed this generalization on the basis of his lifetime lived in the west, in his essay "Thoughts in a Dry Land" he wrote, "Aridity, more than anything else, gives the western landscape its character."[79] Drought is not an aberration on the Canadian prairies, it is the norm. Indeed it is one of a complex of factors creating and maintaining grassland ecosystems.[80] However, the records also show that rainfall receipts are highly variable; typically, cycles of wet years give way to periods of drought. In general terms, the decade of most rapid immigration to western Canada, and the most rapid expansion of farm settlement, coincided with a period of generally moist years. Bumper crops were harvested in 1905 and again in 1915. After 1915, drought hit the Canadian prairies, with widespread crop failures occurring from 1917 to 1920, and continuing through the 1930s.[81] Farmers who had experienced the wet cycle naturally thought of it as being the customary situation and waited in vain for the return of "normal" rainfall patterns. Some would argue that we are still waiting, and have yet to fully accept and internalize the fact of aridity.[82] The devastating impact of the "Dirty Thirties" and the human suffering which scarred a generation was partly due to the fact that the countryside was more densely settled than it had ever been, and neither the homestead families nor the various levels of government had any experience in coping with prolonged drought.

Those of us who live on the prairies are well aware of the fact that drought can strike one area while leaving another unscathed. As I write, central, northern, and eastern Alberta are suffering the third year of bitter drought, while the southern and southeastern parts of the province have enjoyed exceptional rainfall this year. Native grasses not seen for years are reappearing and barns are full of hay. The Saskatchewan geographer, A.K. Chakravati, used data from sixty-three climatic stations over a fifty-year period to produce a series of maps showing where drought struck and which regions enjoyed average precipitation. He concluded that "even during the years of most deficient precipitation, the agricultural areas of the three prairie provinces were affected unequally."[83] For example, in 1929, Alberta and Manitoba suffered widespread drought, while Saskatchewan escaped. In 1936, on the other hand, Saskatchewan was devastated, as were the foothills and southern plains of Alberta.

8.9 A dust storm looms over a village in southern Alberta during the "Dirty Thirties." Glenbow, NA 2685-85

This regional variation in precipitation receipts helps one to interpret the comments of government officials whose reports often seem contradictory – focused as they were on what they saw of local conditions. For example, in 1929, the Livestock Commissioner stressed the difference between conditions in the foothills and out on the mixed-grass plains to the east. He wrote:

> ... owing to the drought over the central eastern part of the province, surplus stock was rushed to market, but in the ranching districts, where most of our cattle are raised, there has been an abundance of grass and feed.[84]

In a similar vein, in 1937 the agricultural agent at Lethbridge described severe drought conditions and high winds which were blowing soil. He reported that farmers east of Claresholm had seeded three times and had still only harvested two bushels per acre; however, west of the railway line from Stirling to Vulcan, wheat crops had responded well to late rains and had given some good returns.[85] The following year things improved dramatically in the south of the province:

> The grazing conditions were the best in years. Native grasses made a remarkable comeback with heavy seed production. Much of this seed germinated during the early fall and made several inches of growth before freeze-up occurred.

> Over large areas native grasses went into winter over a foot in height.[86]

It is clear that during the generally dry years of the Depression the detailed pattern of crop yields and grazing conditions was complex in the extreme. Just as some individuals and classes "breezed through the hard times without a care,"[87] so too some regions did not suffer as badly as did others.

The ranchers of the upper Highwood River were spared the worst of the droughts that devastated the plains. Even in dry years some rain fell and snow melt from the Rocky Mountains fed eastward-flowing streams. However, during the "Dirty Thirties" the usually unpredictable foothills climate became even more capricious. For example, in 1933, a snowstorm in mid-April was followed by a period of heavy rain that prevented farmers from getting onto their land. "Spring work delayed: Sets a New Late Record," was the headline in the *High River Times*. By mid-July, it was drought that was causing concern for both hay and grain crops. On 27 July the paper's editorial was captioned "Unfathomable Weather," it went on to explain:

> There has been something cruelly disheartening in wheat developments over the past week. On top of the season's drought came frost, an unusual and unanticipated blow in this part of the world. True, in 1918, a year very similar to this, there was a July frost … but the calamity of frost has come in 1933, after 3 years of poverty prices, when farmers have been drained to the limit of their resources putting in the present crop. This has been the black week of a black year. It has laid the heavy hand of the depression on all of us.[88]

The following year was something of a reprieve, Professor Carlyle wrote from the E.P. Ranch, a neighbour of the Bar U:

> We have had a delightful autumn until the present week. I do not remember any year in the past twenty in Alberta where we have had as favourable an autumn season, mild and bright with an abundance of sunshine. Stock have done remarkably well. This means much to stockmen as feed supplies in many sections are very low …[89]

The following winter, in complete contrast, was considered to be the hardest since 1907 in the High River district.[90] For more than six weeks the chinook failed to blow. Rivers froze, wells dried up, feed supplies dwindled and cattle shrunk.[91] In May, a dust storm thundered into High River from the north, the wind "sweeping cultivated soil before it, the air was thick with dust and visibility was obscured."[92] In June, the temperatures ranged from several degrees below freezing to more than a hundred degrees Fahrenheit in the shade. "Farmers in the district with 30 years experience say they have never seen drought until this time."[93] Pekisko Creek ceased to flow round the Bar U buildings for the first time since 1881. In mid-August, the editor remarked, "About the only things we will remember this year for is that it has been the coldest, the hottest, the driest, and so on, in the history of settlement."[94] Thus stockmen, even in one of the most favoured areas of the prairies, were hit again and again with natural disasters. On balance, however, it was probably the vagaries of prices and markets which brought some of them to the brink of despair.

After a period of high demand for cattle and exceptional prices for beef during the late 1920s, the bottom fell out of livestock prices in 1931. The Livestock Commissioner wrote:

> The year of 1931 will go down in history in Alberta as recording the lowest prices of livestock and livestock products for the past 30 years.[95]

Little did he dream that stockmen were facing a cyclic downturn in prices which would last until 1939.[96] By 1933, steers off the range were sold for less than $2 a hundredweight.[97] In 1936, low end cattle were bringing in one cent per pound in the Lethbridge area, while 'canner cows' were selling at 50 cents per hundredweight. Indeed, low quality animals were worth more for their hides than their meat.[98] Some cattle shipments actually brought in less than the cost of transport. One of Barry Broadfoot's correspondents told the story of his mother, who shipped a big prime steer from Saskatchewan to Winnipeg because she was advised that she would get a better price for it. "But with the prices then and the weight loss and the trucking costs and the selling commission and what have you, she got a letter back from the agent saying that she owed them $6.00. Without the word of a lie." The woman paid her debt and framed the receipt to remind the family of the hard times.[99] When cattlemen came together for their annual meetings their anger and resentment over prices far outweighed their anxiety concerning the drought.[100] Prices slumped to new lows in 1933 and 1934, and although they rallied somewhat in 1935, they fell back disappointingly in 1936. It was not until 1937 and 1938 that prospects began to improve and a spirit of guarded optimism became apparent once more among cattlemen.

The slump in prices for cattle in Canada was partly due to the imposition of a prohibitive tariff on cattle moving into the United States. This market had always played an important role in the Canadian cattle business. Although the volume of exports from Canada was not great on a continental scale, and amounted to a tiny fraction of the turnover in the Chicago stockyards, nevertheless, the outflow of prime stock tended to have a beneficial impact on prices in Canada.[101] In 1929, the United States acted to tackle their own problems of oversupply, and the tariff on Canadian imports was raised. An observer recalled:

> In 1929, the Hawley-Smoot tariff between Canada and the United States came into effect. This tariff amounted to three cents a pound and resulted in shutting our cattle practically out of the market … It was unfortunate for the cattlemen of Alberta that the depression started at just this time, and between the tariff and the depression the result has been a very large surplus of cattle, with ruinously low prices, which has put the cattle business in a deplorable position. Ranchers and farmers have been producing at below the costs of production, and

8.10 This picture of a cattle drive conveys the characteristics of the mixed grass prairie lands to the east where Burns' Circle Ranch and "76" Ranch were located. Huge areas remained available for extensive grazing until the Second World War, but grass was scarce during the severe droughts of the 1930s. Glenbow, NB (H) 16-495

> unless something is done to remedy the situation, whereby higher prices will be obtained, the largest and best cattlemen in the province will be put out of business.[102]

Shipments of cattle to the United States from Canada dropped from over 160,000 head in 1929 to fewer than 10,000 head in 1931. In Alberta, exports slumped from 28,000 head to just 48. Prices dropped by $3 to $5 per cwt. immediately after the tariff was established.

With the US market closed, Britain was explored as a possible destination for Canadian cattle. The federal government was particularly keen on this solution, for it was already engaged in trade negotiations with the United Kingdom based on Imperial Preference.[103] In 1930, an experimental shipment of six thousand head of finished steers from ranches in Alberta and Saskatchewan was sent to Liverpool. The results were encouraging:

> While ranchers did not realize high prices in the old country, the effect it had on the home market and the prices received for the balance of cattle they had to sell made the shipment well worthwhile.[104]

The following year almost thirty thousand head were shipped and returns were good during the summer. By autumn, however, exchange rates had fallen and it was clear that grass-fed Canadian cattle could not compete in British markets. The costs of getting an animal from Alberta to Britain in 1934 were six times what they had been in 1905, whereas the prices received were actually lower.[105] Gone were the days when the western ranching industry could rely on this distant market as a foundation of its prosperity. With the advent of sophisticated chilling techniques and a well-developed distribution infrastructure, the competitive advantage of freshly killed beef over "frozen meat" had disappeared. Wholesalers in London and Liverpool could draw on beef from the estancias of the Argentine and the outback stations of Queensland, as well as a flood of live cattle from across the Irish Sea. Only the very best of Canadian live cattle could, at times, find a niche in this highly competitive and volatile market.

Nor was there any obvious opportunity to expand the domestic market. Per capita consumption of beef had fallen during the 1920s and pork had become the country's favourite meat. Moreover, the dismal prices received by cattlemen did not translate directly into cheap beef at the butcher's.[106] As unemployment soared to unprecedented levels and nearly half the population was on some form of relief, meat of any kind became something of a luxury. In Montreal in 1933, a family of four received a weekly relief voucher worth just $4.58.[107] All that they could realistically hope for would be some beef bones to make soup. Above all, local markets were too small to absorb surplus cattle production. There were not enough people to provide the demand or the purchasing power to drive up prices. The Calgary market could be "broken" by the sudden arrival of two carloads of twenty cattle, while Edmonton's Swift Canada plant ran at half capacity because the local market could not absorb its products.[108]

The federal government made some tentative moves to try to assist cattlemen. Freight rates were heavily subsidized to encourage the movement of cattle to areas in the northern prairies or eastern Canada where there was feed. Likewise, forage and feed grains could be moved free by rail to sustain producers' foundation herds.[109] The government also organized the slaughter and disposal of thousands of head of low-quality stock to try to rid the prairies of the glut. It was hoped that this would reduce the demand for "feed relief" in the short run, and would lead to better prices in the longer term.[110] These measures were recalled by F.M. Baker in a speech he made to cattle buyers during the 1950s. He said:

> You will remember that in 1936 and 1937 you processed train loads of cattle from southeast Alberta, south Saskatchewan, and southwestern Manitoba, at a price to farmers of one cent to one and a half cents a pound, to save them from dying on farms. Many more were sorted out and sold at correspondingly low prices for further feeding in eastern Canada. Several times similar conditions over wide areas forced liquidation of valuable breeding stock. More times, severe local droughts caused similar action on a less extensive scale.[111]

In spite of these efforts the plight of cattle producers remained very grave. The ongoing droughts meant that the cost of feed spiralled upwards, while the dislocation of international trade led to a glut on the market and very low prices. While all cattlemen were squeezed between rising costs and decreasing returns, different regions in the west and different sectors of the cattle business were affected to varying degrees. The particular experiences of three of Burns' ranches illustrate some aspects of the general story which has been outlined. The following paragraphs trace the fortunes of the Bar U Ranch in the foothills, the Circle outfit on the Red Deer River, and the "76" Ranch in southwestern Saskatchewan.

No region of western Canada escaped the pervasive effects of the Depression; the twin blows of drought and low prices devastated both urban and rural communities. However, the foothills of southern Alberta survived better than most areas. This was a well-established region of comfortable homes, good farm buildings, adequate equipment, and productive fields. Churches, schools, and sports clubs provided social supports. Moreover, the development of the Turner Valley gas field pumped hundreds of thousands of dollars into the region and provided a variety of short-term construction jobs.

The stock books of the Bar U Ranch do not indicate a wholesale reduction in stock inventory during the 1930s. When the ranch was purchased in 1927 there were 5,025 head of cattle on it. In 1934, the total was 5,416; this figure dropped to 4,187 in 1935 and 3,877 in 1936, before rebounding to 5,151 in 1937 and 5,603 in 1938.[112] Moreover, the calf crop remained quite constant at between 1,600 and 1,800 per year. These figures suggest that the stocking ratio at the Bar U was such that herds could be supported on available grass even when there were severe droughts. Actual stock losses seem to have been quite small, amounting to eighty head after the severe winter of 1936–37, and 101 head after the succeeding winter. These losses amounted to only 1 or 2 per cent of the total herd.

As the grip of the Depression tightened, so every effort was made to reduce expenditures, or at least to hold the line. "General expenses" for fencing, repairs, light, heat, power, and gas were actually reduced from $4,266 in 1934 to $3,452 in 1937.[113] So were expenditures on "board," or purchased groceries. The wage bill was the only area to show a considerable increase. This is interesting because individual wages remained stationary, and the minimum wage paid to occasional workers actually decreased. Probably additional hay crews were engaged to put up wild hay in the outlying areas of the ranch to make up for the low productivity of the more accessible hay meadows.

The financial accounts for the Bar U Ranch during the early 1930s have not survived, and only the figures from 1937 onward are available. During this year the ranch recorded a loss of some $3,000 before depreciation, or $5,000 if this item is added.[114] The figures show that the operation was being squeezed by diminishing returns and increased expenses. Returns from stock sales amounted to only $27,000, less than half the figure for 1939, when more encouraging trends were becoming apparent. In the debit column, purchases of feed grain amounted to $8,000; this was quite exceptional, as in normal years there was a surplus of grain which contributed to ranch income. On the mixed-grass prairie to the east, things were even worse.

The Circle Ranch comprised eighty-one thousand acres of land leased from the Canadian Pacific Railway, and about five thousand acres of deeded land. The main function of the outfit was to fatten young cattle received from other Burns ranches. In 1937, there were twenty-five hundred head of cattle on the ranch.[115] The range at the Circle was much drier than on the foothills ranches even during 'normal' times. It was located in the region which bore the full brunt of the droughts. Apparently, all was well during the summer of 1929, although the mixed-grass region had already experienced two years of drought. Burns Ranches paid their half yearly rental with a cheque for $2,447.35, and deposited a further $2,800 as the first payment on some deeded land they were purchasing.[116] However, by the spring of 1932, Don Bark of the Canadian Pacific's Irrigation Development Branch wrote to Burns pointing out that he was "considerably in arrears on your grazing lease in this district." Bark went on:

> We realize that grass was short last year and that livestock prices were comparatively low, but in spite of this we need the money so badly that if you pay 1931 in full at an early date we are willing to give you a free lease for 1932. This, in other words, will in effect amount to three cents per acre per year for those two years, which you will have to admit is a considerable concession on our part, as it will amount to less than taxes.[117]

Burns and Company was swift to accept this offer and a cheque for $4,894.70 secured 80,000 acres of grazing until the end of 1932. In acknowledging this payment, Bark was able to pass on some encouraging news:

> I wish you would tell Mr. Burns that there has now been two weeks of almost continuous rain here and while it has not rained heavily at any time, the moisture and humidity have been so continuous for such a long period that the range prospects now look very favourable indeed. With any reasonable precipitation from now on, Mr. Burns should have more grass on this lease than he has had for years.[118]

Unfortunately, the spring rains gave way to another blistering summer, and major concessions on the lease were maintained until the end of the decade simply because there was so little grass on it. Unless the herd size was to be slashed, the shortfall of range grass had to be made up with purchases of hay from local farmers and by making arrangements with them to winter over bunches of cattle. Expenses for feed rose from a few hundred dollars in 1929 and $1,000 in 1930, to $6,000 in 1933 and $13,000 in 1936. Instead of representing 5 or 10 per cent of total expenses, feed accounted for 60 per cent of the costs of running the ranch from 1933 to 1939.[119]

In 1945, Tom Farrell, Ranch Superintendent for Burns Ranches, wrote to the manager of the Eastern Irrigation District and reminded him of this period:

> Throughout the distressing period in the livestock industry we maintained our herds in the territory and fed heavily during the many trying winters to the distinct advantage of a great many farmers in the district.[120]

In spite of persistent drought which completely undermined the profitability of the ranch, the Circle continued to operate as a beef ranch, fattening young cattle, throughout the Depression. It was to contribute to the profits earned by the company during the heady days of expansion during the 1940s.

The "76" Ranch was located on 100,000 acres of leased land stretching north and east from the Whitemud River in southwestern Saskatchewan.[121] This region was infamous for the severity of its winters and for the frequency and intensity of the droughts it experienced. And yet, from 1929 through 1937, the herd size at the ranch was maintained at between two thousand and twenty-five hundred head. The stock on this ranch were divided into a breeding herd and a beef herd, and entries in a ledger book detail the day-to-day operations of the ranch.[122] The seasonal round of activities proceeded in spite of the economic and environmental storms which raged about the operation. For example, in the fall of 1928, thirteen steers, twenty-six cows, and a bull were sold for $2,731.98, while the following day, C.O. Robinson purchased 317 steers and 199 heifers for $38,595.57. A carload of salt was purchased, Blackleg serum was bought and administered, and lumber was obtained from the De Wolf Company for some construction work.

The same kind of prosaic entries can be found through the 1930s. It is only by making comparisons with earlier years that it is possible to discern the difficult times through which the ranch was passing. For example, at the end of November 1933, a cow was butchered to feed the men; it was valued at $15; only four years previously in a similar entry the cow was valued at $60 and a steer at $100. In 1935, fall sales of stock to Swifts of Canada and Canada Packers averaged only $28 per head as compared to $75 in 1928. The ledger contains a growing number of references to the purchase of feed, for example on 29 February 1936, $842.50 was spent on oil cake; a month later a further $825 was expended, and so on until a total of $3,707. had been laid out. This emergency feeding was only partially successful, for two hundred head of cattle died during the same period. Taken together, the mass of detailed entries paint a comparatively simple picture of increased costs for feed, haulage, and labour, of stock losses, and of low prices. Retrospective comments on the situation by the directors of P. Burns Ranches reached the same conclusions:

> The years 1937, 1938, and 1939 are not representative of our normal ranching and farming operations. In the year 1937 following the drought of the summer of 1936, large quantities of feed had to be purchased at exorbitant prices. Prices of livestock were also at an exceptionally low ebb owing to a combination of circumstances which are not of a natural or reoccurring nature: an extreme lack of purchasing power; excessive drought necessitating a government programme of purchasing cattle at low prices and of slaughtering livestock from drought areas and resultant drop in prices generally.[123]

The ledger book and other accounts of the "76" Ranch illustrate another problem which plagued ranchers during this time of falling prices, namely the difference between the "book" evaluation of stock as recorded in the inventory, and the real world market price which could be obtained in Calgary or Winnipeg. If, for example, you had a hundred steers on your books valued at $70 per head, and a falling market meant that you could only sell them at $40, you incurred a $30 loss on each animal even if your production costs were still less than the selling price. In fact, the more stock you sold the greater your loss in your accounts. At the time of Pat Burns' death in 1937, the inventory value of stock on the "76" was $82,000, while the appraised value was only $34,000.

The audited accounts of the "76" Ranch from 1929 until its sale in 1939 show that yearly losses contributed to a mounting cumulative deficit.[124] The largest losses on the sale of livestock occurred in 1929 and amounted to $31,000. The losses were smaller in the following years and small profits were made in 1933 and 1934. However, the cumulative deficit rose to $125,000 by 1939.

It would be hard indeed to exaggerate the economic impact the Depression had on the cattle industry. For a decade, well-established practices and patterns of trade were disrupted. The total value of cattle in Alberta, which had amounted to more than $18 million in 1929, fell to less than $4 million in 1933. Pat Burns died at the height of the Depression. The lands and buildings on his twelve major ranch properties were valued at $858,597, and the stock on these ranches at some $300,000.[125] Some indication of the slump in cattle values is provided by the fact that a very similar figure, $300,000, had been paid for the stock on just two ranches, the Bar U and the Flying E, in 1927. A brief submitted in 1943 mentions the depletion of capital invested in the company:

> On December 31, 1928, the capital employed by P. Burns Ranches Limited was $2,277,548, but by December 31, 1938 this had fallen to $1,690,244, a decrease of $587,304 or 26%. Operating and other losses, less capital advances for the parent company in this period account in large measure for this depletion in capital.[126]

The same brief provides an analysis of combined profit or loss, for the years 1928–42, of the twelve ranches owned by the company.[127] After posting a modest profit in 1928, nine of the next ten years show losses which averaged $100,000 from 1928 to 1932 and accumulated to total $614,637. They were balanced by profits which grew steadily from $90,000 in 1939 to $202,000 in 1942.

Yet in spite of drought and economic chaos, the big Burns ranches continued to operate. Expenses were pared down and even necessary maintenance went by the board, but no ranches were disposed of nor were any herds liquidated. The management experience which had been built up over the years, coupled with Pat Burns' determination and his financial reserves, enabled his enterprises to survive. His managers and employees had good reason to agree with the senator's terse admonition: "Hang on to the cow's tail and she'll pull you through!"[128] Better times lay ahead.

9

P. Burns Ranches Limited, 1937–50

"It all unfolded as an unlikely series of unforeseeable accidents."[1] – Terry Jordan

Pat Burns had put together a formidable management team consisting of lawyers, accountants, cattlemen, and farm managers. These men had already spent long periods working for Burns and had matured and gained experience under the watchful eye of the "Old Man." As a more optimistic economic climate began to emerge in the late 1930s and early 1940s, they had the opportunity to implement their ideas for overhauling and modernizing the company. Reading the correspondence and weighing the accounts of the Burns ranching empire after the death of its founder in 1937, one cannot fail to detect a mounting momentum of change.

John Burns, Pat's nephew, was chairman of the board, while Alex Newton and J.H. Kelly, both lawyers, were active directors. W. Train Gray was treasurer, while Tom Farrell and Neil Duggan were ranch superintendents. Finally, at the individual ranch level, the company was able to rely on the services of a group of experienced cattlemen: Raymond Clifford at the Bar U; J.L. Sexsmith at the Flying E; D.M. Campbell at the Circle; Bruce Watt at the Walrond; Hope Oliver at the "44"; Oliver Deschamps at the "76"; and Owen Doyle at the Minto. Several of these ranch managers moved from one outfit to another as the years went by, but together they made up a reliable and knowledgeable group which served the company well. This management tradition was summarized in a review of P. Burns Ranches completed in 1945. The author wrote:

> Those associated with Mr. Burns in his lifetime, and at the time of his death, constituted the management. Practical operations including the buying and selling of livestock and other farm products are under the direction of Thomas Farrell who acted in that capacity during Mr. Burns' lifetime. His assistant, Raymond Clifford, has been a practical farm and ranch man for the greater part of his life. There is a foreman in charge of each farm or ranch, who conducts the operations under the supervision and direction of Messrs. Farrell and Clifford. Most of the foremen of the various ranches and farms date back anywhere from 10 to 35 years. The men in positions of responsibility and those under them have grown up in the employ of the company and have a high efficiency record as a result of experience and training.[2]

During the dark days of the Depression, gradual progress had been made in updating infrastructure, and a start had been made on improving the herds. A retrospective comment on the situation at the Bar U makes this clear. Alex Newton wrote:

> When acquired [1927], farm equipment, buildings, fencing, etc., were in extremely bad

> condition. The livestock were in a deplorable state both as to condition and breeding. It took some years to bring the livestock up to the present standard. During this period the livestock purchased was gradually written down and disposed of, fencing, buildings, etc., repaired and losses taken.[3]

Rebuilding was a long-term process, and plans had to be implemented in the context of the economic dislocation which gripped the country during the 1930s. Survival was a real achievement and the pace of change was necessarily muted.

Now, however, the tide had turned. By 1940, the value of cattle and calves in Alberta had recovered to pre-Depression levels, and during the ensuing war years it doubled and doubled again. Prices for cattle followed much the same encouraging upward cycle. In general terms, Burns Ranches recovered from the losses sustained during the 1930s and recorded handsome profits. At the same time, land values rebounded, and the total value of the Bar U in 1947 reached $860,000. Modernization of ranch infrastructure and practice could proceed more aggressively, sustained by these healthy returns.

Integration, Rationalization, and Modernization

The ranches owned by Pat Burns had always co-operated closely. Cows, calves, horses, feed, workers, and managers had been moved from one outfit to another as the need arose. Indeed, the general strategy of using the sheltered and better watered foothills' pastures for breeding calves, and the mixed-grass prairie, with its superior nutritional properties, to fatten stock, can be traced back through George Lane's multi-ranch operations to the days of the open range and the great leases. However, in the years after 1937, the functions of several of the company's ranches became more specialized and seasonal flows of stock from one to another became habitual. It would not be going too far to say that the ranches were welded into a single production system. For example, in 1938, the cow herd at the Walrond Ranch was moved to the Bar U. It consisted of 405 cows and 211 heifers. In exchange, 794 yearling steers were dispatched to the Walrond.[4] Thereafter, the Bar U, Flying E and "44" were primarily cow-calf operations, while fattening of beef herds was undertaken at the Walrond, the Circle, and later, the Lazy H on the Little Bow River. Commenting on the Bar U in particular, Alex Newton wrote: "We have made a practice of wintering the steer calves at the ranch and disposing of them to other ranches in the spring or early summer."[5] During the early 1940s the Bar U had a calf crop of about two thousand head, of which approximately half would be steers. About five hundred heifers were kept on the ranch to inject new blood into the breeding herd, while the remaining five hundred would join a similar number of culled cows in journeying off to one of the other ranches to be fattened. Most Bar U stock were shipped from Cayley to Princess, the nearest station to the Circle Ranch, in April or early May.[6] In a similar way the Flying E and the "44" produced between eight and nine hundred calves between them and sent some 450 young steers to the Walrond each year. Thus the Walrond would have a stock complement of 450 yearlings, 450 two-year-olds, and 450 three-year-olds, for a total herd of between 1,300 and 1,400 head.[7]

The relative isolation of the "76" Ranch made it difficult to integrate into this inter-ranch system, and it may have been this fact, as well as the ranch's mounting deficit, which prompted the board to divest themselves of the outfit. The beef herd was sold soon after Burns' death, and the breeding herd was drastically reduced. In 1938, the total herd was only 278 head. Finally, the company sold "all of its deeded land, leasehold rights, buildings and dams, which were valued on the books at $23,081.20, to the Dominion Government for $44,200."[8] The Lazy H Ranch, situated south of the Blackfoot Indian Reserve, between the Little Bow and the Bow Rivers, was much more accessible. It consisted of about 34,000 acres of leased land and some five hundred acres of deeded land, and was acquired by the company in 1943 to provide more stock-fattening pasture.[9]

Rail links between the Bar U and the Circle Ranch played a major role in facilitating stock movements. The availability of cars, the timing of trips, and the condition of corrals and loading ramps were all matters of concern to Burns' management. In 1941, Tom Farrell wrote to Canadian Pacific Railways at Lethbridge and Medicine Hat:

> May we draw your attention to the condition of your livestock pens and loading chutes at Cayley, Nanton, Claresholm, and Lundbreck, Alberta. These are in a bad state of repair and as we do a considerable amount of loading at all of these points we experience considerable difficulty which we feel sure you will be prepared to remedy. We trust that you will give this matter your earliest attention.[10]

Trail herding remained common practice among the big foothills ranches, for example moving yearlings from the Flying E to the Walrond. It was also an alternative way of moving stock to the Bow Valley Ranch and the fattening pens of Calgary. Hope Oliver recalled a drive in the 1930s which started at the Walrond and picked up additional steers at the "44" and the Flying E. The herd numbered about two thousand head and was accompanied by eight riders and a remuda of horses. The drive took eight days to complete.[11] In a similar fashion, a thousand head of yearlings were trailed from the Bar U in April 1944, probably because of a shortage of box cars.[12] A year or two later, a similar drive attracted countrywide attention, and a newsreel and the newspapers gave the herd star billing. This herd was made up of eighteen hundred head from the Bar U and the Flying E Ranches. The editor of the *Canadian Cattlemen* described the scene with some nostalgia:

> Eleven cowpunchers successfully trailed the cattle over the 67 mile route. The herd made about 12 miles a day.... The route to Calgary had been previously planned by Thomas Farrell, manager of P. Burns Ranches Ltd., and Raymond Clifford, manager of the Bar U Ranch. It was ideal weather for a cattle drive. There was plenty of feed along the road allowances, and there was water at every stop. The herd traveled parallel to the main south highway until they reached the

> Round T Ranch at High River. They crossed the highway at Aldersyde and they were then herded into P. Burns Ranches Limited property east of the highway, and from there the rest of the trail to Calgary feedlots was accomplished with ease.[13]

Such scenes recalled the history and traditions of the cattle industry, but it was an industry which was transforming itself to meet new needs and to benefit from new opportunities.

The management of P. Burns Ranches Limited was closely attuned to changes in the marketplace. For example, John Burns circulated to his board of directors and his ranch superintendents information about the swing toward "baby beef" and innovative feeding systems which were being introduced in the United States.[14] He asked for comments, and it is clear that he expected his company to initiate change in Canada rather than to follow the lead of others. A year or two later Alex Newton reported:

> The trend of public demand for beef has been consistently for smaller cuts and I am of the opinion that it will result in a diminishing number of three year old steers being raised by ranchers. I am inclined to think that disposing of cattle as calves and yearlings will become a more generally adopted practice.[15]

He was looking ahead to times when young stock would leave the ranches to be fattened in feedlots to the exact requirements of the market. Indeed, the company requested the federal Department of Agriculture to send information on "the care and feeding of cattle in feedlots." They were referred to the work of Professor J.P. Sackville in Edmonton.[16] They also contacted Grant MacEwan, who was Professor of Animal Husbandry at the University of Saskatchewan.[17] Later, the company was able to report that "cattle and hogs passing through the feed lots are fed on the basis of scientific formulae."[18] In rather a similar fashion, when a decision was made to expand the dairy herds at the C.K. Ranch and the Nielsen place, the best available advice was sought on the layout and construction of dairy barns.[19] Nor was the flow of information all one way. The Lacombe experimental station wrote for advice from the company concerning the manner in which they should identify their horses. Farrell explained the way in which the Bar U Percherons had been branded with a small iron under the horse's mane.[20] Taken together, the evidence suggests that the senior management of the company was well informed and innovative. One report boasted:

> The development of the herds and the ranches was approached in a scientific manner.... The system included judicious selection, careful breeding, proper fencing and cross fencing, the use of mineral feeds where minerals were lacking, a large increase in the cultivation of hay, feed and grain crops, the separation of ranches and farms into proper economic units and into proper summer and winter units for cattle.[21]

The company was also prepared to pay top dollar to obtain the best stock available.[22] The lengths to which they were prepared to go to sustain a vigorous gene pool is indicated by the fact that Tom Farrell was dispatched to Colorado during the war to obtain new Hereford strains, in spite of currency regulations and travel difficulties. Mr. Gray, the company treasurer, arranged with the Royal Bank to provide American funds:

> It is necessary in the interests of our livestock operations to acquire bulls for breeding purposes on our ranches in Alberta, and it is planned to purchase these bulls in Colorado, for which we estimate that we shall require approximately the sum of $10,000 in American funds by way of letter of credit in Mr. Farrell's name.[23]

Great strides were also made in water management on the ranches. Small catchment dams were constructed wherever practical, and this strategy brought into production for the first time large acreages of natural grazing which had been neglected by stock because they were too far from water.[24] The Bar U was also blessed with a large number of springs. Allan Baker reported that he had forty-eight springs and thirty-nine of them ran year round when he owned a large part of the Bar U Ranch during the 1950s.[25] It was part of the riders' job to patrol these springs and keep them open.

The management of Burns Ranches was exploring innovations in the 1940s which were not to become commonplace in the foothills for another decade. It was ironic that progressive modernization was made possible by the global conflict which had such tragic or at least disruptive consequences for the lives of so many Canadians.

Canadians faced the prospects of another war in a sombre mood very different from the flood tide of imperial patriotism which had heralded hostilities in 1914.[26] However, the Battle of Britain and the heroism of Dunkirk focused public sentiment on the beleaguered British Isles. By the time victory was won in 1945, more than one million Canadians had seen service in the armed forces. On the home front, the entire economy was pre-empted by the war effort. Farms and factories recovered from the Depression and produced as never before. Price controls and rationing were instituted so that surpluses could be exported across the Atlantic.[27] Prairie farmers and ranchers benefited enormously from a period during which they could sell everything they produced at stable prices. However, they lost even the illusion of freedom of choice provided by the free market economy: they could only sell products within Canada and at prices fixed by the government. Because of the long production cycle involved in the beef cattle industry, it was not until 1943 that Canadian cattle herds had rebounded from the low inventories of the Depression. But the period from 1944 to 1950 was one of unprecedented increases in productivity. In retrospect, the war period and its aftermath played as important a part in shaping the Canadian cattle industry as had the Depression.

The broad priorities of the Canadian government were to encourage food production to ensure that Canadians were adequately fed, but at the same time to control consumption to maximize exports to Britain. The "Bacon to Britain" program was the foundation of these efforts. The first trade contract was agreed to in 1939 and called for the export of 291 million pounds of pork; this figure rose in successive contracts to 900 million pounds in 1945. The actual shipments exceeded the contracted figures in every year until the end of the war.[28] The government expected the cattle industry to play an ancillary role by filling the gap in domestic demand for meat caused by these huge exports of pork.

The number of cattle marketed in Canada increased rather slowly during the first four years of the war. In March 1942, the Wartime Prices and Trade Board prohibited further exports of beef cattle to the United States, thus making available some 150,000 to 200,000 head which had been flowing southward. Nevertheless, the low ceiling prices set by the government did little to encourage cattle sales and marketings actually declined in 1942 and 1943. Meanwhile, the demands of the civilian population rose sharply, partly because pork was not available and partly because of full employment and increased purchasing power. In March 1943, the government introduced meat rationing to curb this trend and to encourage a fair distribution of meat products. Finally, in 1944, the long-delayed flood of cattle began to flow off the ranches and farms. "More cattle, hogs, sheep and lambs were received at stock yards and packing plants than in any similar period in Canada's history."[29] Meat rationing in Canada was suspended, and the first two-year "British Beef Agreement" was filled by the end of the first year. Shipments to Britain in 1944 amounted to 147 million pounds, the equivalent of 300,000 live cattle. The following year sales of cattle passed the two million mark, and exceeded the record set the previous year by half a million head.[30] Exports of beef and beef products in 1945 were the equivalent of 543,000 head. Viewed from Ottawa, wartime policies had been vindicated by the performance of the Canadian cattle industry in the last years of the war.

Western cattlemen had a very different perception of government controls: they resented the attempts of outsiders to manage their industry. Above all, "Canadian farmers and ranchers were shocked and alarmed by the closing of the traditional and economically attractive American market to Canadian cattle."[31] They were convinced that selling the best of their young cattle in the Chicago market was the obvious and natural way for their industry to develop. They were prepared to make short-term sacrifices in the interests of the war effort, but they were afraid lest their place in the U.S. market might be usurped by Mexico.[32] Moreover, they argued that the flow of cattle south of the line had a beneficial effect on prices in Canada and kept Canadian packers honest by providing competition. The price ceilings introduced by officials of the Food Administration of the Wartime Prices and Trade Board referred to meat at the wholesale and retail levels rather than to live cattle.[33] This put producers squarely under the thumb of the packers, who could claim that they could only pay such-and-such a price for live animals because they had to operate under government ceilings when they sold their products. Cattlemen claimed that they did not receive their fair share when ceiling prices were raised. Indeed, evidence presented to the House of Commons Agriculture Committee by John McLean of Canada Packers and the Canadian head of Swifts made it obvious that the packers had bought as low as possible and sold at the highest profit. They had often refused to buy live cattle at the "floor" or lowest permitted price.[34]

Canadian cattlemen had responded to the extraordinary demands of wartime and had achieved enormous increases in productivity.[35] They faced the post-war period with considerable anxiety.[36] Collectively, they remembered the deflation which had occurred in 1919, and feared that a similar sharp drop in prices could occur. They were therefore increasingly insistent that their right to export cattle to the United States, "their close and natural market," should be restored.[37] They pointed out that the United Kingdom market, shorn of sentimental considerations, was an uneconomic one for Canada.[38] It could only be regarded as an occasional "salvage or liquidating market."[39] The industry recognized that it could not compete with lower cost beef producers in Argentina and Australia and was also aware that Britain would prefer to purchase beef in Sterling funds when international trade was re-established. At the same time Canadian producers were presented with an opportunity to re-enter the Unisted States market because their competitor, Mexico, was dealing with an outbreak of foot and mouth disease.

The Canadian government, in the person of James Gardiner, Minister of Agriculture, seemed to be wedded to an outdated dream of Imperial Preference. In an extraordinary speech to the Western Stock Growers Association, Gardiner talked of:

> ... our economy, our political institutions, our place in the world is going to be established alongside of the mother of nations, a land which has led us all and which can still lead us all ... a nation which has led the world in the establishment of democratic institutions ... freedom ... and sound finance based on sterling.[40]

Events were to demonstrate how out of touch with post-war reality such a position was. In August 1948, Britain cancelled her beef contract with Canada, and the United States' market for live cattle was reopened three days later.

"The removal of the embargo did change the outlook of the beef cattle industry from one of uncertainty and despair to one of opportunity and hope," wrote the editor of *Canadian Cattlemen*.[41] In the four-and-a-half-month period following the reopening of the US market, 310,000 head of cattle were moved south and earned some $60 million US dollars.[42] The *Annual Market Review* observed that the removal of the embargo "gave a great fillip to the whole market structure" and was reflected "very clearly in the tonic effect on Canadian prices."[43] The United Nations General Agreement on Tariffs and Trade signed at Geneva in October 1947 established important tariff reductions on livestock entering the United States for the next three years.[44] Kenneth Coppock, secretary of the Western Stock Growers Association , expressed the satisfaction of ranchers when he observed: "The industry has now returned to logic and fact.... The future outlook has changed, the fundamentals of the industry are now sound."[45]

Statistics underscore the facts presented in this brief narrative of the transition from the Depression to the controlled expansion of the war years. The figures are reviewed at several scales: for Canada; the province of Alberta; P. Burns Ranches; and for the Bar U Ranch in particular. It took the first four years of the war to build up Canadian cattle herds to the numbers which they had reached in the early 1930s.[46] From 1936 onward, a substantial flow of young cattle from Canada to the United States developed. This proved to be both a lucrative and convenient market for western cattlemen during the early 1940s. The Canadian embargo disrupted this flow for four years. Exports of frozen beef to the United Kingdom are shown in terms of the numbers of cattle the shipments represented. Finally, the flood of cattle moving toward Chicago when the embargo was lifted in 1948 demonstrated that the United States was the market of choice for western ranchers. To sustain these high volumes of exports to both markets without substantially reducing the number of cattle on Canadian farms was a considerable achievement. In Alberta, too, the total number of cattle on farms remained relatively constant at about one and a half million head. However, the value of this stock showed remarkable growth. The provincial herd was worth only $3.8 million in the depths of the Depression, but cattle and calves were valued at $15 million in 1939, and $49 million by the end of the war.[47] The removal of price controls and the embargo on exports to the US had a marked effect on prices in 1948. P. Burns and Company sustained heavy losses for a decade from 1928 to 1938, which amounted to a cumulative total of $0.8 million on net taxable income. These losses were offset by profits earned over the next eight years which amounted to some $1.7 million.[48] At the Bar U Ranch, the total herd size grew modestly during the first years of the war to peak at over seven thousand head in 1945. More importantly, the "calf crop" swelled to more than two thousand head in 1945, 1946, and 1947. Both gross income and profits reflected the unflagging demand for beef at home and abroad and rising prices. The fact that profits grew more slowly than did gross income reflected the fact that changes were taking place in the relative importance of each of the factors of production.[49]

Labour Costs Go through the Roof

The outbreak of the Second World War in 1939 drew thousands of westerners from unemployment lines into the armed forces or the factories. Over a relatively short period there was a profound shift in the balance among the factors of production, land, labour, and capital. As labour costs rose, so mechanization became increasingly attractive. Moreover, as prices rose, farmers had the wherewithal to repay debts and purchase machinery.[50]

The same significant and irreversible trends affected the ranching industry. In an analysis of the costs of raising beef written in 1938, Thomson did not even include labour costs as a significant variable in his calculations. His major concern was with the growing investment necessary to control range lands.[51] However, Chattaway, writing only a few years later, drew attention to the way things were changing:

> ... years ago, labour costs were relatively small because there was little farming done for feed production. The cattle were handled in large

> numbers on the range, and much of the labour was of a cooperative nature, each rancher helping his neighbours on roundups and the like.... Today, his labour costs have risen until they represent the largest item on his expense sheet.[52]

Ranch wages rose from a base level in 1939, when they were lower than they had been in the late nineteenth century, and increased 200 per cent by 1946. This dramatic rise in labour costs prompted ranchers to experiment with new ways of doing things. As cattlemen struggled to attract and keep competent hands, men were replaced with machines wherever possible. The competitive advantage of larger outfits was challenged by highly capitalized family ranches on which the operator and his family did all the work with the help of machinery. In general terms, the struggle to contain rising labour costs has continued to the present. Knud Elgaard published the results of a survey of ranch practices in 1965 and reported that:

> The total amount of labour used per ranch has not changed since 1940 in the foothills region, even though cultivated acreages have doubled and stocking rates increased during this period. The use of hired labour decreased by about 20% during the 25 year period [1940–1965] with the operator and his family making up the difference. In the shortgrass region total labour expended per ranch has decreased by one third since 1940; most of the decrease took place in the hired labour category. The more effective use of labour on ranches in both regions is no doubt due to increased mechanization of ranch operations that occurred during the 1940–1965 period.[53]

The crucial and dramatic changes in this ongoing story took place during the war years, and nowhere were they more far-reaching than at the Bar U Ranch.

In April 1939, Tom Farrell circulated a notice to his ranch managers that wages for the men were to be $25 a month.[54] This was comparable with what the government employment office was paying, roughly equivalent to today's "minimum wage." It was less than had been paid to riders in the 1880s and 1890s, when a dollar a day had been the going rate. However, during the Depression, the prospect of a roof over one's head, hearty meals, and some remuneration, must have made the Bar U cowboys the envy of the rural work force. Much the same rate of pay applied to cooks; for example Ethel Rothwell was offered $30 a month for cooking for the haying crew at the "44" Ranch.[55] However, it was not enough to keep Mrs. Dobson at the Circle Ranch. After only two weeks there she wrote, "I don't find it very satisfactory, for only one reason. So far, it is rely [*sic*] lonesome, my wages are not enough to be able to get a car...."[56] She requested a raise to $45 but was turned down.

Wages were raised on all the ranches in 1942, and an even more generous pay scale was introduced in 1943. This resulted from a conference between J.H. Kelly, Alex Newton, and John Burns which was held in response to a telephone call from Farrell. The latter must have told the senior managers that it was becoming impossible to find and retain competent hands at the current wages. Riders, teamsters, and general labourers were to receive $70; tractor men $85; threshing gangs 60 cents an hour; and foremen $100.[57] This pay scale must have proved competitive, for it remained the same until the end of the war.

As enlistment drew more and more men from the countryside, it became difficult to retain an adequate labour force. Farrell wrote to the Department of War Services on behalf of Tom Feist, who had run a hay crew at the Bar U for several years. "It is a matter of extreme difficulty to secure experienced men to carry on our extensive operations and we believe that there is ample justification for the postponement of this man's military service."[58] In another letter on behalf of Henry Hughes, Gray wrote, "… we would appreciate having the privilege of retaining this experienced engine man."[59] In spite of some favourable responses from the authorities, key men remained in short supply. Kelly commented in November, "We are in the throes of a belated harvest with incompetent help…."[60]

Of course, the social upheaval of the war years sometimes brought men or families to the attention of the company. Roderick Mackenzie wrote looking for a job tending sheep as soon as he was discharged, and was soon at work on the C.K. Ranch. Douglas Hatt applied for a job from Oxford, Nova Scotia, and cited his experience as a rider in the Yorkton area of Saskatchewan before the war. William Bull had worked under Mr. Mayhew, the gardener at the Bow Valley Ranch, and on the E.P. Ranch. He now wrote, "If possible and you can give me a place there, I would like a job with Raymond Clifford at the Bar U."[61] Sometimes, mistakes were made with hiring. Farrell received one letter from the provincial jail in Headingley, Manitoba, requesting him to send the ration book of an ex-Bar U man who had become a new inmate.[62]

Rising wages for individuals meant huge increases in the wage bill for the ranch as a whole. In 1934, the total annual wage bill amounted to $5,409; in 1949 it was $53,974. Moreover, the costs of housing and feeding the ranch hands increased substantially too.[63] The number of those living permanently at the ranch rose from thirteen or fourteen during the Depression years to around twenty during the war. The busiest seasons were, of course, the summer and fall, during which hay was put up and ever-increasing acreages of grain crops were planted. During these seasons the community at the ranch grew to around thirty people, and this figure did not include the contract hay crews which provided for themselves, nor the Native people who provided temporary help. From 1940 onward more and more emphasis was put on the production of grain and tame hay. This led to a considerable rise in the number of hands required.[64] Threshing the grain crops in October extended the busiest period of the year, as it involved a large number of temporary workers. It seems that the potential savings in labour costs which might have been expected because of the mechanization of some operations (ploughing, sowing, and swathing, etc.) were more than offset by the expansion in the scale of farm operations which mechanization encouraged. The greater the acreage cropped, the more men were required for stooking and threshing.

During the 1940s, the Bar U finally set aside the last vestiges of the open range methods which had lingered so long. Valuable stock were nurtured with grain feed and tame hay, and breeding and calving were carefully monitored to maximize production. The huge spread had become a vast stock farm. More labour was required to expand farm operations, but costs were rising rapidly and experienced men were in short supply. It was in this context that the contribution of Native people to the success of ranching in the foothills, and the Bar U in particular, should be evaluated. [Plate 7, colour section]

Help from the Stoneys

During the Second World War, when skilled labour was at a premium in the foothills, many ranches benefited from their historic connections with local Indian bands and with specific Aboriginal families. Indeed, one rancher expressed the opinion that, during this period at least, the Indians "carried a lot of the ranches."[65] It is perhaps convenient to look back from this period when the Native contribution to the Bar U Ranch was so overt to review the decades during which their presence was just as real, but was largely taken for granted and unremarked. The roots of the relationships between Indians and ranchers reached back several generations, for Indians had provided a pool of temporary but skilled labour to help with branding, fencing, and hay making since the first decade of the twentieth century. These activities were compatible with many of their traditions, although their hunting and collecting life had changed for ever. Above all it meant that the family and the group could stay together, and the heads of families did not have to leave for alien jobs in urban centres. For their part, the ranchers had provided a secure place to camp, an opportunity to make a few dollars, and the occasional lump-jawed cow. Sometimes too, the ranchers were able to act as "go-betweens" to smooth relations between Natives and the authorities

Many ranch families had close personal ties with Indian families. Both groups were deeply attached to the foothills country and shared common memories of growing up together. Nowhere was this more evident than at the Mount Sentinel Ranch, where at least two generations of Gardners grew up with the Dixon and Bearspaw families of the Stoney tribe.[66] Mrs. Sarah Minnie Gardner started taking photographs of Stoney visitors and workers during the first decade of the twentieth century, and her wonderful collection spans about forty years. The Wilson and Dixon families are another case in point. After a lifetime of working together and mutual respect, Peter Dixon was buried in one of Stan Wilson's suits.[67] In a similar fashion, Allen Baker's family grew up alongside the family of Webster Lefthand, who lived in the log cabin at the east end of the Bar U headquarters site during the 1950s. Indeed, Webster named his children after Baker's, which caused some confusion as the genders did not always match![68] Josephine Gardner explained that, although Indians would move from ranch to ranch as work dictated, each family tended to have special ties with a particular ranch which were sanctioned by the band. The Rider family were associated with the Bar U, the Bearspaws with the Bedingfelds, and the Dixons with the Gardners.[69] These special associations lasted into the modern period. However, extended families of the Stoneys were large, and this meant that some branches of a family with the same name might be found at more than one ranch.

Employment records from the 1940s show that the management of the Bar U was able to draw on the services of the same group of Indians to provide a wide range of tasks.[70] While the non-Native labour force displayed a rapid turnover, the same Stoney family names occur again and again. Jonas Rider was the undisputed long-time leader of the Stoneys in their dealings with the Bar U Ranch. For more than thirty years he provided skilled and responsible help with cattle and horses. He led by example and taught new generations of young Indians all he knew of roping and managing cattle. Much of the work contracted to the Indian community through the 1930s was managed by Jonas. The other family head who received a number of fencing contracts during the 1940s was Peter Dixon, and family members Neil and Judea Dixon were faithful workers throughout the period. Webster and Eddie Lefthand were representative of a third family which made a major contribution to the ranch, while the Bearspaw and Powderface families were well represented. Peter Pucette, Fred Satler, and Wallace Ear rode for the ranch from 1946 through 1949, and Peter introduced his son to the work with cattle during the latter year.[71]

The continuous presence of Indians on and around the Bar U Ranch is attested to by the fact that two respondents to an oral history project undertaken in the early 1990s mentioned the fact that Fred Stimson, who left the ranch in 1902, was a "good friend to the Indians."[72] This suggests that Stimson's interest in Indian culture, and his championship of their cause, had become a part of Stoney oral tradition. Certainly, Mrs. Leitch, who visited the ranch with her parents around the turn of the century and stayed with the

9.1 First Nations people made a great contribution to the success of ranches in the foothills. They provided a pool of skilled labour for specific tasks and were on call when additional hands were needed. Facing page, left shows a Stoney encampment on the Mount Sentinel Ranch c. 1910. Parks Canada, R.W. Gardner Collection, 95.05.01.04

9.2 Facing page, right shows Webster Lefthand and his children at the Bar U, c. 1947. He became Allen Baker's foreman after the sale of the ranch in 1950. Parks Canada, Joe Hayes Collection

9.3 Cows and their calves held outside the branding corrals at the south end of the Bar U on Stimson Creek. Of the more than twenty riders, more than half would have been First Nations people. Parks Canada, Alwood Rousseau, Rosettis 6397-50, 94.12.01.17

9.4 Branding cattle at the Riggs Place, seven miles north and west of the headquarters site. Jonas Rider was still roping cattle in his old age and several of his family were in the crew. The scene in the corral looked much the same in 1943 as it had in 1919 and in the 1890s. Parks Canada, Joe Hayes Collection, 96.06.01.05

Stimsons, remembered a family of Blackfoot Indians who lived in a tipi quite close to the main ranch house.[73]

The first incontrovertible evidence of Indians playing an important role in the Bar U labour force comes from a series of photographs taken in 1918. These pictures of branding show Native people performing a variety of tasks. A rough count suggests that Indian men made up about half the work crew on this occasion. It was not long after this that Johnny Lefthand visited the Bar U for the first time. His father was working on the roundup and Johnny remembered that they used a chuckwagon out on the range, which he said was more or less unfenced all the way to Milk River.[74] Mark Lefthand remembered being hired and paid by Neils Olsen, who was ranch manager until 1926, while Hugh McLaughlin reported Stoney Indians were camped close to the Bar U at the end of the Lane era.[75]

Mary Wildman, daughter of Jonas Rider, was born on the ranch and grew up there during the 1930s.[76] Thus the pictorial record and oral history hint at the continuous presence and participation of Indians in the life of the Bar U in the interwar period. Elizabeth Lane was touched that "many Indians who had known him a life time" came considerable distances to attend her husband's funeral.[77] In its front-page coverage of Lane's funeral, *The Albertan* mentioned the hundreds of people who had braved the blizzard to attend. The account continues, "on the fringe of the crowd near the entrance of the hall were noticed several members of the Blood Indian tribe, for George Lane was well known to the Bloods and a prime favourite with the older men of the tribe."[78]

During the 1940s, employment records show that Natives were working on the Bar U almost year round. Fencing and brush-cutting contracts were taken up in April as the snow melted. For example, in April 1942, Peter Dixon and his family were paid for "70 loads of brush cut and hauled to the dam," and for two hundred fence posts hauled to the Riggs place.[79] Later, the postholes were dug, wire was stretched, and gates were built. The tempo of work increased as preparations were made for branding. In early June, Jonas Rider and thirteen men put in a day working on the branding corrals and adjacent fences.[80] If the weather in fall and early winter permitted, fencing contracts might continue until Christmas. December 1942 found Jonas Rider and a party fencing haystacks and building a fence around the farm buildings.[81]

Branding was the high point of the ranching year, and skilled help from the Stoney Indians was vital to its success. Lee Alwood recalled one day during the

1940s when they were shorthanded. He sent a rider galloping over to the Emerson Flats, a few miles up Pekisko Creek, to the Stoney encampment. He returned in short order with sixteen men and the branding got under way.[82] Jonas Rider was a very skilful roper and became a fixture at Bar U brandings.[83] When he was quite an old man, while still recovering from a cruel bout of pneumonia, he continued to employ his uncanny skill at heeling calves, even though the young men had to help him hold them. Mark Webster and Johnny Lefthand also worked many brandings at the Bar U. Often, branding was done at the Rigg place, some seven miles northwest of the headquarters site. Here, the Indians would set up camp at the "Red House Ranch," as they called it. While branding involved more men for a longer period than did any other assignment, the Indians relished all jobs which involved riding.[84] They are mentioned as "moving bulls to Flying E Ranch;" "working with cattle;" "moving horses;" and "handling calves." A cattle drive in November 1946 involved four men for thirteen days.[85]

The records show very few cases of Indians being paid for work done during August and September. The 'tradition' of attending the Calgary Stampede and then spending the summer hunting and fishing in the mountains may well have carried over from the 1920s into the war years. This summer period was when several contract haying crews sweated to cut and stack thousands of tons of wild hay on the Bar U. There is no evidence of Native hay crews working on the ranch during the war years, although Tom Feist, a regular hay contractor, said that he hired individual Natives from time to time.[86] At a neighbouring ranch, the Gardners used Natives to put up hay, and Dick Gardner remembered that one of his best haying crews was made up of Indian women.[87] The Stoneys returned to the Bar U in October and helped with stooking and threshing.

Until I had an opportunity to work with the detailed accounts in the Clifford/Alwood Collection I had assumed – perhaps with a measure of cynicism – that Natives were given "pocket money," food and camping privileges, rather than being paid for each hour of work rendered. Such was not the case. Contracts were worked out to the last 6-cent fence post, and the day wage rates for whites and Indians were much the same. However, non-Natives received bed and board.[88] Of course, the advantage for the ranch was that Indians were only paid for work done, they could be called up when the weather suited. Sometimes whites enjoyed their board and got little work done. As wage rates rose rapidly during the war years, the amounts paid to Natives kept pace. In 1942, collectively they earned a total of $1,411.77 at the ranch, and this figure had risen to $2,297 in 1947.[89]

9.5 The group of tents in the middle distance show the Stoney camp at the Bar U's Riggs place about seven miles north of the headquarters. The cattle held in the foreground had been collected and held ready for branding, c. 1943. Parks Canada, Joe Hayes Collection, 96.06.01.07

9.6 These young men are typical of the skilled and highly motivated riders who helped the Bar U Ranch during roundup and branding. Parks Canada, R.W. Gardner Collection, 95.05.01.07

The manner in which Indians were paid changed during the 1940s. In 1942, most money was paid to Native foremen, particularly to Jonas Rider and Peter Dixon, and it was their business to distribute wages to other heads of families.[90] In all probability, this was the way matters had been arranged for the previous two decades. However, by 1947, Indians were treated as any other employees; their names were listed in the time books and individuals received advances on their wages when they requested them.[91]

Since Stoney Indians played a significant role in the life of the Bar U Ranch for such a long period, it is not surprising that people's memories of them provide a somewhat confused picture. One informant will suggest that their camp was in one location while another will point to an alternative location. Most of the vivid images evoked by David Finch's prompting during the Oral History Project can be reconciled with one another by recognizing that they are like still frames in a moving picture. Indian culture was as dynamic in its own way during this period of time as was "white" culture.

For most of the forty-year period ending in 1950, the Stoney Indians who worked on ranches in the foothills maintained winter homes on the Morley reserve. Some families would appear on the ranches in March and April to help with calving. Then the busy roundup and branding season would begin, involving the greatest numbers of Indians. Some would return to Morley toward the end of May to put in gardens, attend the local rodeo, and prepare for the Calgary Stampede. Families would come back to the ranches to take up fencing contracts through the fall, but it was not until the 1940s, prior to the establishment of the Eden Valley Reserve, that some families wintered over close to the Bar U.

The journey backward and forward between Morley and the Bar U would take three or four days, and the route followed the general trend of the present road to Bragg Creek. Families would move by horse, packhorse, and wagons. George Zarn recalled:

> It was quite a sight to see a bunch of Indians moving camp, several democrats and pack ponies, kids from two years up riding little ponies, dogs of all sizes and breeds, but mostly coyote hound type.[92]

The Indians camped on the Emerson Flats while working for the Bar U and the E.P. Ranches.[93] This area, two or three miles upstream from the ranch, comprised a number of open terraces along Pekisko Creek, and got its name from the fact that the famous pioneer cattleman, George Emerson, had homesteaded there during the 1880s. Over the years their occupation of this area became more formalized. Lee Alwood spoke of the Stoneys being "allowed to keep 80 head of horses there."[94] When particular families were working for the Bar U for shorter intense periods, they moved closer to the ranch and camped just across the bridge to the west of the headquarters site on both sides of the road. Jonas Rider and his family had a favourite campsite on the top of the bench where McPherson now has his ranch house.[95] Other respondents mention a camping place downstream from the Bar U in the "bull pasture."[96] It seems unlikely that Stoney Indians camped at the eastern end of the ranch site, where the visitor centre now stands, during the Lane or Burns eras. This area was occupied by a corral complex for exercising and schooling Percheron horses. While they were working at the Bar U the Stoneys did their own cooking, but they were provided with some staples, particularly flour, sugar, and meat. At particular times they shared the produce of the ranch garden and eggs from the chickens. They also ate wild meat and fish from the creek. Hugh McLaughlin remembered the Indian camp as being made up largely of tipis with only a few tents. The photographic record also provides evidence that tipis were often used.[97] However, later observers remembered the Indians living in walled tents, sometimes with ingenious arrangements for stove chimneys. Johnny Lefthand explained that his family used tipis in the mountains when they were smoking meat, but preferred tents at the Bar U.[98] Tipis were retained for ceremonial occasions like the Calgary Stampede long after their general use had disappeared.

The Stoney presence had a marked effect on game and predator populations in the foothills ranching country. They reduced the number of deer and elk, and shot coyotes for their skins. Native winter traplines brought in beaver, mink, and weasel.[99] It was King Bearspaw who, with Oscar Brandham the head rider at the Bar U, finally tracked down the huge black wolf which had been preying on cattle on the lease to the west of the Highwood River. It took the two men a week to trail the animal and get a shot at him. He measured about nine feet when he was stretched out, and earned the pair a $90 bounty.[100]

Some of the Indians associated with the Bar U Ranch achieved fame for their cowboying skills at national and international rodeos. Jonas Rider had a successful career in competition, and was calf-roping champion at the Calgary Stampede in 1924. Johnny Lefthand competed from the late 1920s onward. At first he was a bareback rider; later he was an outrider and chuckwagon driver. In 1932 and again in 1944 he was champion wild cow milker. In 1939, he accompanied a group of cowboys from Alberta on a trip to Australia to an International Rodeo in Sydney.[101] Johnny took the honours as all-round cowboy, but broke his arm doing so. He remembered the voyage via Hawaii, Fiji, and New Zealand with nostalgia, but said he quickly tired of the ocean, "everyday nothing but water." Johnny said that he used to train for bronc riding rather differently from today's cowboys. He and his friends practised by pulling down a clump of willow brush, strapping a saddle on it, whereupon several young men would pull it down and let it rebound upward while he tried to ride it! He got into shape by pounding willow fence posts with a sixteen-pound post maul.

An employer is likely to evaluate the potential labour force of a particular town or region using five criteria. He will ask himself: what is the potential availability of workers? how reliable are they likely to be? what skills and experience do they have? what competing jobs are available and what is likely to be the turnover among workers? and finally, how flexible will the workforce be? will it be possible to shift easily from one task to another? Judged in these objective terms, the Stoney Indians who served the Bar U for four decades must receive a very high rating. Because of their desire to maintain some elements of their culture they were drawn to ranch work. They made available a pool of temporary labour that could be deployed quickly to meet a variety of needs. Over a long period of interaction, Native 'foremen' and ranch managers got to know and trust each other. Men like Jonas Rider and Peter Dixon could be relied upon to carry out tasks with a minimum of supervision. The Indians brought to ranch work skills which they had acquired while growing up and an empathy toward horses and cattle. The records demonstrate a very low turnover, and one generation learnt from their elders the nuances of what was expected. Native workers were flexible. They preferred working with stock, but fencing contracts had the advantage of involving the whole family and were carried out independently. Working in the harvest fields was the least popular activity, and there were sometimes difficulties due to different cultural attitudes as to time.[102] Extant accounts show that Indians were paid at much the same rates as non-Natives, but the attributes they brought to the work meant that they earned every dollar they were paid. During a volatile time of labour shortages and soaring labour costs, when the factors of production began to favour smaller family ranching outfits, the Native labour force did much to ensure that the last decade of the Bar U Ranch's corporate existence was one of triumphant profitability.

Hay, Crops, and Horse Power

As we have seen, haying was part of the seasonal round at the Bar U from the very beginning. George Lane used the ranch primarily as a cow-calf operation and thousands of tons of hay was fed to weaned calves and weak cows. In addition, as the Percheron stud was developed, the stallions, mares in foal, and colts required a balanced and high-energy diet. Tame hay, green feed, and grain were used in increasing quantities. We have already described Pat Burns and Cornelius Duggan's innovative methods of winter feeding steers for the spring and early summer markets. These cattlemen were never satisfied that they had enough hay put up! By the 1940s the Bar U routinely stacked forty-five hundred tons of wild hay in a given summer. Moreover, ever-increasing quantities of tame hay and grain crops were grown for feed. This meant that the entire cow-calf herd could be fed through the winter should it be necessary.[103]

We are used to regarding hay as a crop like any other, part of a planned rotation of grain crops which might include wheat, barley, and clover, alfalfa or timothy. It comes as something of a shock to us to comprehend that almost all the hay put up at the Bar U throughout its long history was native hay. Huge acreages of flat land were protected from heavy grazing and the natural grass was "harvested" every second year. A wet year might provide a thick hay crop two years running and, where irrigation water was available, hay could be cut every year. The rough fescue, or 'upland' grass, made the best hay. It cured on the stalk and could be cut and moved to the stack within a few days. "There has never been better hay cultivated or wild. Cattle could fatten on it; thin work and saddle horses turned out in the fall to rustle during the winter would come in 'rolling fat' by spring...."[104] Sloughs and wet river bottoms produced a far greater volume of grass, but they had to be cut at precisely the right time and the grass they produced did not have the nutritional properties of the rough fescue. Hay fields between Pekisko Creek and Stimson Creek, lying to the east of the ranch headquarters and surrounding the Bar U farm, were cropped for forty years with no diminution of their bountiful returns. In the Big Hole country of Montana ranchers still rely on wild hay, and the following brief description would have applied equally well to the Bar U.

> Most of the hay meadows in the Big Hole have neither been tilled or sown in decades.... Year after year, the soil carries timothy, Redtop, blue-joint, nut grass, a couple of fescues, creeping and meadow foxtails, clovers, and bromegrass, as well as some miscellaneous rushes and sedges.... During Russ's fifty-something years in the Big Hole, the hay crop has never failed.[105]

Haying was one of the most traditional niches in the wide spectrum of agricultural activity. It was a labour-intensive activity which involved a long-lasting partnership between men and horses, for haying used horse power, wood, ropes, and tackle and a lot of human ingenuity. The operation consisted of four linked tasks: mowing, raking, gathering, and stacking. Mowing was the first to be mechanized. Fred Ings described haying on his OH Ranch during the 1880s:

> The grass was so good in those days that the cattle grazed out on the open range all winter and it was possible to roundup prime beef any time. We didn't wean our calves then and only cut what hay our saddle horse needed. To cut this we used an American make of mower, the Walter A. Wood, a high geared machine and the only one in those days capable of cutting the high, thick, wild grass.[106] We used basket racks. The hay was pitched by hand and stacked by hand.... Even later, when more was put up, we used the same primitive methods.[107]

Mowing was not a job for the faint of heart. The horses at the beginning of the season were virtually unbroken, the terrain rough and pockmarked with gopher and badger holes. The driver's seat was mounted on a four-foot length of spring steel, and the ride was like straddling a frisky bronc. Heavy grass frequently jammed the mower and the viscous sickle blades had to be cleared. Moreover, it was the driver's responsibility to grind the blades at least once a day. The cut hay was whipped into windrows using dump rakes. These rows were then collected with an ingenious homemade buck rake or "sweep." These had teeth about ten to fourteen feet wide attached to a backstop some four feet high. The driver sat at the back and controlled the angle of the teeth. Horses were harnessed at the ends of the rake and walked on either side of the windrow picking up the hay which curled up against the backstop until a full load was obtained. Then the driver raised the teeth and drove the team slowly to the site of the stack. Some buck rakes slipped sledge-like over the ground, others had small wheels. The final stage of haying involved stacking the hay so that it would be more or less protected from the weather and the depredations of cattle or elk and deer. Well-stacked hay would retain most of its food value for at least two years. During the 1930s, stacks were often topped with willow brush to keep the hay from blowing. The higher the stack, the less hay was exposed to the elements, and some of the stacks made each year at the Bar U were monsters. The

9.7 Using a push-pole stacker to build a hay stack at the Bar U farm. Notice the hay sweeps with a Percheron horse at each end and the driver standing on a plank. Parks Canada, Olsen Family Collection, 96.03.01.53

men who worked up on the stack were the most skilful and experienced in the whole operation. It was their job to spread and trample the hay in such a manner as to preserve a firm coherent shape as the stack grew and then to build the top thatch to shed water.[108]

A variety of techniques of increasing sophistication were used over the years at the Bar U to lift the hay to the top of the stack. Indeed, Jordan suggested that "hay-related material culture represents the major innovative western component of the built landscape...."[109] The simplest labour-saving aid was the slide, constructed of planks laid on a frame at a relatively slight angle up which the buckrake could be pulled to dump its load. This method could build a long, low stack quite quickly. It was replaced by the push-pole stacker, which used a team of horses to push a "plunger" up a steeper incline. This allowed higher stacks to be constructed, and was used at the Bar U during the 1920s and 1930s. Tom Feist, who put up hay on contract for the Bar U from 1942 to 1950, remembered using a push-pole stacker for the first year before changing to an overshot stacker, which emulated the lifting action of a man with a fork. These overshot stackers required fewer men to manage them and came into general use as ranchers responded to the labour shortages resulting from the Second World War. About a month after the stack had been completed it was carefully measured. The volume was calculated by measuring the length, width, and height of the stack after it had had time to settle. Tom Feist commented, "Tame hay was divided by 480 cubic feet to the ton, and Burns had its own measurement of 512 cubic feet per ton. Burns was getting its tonnage...."[110] Finally, the stack was fenced and a fire guard was ploughed around it.[111]

Enough has been said to establish that haying was a labour intensive activity. Big ranches could either hire seasonal hay crews or draw up a contract with a trusted local man for hay at so much a ton. The Bar U used both strategies. The labour force more than doubled during the growing season and the "dormitories" at the headquarters and at the Bar U farm were full. In addition, at least one contract crew would be working independently somewhere on the hay meadows. Tom Feist's crew during the 1940s was made up of a dozen or so workers: four men on mowers, two on rakes, two 'sweeping,' two in the stack, and one man sharpening mower sickles. Feist himself worked the stacker. The crew would use twenty or more horses at one time, and those pulling the mowers had to be changed frequently. Feist built up a horse herd of his own as well as borrowing animals from the Bar U. He needed at least thirty horses during a busy summer. The crew lived in a mobile bunk car out where they were cutting, and were fed from a cook car. The long day started at 7 a.m. and went on until about 6 p.m. One year Feist and his crew put up twenty-three hundred tons of hay, but the average figure was between fifteen hundred and two thousand tons.[112]

Although an efficient hay crew depended on the skill and experience of a few key men, there was room for any willing help. Many young men earned their first wages driving a rake or managing a team at the stack.[113] Horse powered haying was a shockingly dangerous occupation.[114] Broken arms and legs from runaways were common. Like the deck of a sailing ship, the haystack was a place of ropes, wires, and tackle blocks, any of which could catch a wrist or ankle or remove a finger. Oliver Christensen, who worked at the Bar U for several years, remembers the excitement of a runaway but not the fate of the driver:

Our haying crew only had one runaway that summer, and it was a good one. A big sweep rake was driven by a young Indian kid named Henry Aker who was about fourteen years old, just a small young kid. He was a nephew of Nat Cayenne, an Indian from across the border somewhere who worked with us that summer. Henry's job was to follow along and sweep up all the loose hay left on the ground after the windrows had been collected and hauled to the overshot stacker. It was dinner time and Henry had the sweep rake down at the end of the field on the east side of the farm buildings, down toward the store run by the Baines family, the Pekisko Trading Post. Other fellows had already unhooked their teams and were starting for home. Henry's team caught sight of them and decided it was time to go too. They knew the routine, and Henry didn't have a chance: those horses started running and he couldn't slow them down, let alone stop them. Away they went, racing for home with that fourteen foot sweep bouncing along behind and Henry hanging on for dear life. I don't know how he stayed on. All the Bar U horses knew that when they went home. They always went into the corral. That was how they were trained. The only problem was that the gate to the corral was only twelve feet wide. The horses hit the gate dead centre, and that fourteen foot sweep just folded into a Vee. I can't remember what happened to Henry, but he probably got thrown off when the rake hit the gate.[115]

A cow-calf operation needed hay in large quantities not to fatten stock but rather to maintain breeding cows and weaned calves in robust health during the winter. Without some shelter and feeding many calves would be lost to exposure, and even those that survived were liable to lose up to one third of their body weight.[116] The system used at the Bar U aimed at preventing weight loss, and during most winters modest weight gains were realized. The amount of hay required in a given season depended not only on the tonnage left over from the previous season, but also on the amount of alternative feed available. The Bar U farm produced large quantities of grain and green feed which was fed to stock along with hay. This expansion of farm operations was another facet of the transition from extensive to more intensive ranching.

9.8 By the Second World War tractors had replaced horses in several crucial operations in the fields. Here a crawler tractor pulls binders at the Bar U farm. Parks Canada, Oliver Christensen Collection, 94.07.01.07

George Lane had broken some fifteen hundred acres of the Bar U in the years leading up to the First World War.[117] The cropped acreage had fallen to less than a thousand acres during the 1930s. However, a crawler tractor was acquired in 1938, and, as grain prices rose due to wartime demand, wheat was sown extensively again and the tilled acreage expanded. There were two thousand acres in production by 1941 and thirty thousand bushels of wheat and coarse grain were threshed. Such was the urgency to break new land that Lee Alwood ploughed all night on occasion, pulling three ten-foot double disks.[118] Land broken one year was seeded the next. Night ploughing was an unusual measure, a response to the relatively short season during which conditions were optimal for cultivation. By the mid-1940s about thirty-five hundred acres were cropped at the ranch. The output of wheat remained fairly constant at between twenty-five thousand and thirty thousand bushels; oats were the single most important crop; while barley and rye played less significant roles. The fact that the Bar U was located close to the margin of the area in which wheat could be grown with some expectation of success is brought home by the fact that in 1943 and again in 1948, there are references to hundreds of tons of frosted wheat bundles available for feed. A letter to the Canadian Wheat Board outlined where Burns Ranches grew grain and the purposes for which it was used. It is worth quoting in full:

> Most of the company's wheat production takes place immediately adjacent to Calgary at which point we are very large wheat and grain producers.... We will very shortly have fed all the grain we have grown in the Calgary area. We have some grain in store at the Bar U, which is a 70 mile haul, however, our practice has always been to keep large reserves of grain on this ranch in order to meet the varying feed conditions of the future.
>
> This company is well equipped and feeds some thousands of head of hogs and cattle annually and it is both willing and anxious to continue to do so during the present period of great demand for home and overseas use. You will, I am sure, appreciate that it cannot continue to do so unless feed can be purchased at the price grain is available to other feeders. The company is presently feeding at Calgary approximately 1,000 bushels a day. The company would be grateful for an early reply as the situation is urgent if we are to continue feeding hogs and cattle on the various ranches.[119]

At the Bar U, calves were wintered in corrals at the headquarters site and fed hay and a ration of grain from the time they were weaned in October. It was usually possible to delay feeding mature stock until after Christmas. Then the hundred or so bulls would be fed hay in the morning and rye or oat bundles in the afternoon.[120] Their feeding grounds were along the flood plain of Pekisko Creek to the north of the corrals. In many winters the heifer herd and the cow herd managed very well on their winter range and needed no supplementary feeding.[121] But then would come a bitter winter when the snow piled up and the chinooks refused to blow. Gordon Davis had good reason to remember the winter of 1936–37 because he suffered the biting cold every day getting feed for the herd. It snowed from Christmas Day until mid-February. The cattle were brought in from the hills and fed in the field to the south of the headquarters site. Each day the stock consumed thirty-eight tons of hay and five hundred bushels of oats. This meant taking teams out to the stacks and bringing back loads of hay or bundles of oats, often the haul was two or three miles one way.[122] In a normal year, feeding was handled by three or four hands picked from the hay or threshing crews, but in an emergency everybody helped including riders, like Davis, and the choremen. The winter of 1948–49 was another memorable one. There was plenty of hay and feed put up, the problem was getting out to the stacks to collect it. Caterpillar tractors had to be used to cut a passage out of the ranch site onto the fields, and then teams of Percherons strained to pull sledges of hay to the sheltered feeding grounds. As the winter progressed and made all movement a battle, Alwood turned the heifers out to graze on some old stacks. He recalled:

> Down south in the old HL I had these heifers in there and they'd eat underneath [the stack] and make a big toadstool. I used to go down there every morning and run my truck into them to push them over. This one day I couldn't get this one to tip over. I tried every way I could. The next morning I come down and seen it was over.

> There were two dead heifers under there – it had fallen on top of them.[123]

The fact that Alwood could refer so casually to using his truck on a daily basis hints at the progress that had been made toward mechanization. In 1940, there was only one half-ton truck and one TD 40 International tractor on the ranch. By 1949, there were four trucks, two of them one-tonners, and four tractors. Horse-powered haying equipment was being replaced with power-driven machines; there were three power mowers, a hydraulic hay sweep, two hay balers, and a power binder. A welding outfit and compressor had been acquired to help keep the new machines running. Power saws, a post hole digger, garden cultivator, milking machine, and grain treater all contributed to a reduction in the time taken to complete necessary jobs around the ranch.[124]

Experience at the Bar U, and with Burns' ranches in general, emphasizes the fact that mechanization was a gradual process which took place surprisingly late. It was not embraced wholeheartedly until the mid-1940s.[125] On most farms, during a long transition period, tractors and draft horses co-existed, each being employed for the tasks at which they gave the best return.[126] Nor was the march toward mechanization irreversible. Changes in the balance of costs and returns produced a swing back toward horse power during the 1930s. As Alberta entered the Depression the demand for draft animals remained strong. Farmers weighed the low price of wheat against the high price of gasoline and continued to opt for horse traction. Horse breeders suggested that: "The proper tractor is one that can be manufactured on the farm. It is better to raise one than mortgage one's soul for gasoline."[127] While the Livestock Commissioner explained:

> Farmers of late have been relying on motor power, and therefore have not been raising many colts. In the past year a great many of them were forced back into using horses and this created considerable demand.[128]

If the adoption of tractors and mechanization in general was a slow process on the farms of the Canadian west, on ranches movement in this regard was positively glacial. Ranches had an ongoing need for horses and the ability to run them at low cost. Saddle horses were required to manage the cattle herds and could be pastured alongside cattle at very little extra expense. The necessary skills for rearing and training horses were not only present on ranches, they were cherished and venerated. To a farmer there were real savings to be made from replacing his teams with a tractor, for the oat field and the pasture could then be used for cash crops. Moreover, tractors did not require rest and could meet the challenge posed by a short growing season. On the ranches of the foothills the same pressures for change were not present.

The hidden cost of maintaining horse herds on ranches was reflected in the labour cost. As long as wages paid to riders and hands remained at the same level or below what they had been paid during the nineteenth century, as was the case during the Depression decade, there was little need to meddle with the status quo. The Second World War changed this. Competent help became hard to find and ranch wages rose rapidly in an attempt to compete with those offered on the oil fields and in urban occupations. Mechanization became desirable on all but the smallest ranches, and the high prices received for meat and grain during and immediately after the war provided the ranchers with funds to make the changeover.[129] The total number of horses in Canada and Alberta fell only slowly between 1921 and 1941; indeed there was an increase in the number of horses on ranches between 1938 and 1941.[130] However, the next decade saw a precipitate decline. The number of horses per ranch dropped from an average of twenty-two in 1940 to only seven in 1965. At the same time investment in machinery had increased two-and-a-half times.[131] As we focus our attention on the rearing, training and working of horses on the Bar U Ranch during the 1930s and 1940s, we should be aware that our particular case has much to tell us about the general process of change in agriculture.

On the Bar U Ranch the vital role played by horses had changed little since the beginning of the twentieth century. When Gordon Davis was hired in 1931, he explained that there was still almost complete occupational separation between "riders," who tended the cattle herd and the men who worked with the Percheron teams on the farm.[132] The former had their own sparse rooms in the bunkhouse, while the latter slept upstairs in the cookhouse dormitory style. Moreover, mirroring the practice of the open range days, each rider had his own string of horses assigned to him which nobody else would ride. Some of the men would have a horse of their own, but most used horses from the ranch herd. Riders had their own gear, saddles, and bridles, but the ranch bought the lariats.[133] A decade later, when Ray Fetterley signed on, nothing had changed. He brought his mount Kitten with him and was assigned a string of six other horses, one for each day of the week.[134] Young horses were 'broken' or 'gentled' by men who stayed over at the ranch for several weeks at a stretch and were paid for each horse they trained. Bert Sheppard came over from the OH Ranch on several occasions to perform this function, the last time being during the winter of 1931–32. He said that he could handle about five horses a week for $5 a head.[135] Earl Clarke and Slim Martens put in several seasons working with the horses during the 1930s, and later through the 1940s Alex Murdock, Dale Millar, and Dave McGregor were all remembered as horse breakers.[136] These man had a whole repertoire of techniques at their disposal.[137] In general terms, the value of saddle horses had increased and the rougher methods, which had often resulted in fatalities, were no longer acceptable. Purebred Percheron colts and light thoroughbreds were treated with the care and respect that their potential value demanded. There was, of course, no way that a professional horsebreaker could turn out a fully finished cow pony. What they did, in the few days available to them, was to introduce the horse to halter and saddle, teach him to stand while being handled and to tolerate a rider. The assumption was made that riders were quite capable of handling a bucking horse and that each cowboy would train the 'green-broke' young horses in their string according to each horse's potential. Ironically, a bad horse breaker could break a horse in such a manner as to prevent it from ever becoming a good cow pony, while a good initial experience of working with people could do no more than set a horse on the right path. The skill of working cattle was learned by a horse on the job. Sometimes the green-broke young horses had benefited little from their early training. Ann Clifford told of a case in point:

> One time one of the boys got bucked off four or five times and the last time he was dumped on his head in a wheelbarrow. Mac said: "I have had it, I quit!" Raymond told Mac to go pick another horse and to give the horsebreaker a little more time with that one.[138]

Ray Fetterley reckoned that Dave McGregor, the horsebreaker in 1948, usually rode a horse a couple of times and called it broken.[139] Most horses settled down to a work routine, but there were occasional outlaws like one horse in Tom Feist's string which ended up in Madison Square Gardens in New York as a famous bucking horse.[140]

When Oliver Christensen was promoted from the Bar U farm to become a rider in 1945, he inherited the string of horses used by the cowboy he replaced. He remembered Annie, Rex, and Red as being decent horses who were reliable and knew their job. The next spring he got some new recruits. He explained:

> In the spring of 1946 the Bar U hired two horse-breakers to rough break about 30 head of saddle horses for the ranch. I got ten of them in my string. I got to name them all too. Some of the horses were good, but some of them were hammerheads. We riders had to take them out and figure out which of those broncs were going to make good saddle horses. Well, we had a stampede every morning for a while trying to sort that out! Sometimes we'd have to front foot one of them to get on. We'd saddle up in the barn then lead the bronc out to the corral. If he wouldn't let you on his back we'd make a loop with our lasso, walk the horse into the loop at his front feet, then when he reared up, we'd jerk it tight, and down he'd go onto his shoulder. Then we could straddle him on the ground, slack off the lasso, and let him up with you already in the saddle. Sometimes a horse would buck after that, and sometimes he would just go to work. You never knew.[141]

The best horse Oliver ended up with out of this bunch was Cliff, a big, dappled iron-grey horse, part Percheron, part Thoroughbred. He proved to be tireless, and was Oliver's choice whenever he needed a horse for some hard riding, for example up in the mountains.

Work horses were trained in much the same way. Earl Clarke in particular had a precious gift of inspiring confidence in young horses:

> This man could go to the corral and throw a rope around one of these wild carefree Percherons … and within half a day they would be following him around all over the ranch.[142]

Once halter-broken, the colt was hitched to a wagon alongside a big black Percheron gelding named Dan, who weighed about a ton. Between the big horse and the human trainer the young animal was soon shown how to perform. Nevertheless, when Bar U teams were loaned to hay contractors like Tom Feist, they were still pretty wild. After six weeks of hard work in the hay fields they were so well schooled as to be unrecognizable. Feist had a real battle with the ten mules he was loaned his first year as a contractor on the ranch, but he came to swear by them.[143] During the haying season, there would be fifteen mowers cutting hay and another ten teams working rakes and around the stacks. By the time the hay was put up, the grain crop was ready to cut with horse-drawn binders; after that came the threshing, and then a respite for the teams until it was necessary to start hauling feed, sometime in the new year.

George Lane had built up a world-class Percheron stud farm at the Bar U during the years before the First World War.[144] The legacy of this achievement lingered into the Burns era. Although the papers of the 1927 transfer from the Lane estate to Pat Burns do not mention any particularly valuable sires, some purebred mares were sold with the ranch, and the work horses on the ranch were of very high quality. For a time, Sandy Crawford, who had served at the ranch under Lane, continued to manage the breeding program. Records were carefully maintained, and some horses, like Pekisko Mark, were shown with success.[145] However, the Bar U was run as one cattle ranch among several. While the production of young work horses was economically justifiable, the maintenance of an elaborate breeding schedule and a man to administer it was not. Sometime in the early 1940s a less labour-intensive regime was introduced, although purebred Percheron mares were still sent off the ranch to be bred with superior stallions, like Charcoal at Nanton.[146]

Although the Bar U was closely linked to Percherons, it was also home to a succession of Thoroughbred stallions. Desmond Star, Burdock, and Markab were all sought-after sires. Crossed with working mares, they produced excellent saddle horses. Ann Clifford remembered the Tennessee Walker 'Rex Stonewall' best, and the lovely Palomino 'Sun Dog.' These horses accompanied the Calgary Stampeders to Toronto for the Grey Cup in 1948.[147]

Most of the young horses reared on the Bar U went to work on the ranch, or were moved to another of the Burns outfits. However, there were often surplus animals, and these were sold. The reputation of the big ranch and the wide-reaching contacts of Pat Burns ensured that there was a steady stream of inquiries for horses arriving at head office.[148] Advertisements offering horses were placed in *Canadian Cattlemen*. A five-year-old saddle mare broken to ride and weighing around 1,050 lbs. was available at $75 to $100 in 1939. One inquiry, for twenty to thirty saddle horses, came from Niagara Falls, Ontario.[149] J.W. Hughes bought young Percherons by the carload in 1939 and 1940 for dispersal among his ranches in the Okanagan Valley.[150] However, it is clear that the demands of the ranch came before sales. Kelly, the general manager of Burns' operations, replied to inquiries about a team in the following terms:

> ... we hardly have enough quiet horses to handle the work. However, when harvest is finished, he [Farrell] feels satisfied that he would have a team that would be suitable for your purposes.[151]

Sales of horses continued into the post-war period; in 1946 they were worth $1,430 and in 1947, $1,576.[152]

The final disposal of the Bar U horses had a traumatic impact on all those who took part in their last drive. More than thirty-five years after the event, it was still hard for Raymond Clifford to talk about it.[153] To all who worked at the ranch, and to the whole neighbourhood, this occurrence signalled the end of the Bar U in a way that all the rumours of sale, and the partial disposals of land, did not. Raymond explained that Burns Ranches had between five hundred and a thousand head of horses. At the Bar U alone they had seventy-two broken saddle horses and a breeding herd of 250, including many purebred Percherons. He got orders to take them to Calgary to Red Top, the meat processor. Four riders trailed the herd northward, while the stallions were shipped by truck. When they got to the corrals, Raymond said to the men, "Peel those saddles off too." The men said, "Aren't we taking these riding horses home?" "No," said Clifford, "They are to go too." Some of the finest horses in Canada were disposed of at a quarter of a cent a pound for fox meat.[154]

9.9 The Bunkhouse and the garden looking eastward in 1947. The small buildings along the crest of the coulee behind the studhorse barn are mobile granaries. Parks Canada, Alwood Rousseau Collection, Rosettis 6397-16 94.12.01.05

A Last Look Around

In 1947, the Burns Company hired a professional photographer to produce a portfolio of pictures which showed off every facet of the Bar U Ranch: buildings, stock, range, and water. These photographs were to be included in a package which was sent to prospective purchasers. Joe Rosettis rose to the challenge, and his magnificent collection provides the most comprehensive visual inventory ever taken of the ranch.[155] Never had the headquarters site looked better;

his pictures conveyed a sense of neat functionality. The trim around the windows and doors was freshly painted. For the first time the dwelling houses, the old post office, and the blacksmith's shop were painted white. The big garden appears to have been carefully planted out. The hay racks are all aligned and all the fences visible are in good repair. The war years had seen a few changes. A new tractor shed had been built east of the bunkhouse; the blacksmith's shop had been relocated back against the coulee and a fuel storage area was created close to it. Buildings made redundant by the scaling down of the Percheron breeding program were used for grain storage, particularly some of the stalls in the U-shaped barn north of the river. The capacity of the chophouse had been increased by the addition of an elevator, while one of the photographs shows some twenty mobile granaries that were used to store grain at harvest time. Finally, the log house, originally sited to the west of the work horse barn, had been moved to a new position at the other end of the 'street' to the east of the new tractor shed. A caragana fence and a shallow depression marked its old position. The promotional literature explained that electric light was available in all the dwellings and in the garages and blacksmith's shop. During 1942 running water, showers, and inside toilets were added to the cookhouse.

Rosettis' photographs of the headquarters are almost completely devoid of human figures. Nevertheless, the neat buildings continued to house a busy community. Ann and Raymond Clifford still lived in the foreman's house close to Pekisko Creek. However, Raymond was rather less of a presence on the Bar U because he was now responsible for overseeing the management of all the Burns ranches south of Calgary. He spent a lot of time on the road and at the head office in Calgary. His place as foreman of the Bar U had been taken by Lee Alwood.

Lee was a local boy. His grandfather had homesteaded just to the east of the Bar U, and his father had worked on haying and threshing crews at the ranch.[156] However, Lee had turned his back on the farm. He had found a succession of jobs in construction and had shown great aptitude for working with machinery. He worked for a drilling contractor sinking an exploratory well on the Bar U, and then on the irrigation ditch from the E.P. Ranch to the Bar U. He married Ann Clifford's sister Alice, and the pair settled at the Bar U in 1937, when Lee started work at the Bar U Farm. As the cropped acreage increased and the process of mechanization began, so Lee's responsibilities increased and his competence was demonstrated. In 1942, Lee was promoted to farm foreman, and in 1944 he took over as 'boss' of the whole ranch.

When they arrived at the Bar U, the Alwoods lived in the log house originally built for the Nashes and recently occupied by Tom McMaster and his wife. The tempo of social life at the ranch picked up after Lee and Alice arrived. There were regular visitors from High River as cars and trucks became more generally available and the roads were improved. Jeanette was born in 1939 and her brother 'Stu' a couple of years later. It was not long before the western end of the headquarters site began to develop something of the family-oriented feel that it had enjoyed a generation before when the Olsen and Paulin families had run wild about the ranch. Although the children were forbidden to hang around the barns and the cookhouse, their presence was enjoyed by the men. George Lawrence remembered that Jeanette Alwood "was the prettiest little thing you saw in your life. Gee whiz! she was a pretty little girl.... He [Lee] was some proud of her too and I don't blame him."[157] The closest other families with children were the Baines and the Gardners, but visits were not frequent because the distances were considerable and the adults were all busy people. Instead, Jeanette and Stu played with the chickens, cats, sheep, and the minnows in the creek. They tried to avoid the attention of the dairy cows that grazed the site and in particular the Ayreshire bull, which had a bad temper. The addition of Sandy the Shetland pony to the family added a whole new dimension to the children's adventures.

Lee Alwood developed an extremely effective management style over the years. He was fair but brooked no nonsense. He carried a cheque book in his back pocket, not only to provide an advance to a cowboy on his way to town, but also so that he could fire a man on the spot who had transgressed too often. The young men who he recruited on the hotel steps in High River and Nanton are now old men, but they remember Lee with respect and affection. Floyd 'Swede' Erickson said, "... he had a nice way about him, he didn't give orders so much as ask 'would you like to do ...' whatever job it was."[158] He was a patient man and taught young recruits the tricks of their new craft. Oliver Christensen remembered, "None of us could ever remember him raising his voice or getting angry about anything. If you needed something he would show you how to do it, and not make you feel like a fool for not knowing...."[159] He was not a distant and aloof leader, but joined the men regularly for meals and to play cards in the evening. On their rather infrequent sprees to High River, Lee might join the "boys" in the bar, and on one occasion he admitted to sleeping it off in the bath at the hotel![160]

9.10 The headquarters site spruced up for the possibility of a sale. This was one of a series of photographs circulated to prospective buyers. Parks Canada, Alwood Rousseau Collection, Rosettis 6397-13 94.12.01.06

9.11 Bar U cattle pass through the site of a fire as they are driven toward summer grazing in the forest reserve, c. 1947. Parks Canada, Ed Peters Collection, 96.10.01.09

Lee's gifts were in organization and leadership; his background and experience was with machinery and farming rather than with cattle. It is significant that I have not seen a photograph of Lee mounted; instead he is shown with a high-pressure hose above the corrals, treating cattle for warbles. Fortunately, he could rely on the "cow sense" of the long-time head rider, Oscar Brandham, and the experience of men like Ben Jarvis, as well as on the enthusiasm and energy of the young men to whom being a cowboy at the Bar U was a dream come true.

During the 1940s the Bar U bunkhouse was home to a group of young riders who inherited the mantle of men like Charlie Millar, Ebb Johnson, and Charlie Mackinnon. They were all young, in their late teens or early twenties. They had the general background skills of farm boys but no training or experience as cowboys.[161] What they brought to the job was a very high level of motivation. This determination and willingness to learn was reinforced by competition with their peers and by the traditions of the old ranch.[162] Indeed, they were lured to the big ranch by its reputation, and by the image of themselves 'riding the range,' rather than mucking out the barn or checking on the chickens of the home place. Joe Hayes was born in Maycroft, Alberta, and grew up on the family farm at Eureka River during the Depression. In 1942, the eighteen-year-old went to the Calgary Stampede in the hope of landing a job. He was lucky, and spent the summer with Cliff Vandergriff's hay outfit on the Bar U. In the fall, he was kept on to haul hay for the cattle and by spring he was a "steady hand." He spent some time as a rider, but his real skill lay in his ability to fix things. He explained, "I drove the cats and other farm machinery and did repair work when needed."[163] Not surprisingly, the company wanted to keep him and wrote to the authorities requesting that he be deferred from active service because he was an experienced ranch hand and engine man.[164] Joe, however made up his mind to join up and 'do his bit.' Floyd 'Swede' Erickson was twenty-three when he left the farm in Kingham, Alberta to go "stooking." He soon made a stake and took the bus to High River. In the St. George Hotel, he came across Jack Crowe, who had been working at the Bar U ranch for three or four years. Jack introduced Floyd to his boss Lee Alwood, and he was hired on the spot. Like Joe, the novice started hauling hay, but he soon got a chance to show his mettle as a rider. In the spring of 1946, when Oscar Brandham, the head rider, left the ranch, Lee Alwood approached Floyd after breakfast and asked him to take over as head rider. Floyd hesitated, saying, "But I've never done that before." To which Lee replied, "Gosh man, you can learn!"[165] He did, and looked after the cattle at the big ranch until early in 1949. Oliver Christensen was raised on a farm at Standard, Alberta. He was drawn toward 'cowboying' from an early age. While growing up, he rode the big work horses whenever he could, and was in heaven when his school teacher allowed him to board her horse during the weekends.[166] His first job away from home was at the Burns feedlot in Calgary, but June 1945 found him working with a haying crew on the Bar U. He worked hard and made a favourable impression. In October, when Jack Crowe and Harry Kerr decided to move on, Oliver realized his dream and became a rider on the famous ranch. His bunkhouse buddies included Dallas Campbell and Buck Cofield. Lee Alwood told a good

9.12 Ed Peters mounted on Jiggs. The crew were up in the forest reserve to the west of the Highwood River pushing cattle onto summer pasture. Parks Canada, Ed Peters Collection, 96.10.01.01

story about one nameless neophyte who might well have been any of these young men. He explained:

> We used to get all them young fellows, they all wanted to be cowboys. So this old fellow Dunc Fraser, an old cowboy, moved up there [to the Rigg place] ... so he took this young guy up with him and they were setting up. They took their groceries and were moving in. He gave this kid an old block of wood, a bunch of coffee beans and a hammer and had him mashing up the coffee beans! There were two coffee grinders hanging on the wall.[167]

Ed Peters was a Manitoba boy. He had a particular reason for pursuing a life in the saddle. He was born with deformed ankles and feet, and in spite of several operations, it always pained him to put in long hours on his legs.[168] He used some of the first money he earned to enrol in a riding course which he had seen advertised in *Prairie Farmer*. This, coupled with his youthful determination, made him an usually good horseman. In the spring of 1946, Ed jumped a train heading westward, and found himself a job with Charlie Duggan on Burns' sheep farm near Crossfield. This was much better than working underground in Timmins, Ontario, but it was still not quite what Ed was looking for. In December 1946, he moved to the Bar U to haul hay for the winter. Finally, in the spring when Oliver Christensen left, Ed took over his job as rider. He loved the place and his job and stayed at the ranch until it was sold in 1950. Then he joined several other hands in the move to the Bow Valley Ranch. Peters took a keen interest in rodeo, and when he noticed that one of the boys driving a dump rake at the farm spent all his spare time riding, he encouraged the young man and worked with him on a bucking barrel.

Lee Alwood had no particular reason to bemoan the departure of Hop Sing when he drove him into High River for the last time in the spring of 1942. The little Chinese cook had worked tirelessly to feed the crew for some fifteen years, but he was set in his ways and not at all awed by authority. Lee had recently become ranch foreman and had his own ideas about how things should be done. However, if Lee had known the problems he was to have getting and keeping cooks during the next three years, he would have surely begged Hop to stay! It was not easy to attract a good cook to an isolated ranch during the war years when many opportunities beckoned. Mrs. Quebec and Mrs. Phillips were the first to try to fill Hop's shoes.[169] They were faced with a formidable task in their attempts to modernize the kitchen and the way meals were prepared. One of the first things they insisted on was the removal of the pack of semi-feral cats that surrounded the cookhouse. Lee Alwood had to shoot about thirty of them.[170] Eileen Watrin took over from Mrs. Quebec in 1944. She was the wife of Eddie Watrin, who was a regular hand on the ranch. They stayed in the cook's quarters and Eddie helped her with the chores whenever he had the time.

Meanwhile, Mrs. Grierson, or Myrtle Farnham as she preferred to be called, had been hired to cook at the Bar U farm. Her story demonstrates eloquently that a sense of adventure, and a willingness to take risks, was not confined to the young men who aspired to become riders. In the spring of 1945, twenty-three-year-old

Myrtle hitched a ride down to Calgary, carrying baby Hazel in her arms. They spent the night in her sister's bed-sitting room. Her memoir continues:

> Next morning, Beulah walked us to Burns' office on 9th Avenue; Mr. Farrell was the supervisor who hired all the cooks for Burns ranches....[171] He looked me over, then said there was no opening for a cook. I wasn't eating but I fed the baby. Well, Beulah went to work. After lunch the phone rang; it was Mr. Farrell saying that he had a cooking job open for me on the Bar U. I was to go down on the bus Monday afternoon, and the foreman, Mr. Alwood, would meet me. By the time I bought my bus ticket for $1.25, I had only .25 cents left. We got into High River, and no one met us. I prayed to God to guide and help me. I'd never traveled, but I did know I could take a hotel room, and didn't have to pay until morning. I got thinking that I could clean the kitchen or restaurant after it closed, and also clean the bar. The baby was a good sleeper. I'd fed her solid food at Beulah's; thus, with the.25 cents I could buy five baby bottles. I hadn't eaten for three days, only a little snack at Beulah's. At any rate I had it all figured out. I sat there in a big leather chair for two hours, then a man came along and asked me if I was Mrs. Grierson; I jumped up and said, "Are you ready to go?" He was a very kind voiced man, and said that his wife was shopping, and that we'd have supper in town. Then he left, I thought, "Oh God, I only have .25 cents," which at that time would buy a piece of pie and a cup of coffee; I didn't realize he would pay for the supper. They came and we went into the cafe. Mr. Alwood got a high chair for the baby. They ordered roast beef. I ordered the pie and coffee. Oh! I was very hungry, and very concerned about my future. Mrs. Alwood fed all her mashed potatoes to the baby. I felt embarrassed, but I couldn't help it.
>
> We then left for a long drive out to the Bar U Farm; it sat back from the road. The first thing I looked at was the stove, which was a Home Comfort. I was happy to see that, as the stove is a good cooker. All of the rest of the cookhouse was very barren, all wood floors, and the bedroom very bare but it had a leather chair that would fold down, with arms on the side that made the baby's bed. In the dining room were two round tables; the centre round moved. I put everything on the centre, even the coffee pot and the tea pot. The kitchen had no cupboards. There was a water tap just for cold water; one had to carry the wood in and ashes out. When I looked the place over, I prayed to God to give

9.13 Bar U cook Myrtle Farnham with Dallas Campbell, 1945. Parks Canada, Dallas Campbell Collection, 98.01.02

> me strength to be able to do this job, and also for God to guide me the right way.[172]

Myrtle's prayer was answered. She coped valiantly with the demands made on her: cooking for forty men, baking pies, cakes, and cookies and dealing with a mass of laundry. She soon became a favourite with the men who called her "Sis." Oliver Christensen and Dallas Campbell vied with each other for baby Hazel's affections and helped to teach her to walk. Alice Alwood and her children arrived at the farm to give Hazel a surprise birthday party, and Eileen Watrin sent a gift.[173] Myrtle finally got a few days off during the Calgary Stampede, and "six boys from my cookhouse" took her on many rides. Her young cousin Ralph Zutter joined the haying crew, and her sister Beulah came down to help out during the busiest time of the harvest season. Altogether it was a happy time in spite of the unremitting grind of work.

When Mrs. Watrin and Myrtle left the ranch, the hands fell on hard times. The replacement cook refused to prepare the usual big roast and all the trimmings which had become the traditional fare for the branding crew. She said that she refused to cook for Indians. Alwood had to buy Burns wieners and beans and the boys cooked them on the branding fires. Lee said, "I fired that cook. I hated doing that 'cause most of the time they would start bawling. The men, it wasn't too bad firing them but the women, I didn't like doing that."[174] Little did he know that his troubles had only just begun. The next cook was a disaster. Oliver Christensen said that they lived on baking powder biscuits all winter, "She was a horrible cook – she couldn't cook for sour apples."[175] Oscar Brandham, the head rider who had been at the ranch since the mid-1930s, complained to the foreman. He said that if she wasn't gone in thirty days he would be. Over the next few weeks, however, things changed. Floyd Erickson takes up the story:

> Oscar got sweet on the cook.... At the same time, another rider on the ranch, Lee Wall, began dating the cook's daughter. Now, all of a sudden, the cooking didn't taste half bad and Oscar forgot about his threat to leave the ranch. But Lee Alwood hadn't forgotten, and one day to the month that Oscar had lodged his complaint the foreman walked up to him and said, "Got your bags packed?" Oscar replied, "What do you mean, have I got my bags packed?" going a little pale. "Well its thirty days since you served notice, and the truck is ready to take you to town, let's go!"[176]

So the ranch lost both an indifferent cook and an experienced head rider. It was enough to bring to mind Bert Sheppard's laconic reply to the question, "How many women were there on the Bar U?" "Too many!" he replied.[177] Fortunately for everyone, Lee Alwood's luck had changed: he hired the reliable Mary McCloskey. She was a wonderful cook and soon gained the respect and affection of the crew. Oliver Christensen remembered:

> ... she fed us well: lots of meat and potatoes, roasts, vegetables, and there was the odd apple pie or something good for dessert. When the fresh apples ran out, she served CPR

strawberries – stewed prunes – or tapioca, which we called fish-eyes. We hardly knew there was a war going on and food was being rationed, we ate so well. I remember thinking Mrs. McCloskey was pretty elderly back then, although she was only in her forties. She sure worked hard.[178]

She continued to cook at the ranch until it was sold, and then accompanied several of the workforce to the Bow Valley Ranch. The men at the farm were also blessed with the arrival of Mrs. George, and her daughter Wanda, who cooked there for several summers. Often they would cook the meals out in the mobile cook car close to the place where the men were working, but the pair lived at the farm.

As well as showing the buildings at the headquarters, Rosettis' photographs also provided a picture of the state of the Bar U range. They show the rough fescue 'prairie wool' which would be cut for hay, and the dense stands of tame hay closer to the farm. Pekisko Creek and Stimson Creek were running clear and full, and showed little evidence of riverbank degradation.[179] While there is plenty of evidence of fenced pastures and branding corrals, the swelling grassy hills to the north and west of the headquarters looked unchanged since they were viewed for the first time by Fred Stimson in 1881. The rangeland was in good heart; for sixty-five years the ranch had sustained its herds of cattle and horses without damaging the grass. The presence of the Bar U, and other ranches like it, had prevented the land along the creeks from being homesteaded, an outcome which could only have led to impoverishment for the settlers and the degradation of the grassland. Partly by design and conscious planning, partly by accident and enlightened self-interest, successive

9.14 Mrs. George, the Bar U farm cook with her daughter outside the mobile cook car, c. 1943. Parks Canada, Joe Hayes Collection, 96.06.01.29

Bar "U" Ranch,
Pasture Lands &
South end of Ranch,
adjoining Wild Hay lands.
6297-69.
June 1947.

managers of the big ranch had protected the complex and fragile grassland ecosystem by not demanding too much of it.[180] They established a tradition of stewardship which is being articulated with increasing vigour by the present generation of foothills ranchers who face unprecedented competition for land from the powerful urban majority. John Cross of the a7 Ranch spoke recently of being "entrusted with the stewardship of a fragile land."[181] Ian Tyson voiced his objection to oil exploration in the Pekisko area by claiming, "This land is intact and pristine because ranchers have made a living from it and protected it...."[182] Such is the honourable legacy of the Bar U Ranch.

The Sale of the Bar U Ranch

Since the turn of the century there has been a marked tendency toward stability of land tenure in the foothills country. Land painstakingly acquired over the decades was not to be parted with lightly, but rather it was to be passed on to the next generation. In keeping with this attitude the Bar U Ranch had maintained its territorial integrity for almost fifty years. Made up of both leased and deeded land, the ranch sprawled over nearly five townships and covered seventy thousand acres. The boundaries of this range had been established during George Lane's ownership when he was able to purchase thirty thousand acres from the railway company and to obtain a closed lease for forty thousand acres of some of the best winter pasture in the province. Thus the rumour that the great Bar U Ranch might be up for sale must have spurred an enormous amount of interest throughout the region. For some, there would be an opportunity to round out a troublesome land border; for others an opportunity to regularize long-standing agreements with neighbours concerning pasture lands; while still others might be in a position to buy additional hay land, which in turn would allow a larger cow herd to be run and a larger calf crop to be expected. However, Burns Ranches held assets and investment opportunities on a scale which attracted interest from outside the region, both from across Canada and internationally. "Moneyed interests" in eastern Canada, New York, and California flirted with the possibility of a major purchase of Canadian ranch land. Such an investment might not yield a very high rate of return, but was regarded as being rock solid and as having the potential for long-term growth. In a world embroiled in a global conflict, with governments making unprecedented demands on the assets of individuals and corporations, land must have seemed an excellent balance to a volatile investment portfolio.[183]

Surviving records suggest that the board of directors of Burns Ranches were willing to consider liquidating some of their ranch assets by the early 1940s. This would not have been possible while Pat Burns was alive (he died in 1937), nor would a sale have been viable until cattle prices began to pick up after the long Depression. It has always seemed rather ironic that the ranch was finally broken up in 1950 after enjoying a five-year period of unprecedented prosperity. Now it becomes clear that the board was prepared to consider offers for the various ranches as going concerns at least as early as 1941.[184] Thus, on the one hand, one cannot fail to be impressed with the organizational structure and the on-the-spot management of Burns

9.15 The range and the riparian zone along Stimson Creek had never looked better than it did in 1947 after 65 years of well-managed grazing. Parks Canada, Alwood Rousseau Collection, Rosettis 6397-69, 94.12.01.23

9.16 Spring runoff along Stimson Creek, showing no signs of accelerated erosion due to heavy grazing. Parks Canada, Alwood Rousseau Collection, Rosettis 6397-47,94.12.01.159.24

Ranches during the 1940s; while on the other hand, one has to realize that the ranches were but one part of a corporate empire. John Burns, Pat Burns' nephew, and many of his fellow directors, were the same men who had made the decision to divest themselves of the ranches in 1926 so that they could concentrate their investment and managerial energies into meat-packing, fruit wholesaling, and retail dairy enterprises. As Canadian industry geared up to meet the demands of war, the opportunities for further investment in established fields and the possibilities for exploring new opportunities must have been very appealing. For example, Calgary was poised for rapid population growth as soon as the troops came home.[185] Burns enterprises owned land which could be developed for commercial and residential purposes, and river gravels and sands which might become valuable resources for building. The directors were very well aware that even the unprecedented profits posted by the ranches represented only a modest return on a large investment. In a confidential memo to John Burns, Newton pointed out that it would be most unwise to send a statement of estimated profits to prospective buyers. He wrote: "I think the reaction of a practical rancher will immediately be that the estimated profits would not in his opinion justify the capital investment."[186]

Thus, the breakup of the Bar U Ranch in 1950, and the increasingly sophisticated and energetic efforts employed to sell off the ranches, must be regarded as the final successful conclusion to a process that had been underway off and on for close to a decade. In the following paragraphs some of the most important correspondence between land agents, prospective buyers, and the board will be reviewed. It is hoped that this will provide a context for the eventual sale of the ranch and its lands.

At first, the company's preferred strategy for the sale of its ranches was to sell them as going concerns, including the stock and all the equipment. While some of the smaller ranches might have attracted local investors, there were not many Albertans in a position to offer more than half a million dollars for ranches like the Bar U and the "44." It was to land agents in Los Angeles and New York that the company looked to put together such big deals. One such man was James Dwyer of Los Angeles. In a spate of handwritten letters and notes to Alex Newton during the early months of 1941, Dwyer pointed out that there was considerable interest in Canadian ranch properties, and that he was the man to bring buyer and seller together. He wrote, "There is good activity in big ranches and we feel confident we can find you a satisfactory buyer."[187] A week later he explained that one likely prospect had bought a 160,000-acre ranch in New Mexico, but he went on to assure Newton that there were plenty of others. He talked of the real estate boom in Los Angeles, and, for the first of many times, he suggested a trade of an Alberta ranch for property in that city. "I believe that this would be less trouble and pay better dividends than the ranch. There are some good trades in large apartment homes, hotels, and office buildings."[188] A week later, Dwyer reported that his efforts were generating considerable opposition from brokers who were doing everything in their power to discredit Alberta ranch lands. He suggested that the way to foil their plans was to grant him an exclusive option on the properties for ninety days.[189] Newton replied that Dwyer's proposals had been discussed by the board, that the company was not interested in acquiring land or buildings in California, that they were not prepared to grant an exclusive option, and that they were in no rush to sell. He concluded, "The ranches are profitable but as they form part of an estate they will be sold over a period of years."[190] Dwyer's rapid-fire letters continued, but nothing came of them, and Pearl Harbour seems to have curtailed Californians' interest in the Bar U.

Canadian land dealer C.J. Wilson, based in Toronto, benefited from his earlier acquaintance with Pat Burns and his ongoing friendship with John Burns. He wrote that he had a potential buyer in Chicago, who he described as a "very reliable fellow." He concluded his letter: "... your uncle always said it was a good time to sell when there was a buyer."[191] Newton sent him the general prospectus on the Bar U and "44" Ranches, and followed up with a more detailed letter.[192] When Wilson queried the fact that the figures did not include any sales of beef cattle, Newton explained that: "we have made a practice of wintering the steer calves at the ranch and disposing of them to other ranches during the spring or early summer."[193] The price being asked for the Bar U as a going concern was $650,000. Once again after a brief flurry of interest no serious buyers emerged. Wilson's earlier enthusiasm was dampened: "I found out there were not many people looking for ranches at the present time on account of the war. People are very nervous about investing money."[194] The possibility of selling the Bar U as a going concern remained a favoured option throughout the 1940s. In April 1949, while other strategies were being actively pursued, John Burns wrote to F.H. Armitage in Edmonton, "We are pleased to have met you this morning to discuss the prospects of the sale by you of the above ranch [the Bar U] as a going concern to American interests." He went on to describe some of the attributes of the ranch and the terms of sale. He even mentioned that a promotional film in full colour had been made to allow prospective buyers to view the property in their homes.[195]

Closer to home, communal groups like the Hutterites were interested in acquiring large blocks of land on which to establish new colonies. This often involved a realtor in complex negotiations with a number of farmers. The possibility of buying at least ten thousand acres at one time was obviously very attractive. Byron F. Tanner of Lethbridge had handled several purchases for the Hutterites and had shown considerable interest in Burns' properties. As he remarked:

> Our office has been in touch with your company for several years now, both by correspondence and by personal contact. We have shown various of your properties several times, including the Bar U, the Forty Four, and the Bow Valley property.[196]

He then introduced a firm offer for ten thousand acres of the Bar U farmland from the Miami Colony of New Dayton. This interest on the part of the Hutterites was not surprising since, on the one hand, demographic and social pressures within Hutterite culture made periodic colony division an important survival technique, while on the other hand, land sales to Hutterites had been forbidden by the Land Sales Prohibition Act of 1942. By 1946, the legality of this act was under review and discussions concerning the Communal Property Act – which was to be its successor – were ongoing. Hutterite leaders and the realtors who worked with them were eager to take advantage of the repeal of one act before another replaced it. In a letter to John Burns from George Waldner of Miami Colony, the leader said, "We are very anxious to determine our next location as soon as possible."[197] In this case the offer came to nothing and the colony established a daughter colony in Montana.[198]

Another group that showed considerable interest in the Bar U Ranch were Mennonites from Manitoba. Under the leadership of Joseph Rempel, they were looking for a big block of land which could be subdivided into three- to four-hundred-acre lots for individual families. This was a strategy which had worked successfully on the Namaka Farms property of George Lane during the late 1920s.[199] Rempel expected to have as many as two hundred families involved, and arrangements were made to show an advance guard around the property.[200] Apparently, nothing came of this overture either.

Alex Newton, a director of Burns Ranches and John Burns' trusted lieutenant, knew very well that a group of Stoney Indians had left the Morley Reserve and had been more or less permanent residents of the upper Highwood valley for several decades. These families provided invaluable seasonal labour to the ranches of the neighbourhood. They were in desperate need of a permanent land base. Newton contacted G.H. Gooderham, the Regional Supervisor of Indian

Agencies, and suggested that the government might want to purchase one of the Burns ranches on which to establish a reserve. The matter was discussed in Ottawa, but the consensus was that it was impossible to purchase a ranch for the Indians at that time.[201] However, it may be that Newton's inquiry planted a seed in the bureaucratic minds in Ottawa, for, when the reserve was finally established, it was located on three or four sections of land along the Highwood which had been purchased from Burns by Frazer Hunt, the American explorer. Moreover the reserve took the ranch name – Eden Valley. Hunt had used the place as a summer retreat but had hopes that his son Bob would ranch there on a permanent basis. In 1944, Bob was drafted into the armed forces and the ranch was sold to Earl G. Garner, who owned it for a few years before selling to Indian Affairs.[202]

In 1949, after almost a decade of trying to sell the Bar U ranch as a single unit, the board decided to sell off the land and assets piecemeal. Neighbours were asked quietly if they were interested in acquiring land.[203] The following year the local press commented on the progress of the sales. The *High River Times* had a headline which read, "First Break-Up in Famous Bar U: 4320 acres sold." The article went on to suggest that this might be the start of a trend toward a higher population for the foothills, with several small ranches taking the place of a few big ones.[204]

It was not until August 1950 that the heart and soul of the ranch, the headquarters site, eight adjacent sections of farmland, and the greater part of the winter pasture, was sold to J. Allen Baker. He bought four thousand acres of deeded land and acquired sixteen thousand acres of the lease, along with the brand, the ranch buildings, and five hundred head of the best Hereford cows.[205] Baker remembered his purchase of the ranch well, as it changed the course of his life. In addition to a flourishing business as an auctioneer, he had been running a farm south of High River. His grain crop had been frosted yet again the previous year, and he had determined to pull up stakes and head for Montana. He knew the country down there well and fancied a ranch near Bozeman, where he felt the climate was rather better, and where he hoped he could raise a herd of prize Herefords. Jim Cartwright, the owner of the D Ranch, the Bar U's western neighbour, persuaded Baker to come over and have a look at the Bar U before his departure. Baker liked what he saw so much that he engaged realtor Alf Tannis to look into the possibility of a purchase agreement. Eventually the deal was worked out.

Neighbouring ranchers were the chief beneficiaries of the break up of the Bar U Ranch. This is demonstrated by Map 9.1 [see colour section] which was drawn using purchase agreements, lease assignments and newspaper accounts.[206] In Township 18, Range 3, for example, which included the northerly sections of the Bar U land reaching up to the bend in the Highwood River, J.A. Hughes, who already owned three or four sections of adjacent land, was able to double his acreage by acquiring both deeded and leased land from the Bar U. Likewise, W. & R. Runciman, the Hogg brothers, and the Rowlands were able to obtain convenient sections.[207]

To the east, farmer and entrepreneur Roy Charles Henderson purchased the Bar U Farm and seven sections of land. But he only owned it for a few years before selling it to Allen Baker's brother-in-law, Harry

W. Hays. It seems likely that the farmland in question was better suited to grow feed grains in support of pastoral activities than it was to wheat production. Perhaps too, Henderson became over-extended with his investments in car dealerships.[208] R. R. McLean was able to round out his land holdings by purchasing some key sections. His letter exemplifies the process by which local ranchers could benefit by acquiring Bar U lands, and for this reason it is worth quoting in full:

> Dear Sirs,
> Your records will show that my uncle, Mr. Dean Pritchard, and myself purchased from you Section 12 and South half Section 13-17-2, West of 5th. We previously owned the North half of Section 13 and Section 24 and since the date of purchasing from you the above described lands I have acquired Section 1-17-2, West of 5th. There are certain portions of the Bar U Ranch referring to the north half of section 36, and 160 acres in section 35-16-2, west of 5th, that would fit in with the holdings of my uncle and myself. If, at any time, there is a suggestion of disposing of these two parcels I would appreciate very much having the opportunity of discussing with you the purchase of the same. Yours very truly, Rex R. McLean.[209]

The "Bar U Flats," the valuable hay and arable land bordering the meandering course of Stimson Creek, was subdivided among a number of local buyers. R.W. Gardner was able to add to his well-established Mount Sentinel Ranch by purchasing two sections and leasing three more. Other members of the Gardner family acquired an additional two or three sections, while the Blades family also bought two sections reaching up into the Porcupine Hills.

To the south and southwest, S. J. Cartwright of the D Ranch made the single largest purchase involving eight sections. This meant that the boundary of his ranch moved eastward down Pekisko Creek from its headquarters at the base of the mountains. The purchase of the E.P. Ranch from the Duke of Windsor in 1961 completed this process of land acquisition in the Pekisko drainage.[210]

To the west of the Bar U Ranch lay the beautiful bench lands along the upper Highwood and the forested summer pastures of the mountain front. Most of the Bar U land was concentrated in Township 17, Range 4, but Burns Ranches had interests from time to time in the Buffalo Head Ranch, Guy Weadick's Stampede Ranch, and the Eden Valley Ranch of Frazier Hunt. A number of transactions disposed of these holdings during 1949 and 1950. Firstly, the Kuck Place was sold to Pieter Tjebbes in a deal which involved 672 acres of deeded land and 11,000 acres of leases. Tjebbes ranched there for a year or two before selling out to Earl Nelson.[211] Secondly, the Buffalo Head Ranch was sold to Fiorvante de Paoli.[212] Thirdly, Hunt sold the Eden valley Ranch to Garner as detailed previously, while the Stampede Ranch, which had been bought from Guy Weadick in 1946, was retained through the 1950s.[213]

Was the sale of the Bar U Ranch lands and assets in a piecemeal manner to neighbours a good business decision? Did the company realize as much as it would have if it had been able to sell the ranch as a going concern? These are important questions, and they are hard to answer given the fragmentary information available. A point of departure is to review some evaluations made during the 1940s. For example, the ranch was appraised in 1947, and the value of land, stock, and improvements was estimated at $860,000.[214] Given that appraisals are usually conservative and that land and stock values were rising between 1947 and 1949, one can reach a conclusion that any sale which did not exceed this evaluation by a comfortable margin could not be regarded as a "good deal." A few years before, the Bar U had been offered for sale as a going concern for $650,000.[215] In later responses from the company to inquiries from possible purchasers, no figures were committed to paper. However, in December 1949, an interested party was told that the company was asking $26 an acre for its deeded land.[216] This would have meant a total of $754,000 for all the deeded land. What about the extremely valuable leased land which constituted such a prize asset? Obviously, it was worth a lot to acquire some sections of prime grazing land from the government. Most of the Burns leases had been recently renewed and were good for another fifteen or twenty years.[217] On the other hand the leaseholder had to pay a not inconsiderable fee based on the carrying capacity of the range in a particular year. In a 1945 evaluation of Burns' ranch lands, leased lands were appraised at $1 an acre.[218] However, an offer for the Buffalo Head Ranch in October 1948, valued the leased lands at $2.50 per acre.[219] Given the fact that the assignees of leases could look forward to more than a decade of secure tenure, and given the location and quality of the pasture, the latter price seems more realistic. Thus the forty thousand acres of leased land would have raised $100,000, for a total return from Bar U Ranch lands both deeded and leased of $854,000.

9.17 The last drive of two thousand head of cattle from the Bar U to the feedlot in Calgary in 1949. Glenbow, NA 2575-6

One way to check this estimate is to use the data that is available on specific sales of particular parcels of land. The "Fifth and Sixth Tenders" show that there was, as you might expect, a marked difference between the price asked for the rolling pasture land in Township 18, Range 3, which averaged $22.89 per acre, and the flat arable lands of the Bar U Farm, which were sold for an average of $36.78 per acre. If we assume that the same prices were obtained for farm and pasture lands in other townships, a complete estimate of returns from land sales can be worked out. A conservative assumption that $30 per acre would be realized for farmland after discounts and expenses, and a comparable figure of $20.00 for rangeland, produces a total return of $860,300. If we use the average figures actually realized in the Fifth and Sixth tenders, then the total return would have been $1,019,928.[220] My conclusion would be that Burns Ranches benefited from rising land values during the post-war period and sold their lands for almost one million dollars in 1950.

It was the cattle rather than the land which had made up most of the value of the Bar U Ranch during its long history. Thus, it should come as no surprise that the sale of stock more than matched the returns garnered from land and buildings. It is hard to establish precisely the size and makeup of the Bar U herd in 1949. The primary role of the ranch was that of a cow-calf operation which provided calves to other Burns ranches where they were fattened. This role is illustrated by the stock returns for 1945, which mention 2,015 calves, 3,778 cows and 100 bulls.[221] But was the whole herd sold? On the one hand, the board was under no compulsion to rid themselves of the whole herd, for some of the other ranches could have absorbed much of the displaced stock; on the other hand, the company was trying to extricate itself from the ranching business and cattle prices were high. These factors would surely have encouraged as complete a clearance sale as possible.

In November 1949, Tom Farrell, the overall ranch manager, produced a memo for the board and for potential buyers indicating the selling prices for various categories of stock. He added, "These prices, I believe to be very close to the present day market values and I would be quite willing to sell at these prices."[222] If these suggested sale prices are applied to the 1947 stock numbers, then the company should have received almost one million dollars from the sale of its herd. However, Farrell's down-to-earth estimates were based purely on market values and did not take into account the reputation of the ranch and the quality of the stock. Nor did he consider the fact that some buyers would be willing to pay premium prices to acquire a part of a prestigious herd.[223] The sale of 309 cows from the Bar U at the community auction in High River brought in $77,000, an average of $249 per head, which was almost double the average of Farrell's estimates.[224] J. Allen Baker had picked five hundred head of the best cows from the breeding herd as part of his purchase. If these cows were valued at even $200 per head they would have been worth $100,000. Another sale in the spring did equally well, a milk cow netted a record $400, while a well-broken saddle horse went for $125, rather than the $50 that Farrell had expected.[225] It would be wrong to make too much of these individual cases, which apply to such a small fraction of the whole herd; however, they do suggest that the estimates based on Farrell's valuations may be rather conservative.

To summarize: after a decade of looking for an individual or company to purchase the Bar U Ranch as a going concern, Burns Ranches decided to sell off its assets in land and stock to the highest bidders. The evidence suggests that the company received about one million dollars for the land and a similar sum for its stock.[226] This was more than double the appraised value of the enterprise. It was a triumphant outcome which would have pleased Pat Burns or even Fred Stimson! Finally, reference was made earlier to the fact that land tenure has been very stable in the foothills. In this context it is interesting to see from the latest available municipal maps of land ownership that much of the original land owned by the Bar U remains in the hands of those that purchased it in 1949 or 1950.[227] Some changes have occurred and some family operations have been incorporated, but continuity of land tenure is still very much in evidence.

10

Postscript and Retrospect

"... *for somehow, against probability, some sort of indigenous, recognizable culture has been growing on western ranches.... It is a product not of the boomers but of the stickers, not those who pillage and run, but those who settle, and love the life they have made and the place they have made it in*."[1] – Wallace Stegner

It is more than half a century since Burns and Company sold the Bar U and the ranch became a family outfit for the first time. During this long period there have been far-reaching changes in the cattle industry. In many ways, these have been as dramatic as those which faced Stimson, Lane, and Burns. In 1950 the industry was poised for a period of spectacular and prolonged growth. The total number of cattle on Canadian farms grew from about three million to nearly ten million head in 2001.[2] The centre of beef production has moved steadily westward, and today the prairie provinces account for more than three quarters of Canada's beef cows. Alberta has increased its share of slaughtered cattle from 30 per cent in 1976 to 65 per cent in 2000.[3] This shift was already underway during the 1970s, but intensified when the Canadian government stopped subsidizing the movement of grain and gradually removed "the Crow Rate."[4] More and more grain which would have been sold abroad stayed in the west, and calves which had in the past been shipped to Ontario for fattening stayed on the farms and feedlots. As the size, capacity, and efficiency of Alberta's feedlots increased, so the meat-packing industry relocated to the centre of beef production. Locally produced cattle were assured of nearby markets.[5] An industry which provided opportunities for reducing the bulk of its raw material during processing was able to conform to the laws of economics for the first time, and located close to those raw materials. The value added stayed in the west and the overall transport costs were slashed.[6] During this period, the links between the ranch and the auction lot or feedlot were also streamlined. Semi-trailer cattle liners which provided "door to door service" replaced the time-consuming cattle drives to a loading point on the railway.

The cattle industry was also shaped by a number of factors acting upon it from outside. Like any other commodity, the price of cattle is cyclical, with a typical wavelength of ten to twelve years.[7] In Canada, cattle numbers grew from 1951 to 1964, and again from 1969 to 1975; they then declined quite steeply to bottom out in 1987; then they grew again to peak in 1997, and since then they have continued to decline, partly due to widespread drought. Superimposed on these cycles have been far-reaching changes in consumer behaviour. Per capita consumption of red meats in general and beef in particular has declined more or less continuously over the past two decades. By the mid-1990s beef consumption had stabilized at the level it had been in the early 1960s. This reduction was partly due to new concerns with diet and nutrition, particularly the links between saturated fats and heart disease. Cattlemen have always been attuned to changes in the marketplace, but they have never before had to confront the question, "Is red meat healthy?"[8] The aging of Canada's population and aggressive competition from poultry and pork have also affected demand. The majority of Canadians who still relish beef tend to purchase it in novel forms. The numbers of meals eaten outside the home continue to rise, and the amount of beef which reaches the consumer in the form of hamburger patties and "prepared foods" is still growing.[9]

Western cattlemen came into their own during the last decade of the twentieth century. Their ranches represented the undisputed centre of beef production, and were served by modern packing plants on their doorstep. It was ironic that when they had so much economic power, their political heft was but a shadow of that wielded by their predecessors. Indeed, the sanctity of their leased ranges was threatened by the demands of the urban majority for recreational space and by the incursions of the oil and natural gas industry.[10]

Even while these changes in structure, location, and consumer behaviour were taking place, a parallel revolution was affecting the very nature of Canadian cattle. The stranglehold of the traditional beef breeds and the quasi-religious adherence to the "purebred aesthetic" were shattered by the import and widespread adoption of new breeds of cattle from Europe.[11] Pioneering work on cross-breeding during the 1930s had challenged the assumption that purebred cattle were superior to cross-breds in terms of hardiness, rustling capabilities, weight gain or any other criteria. The majority of cattlemen, however, held fast to their chosen purebreds and little notice was taken of these important findings until the 1950s. By that time the results of Canadian breeding research were widely disseminated, and experiments in Texas, where classic British breeds had been crossed with Brahman cattle from India, were watched with interest. Synthetic breeds like the Santa Gertrudis, Beefmaster, and Brangus proved to be better adapted to the semi-tropical conditions of the Texas range.[12] During the 1960s more and more Canadian cattlemen were cross-breeding their herds in search of "hybrid vigour." They hoped to combine the best features of several familiar breeds. Bulls were evaluated not so much in terms of how they looked in the show ring but rather by how they performed.[13] Finally, the intellectual climate was ripe for more radical change. In 1965, arrangements were made for a quarantine station to handle European cattle imports to be established on the St. Lawrence. The next ten years saw the introduction of more than a dozen "exotic" breeds, the most common being Charolais, Simmental, and Limousin. The objective was to benefit from the size and rapid weight gain of these breeds while retaining the hardy rustling qualities of the familiar breeds. By the 1980s, two thirds of the cattle marketed in Canada were cross-breds, and livestock markets consistently rewarded them with premium prices. The term "cross-bred" had become an accolade rather than an epithet.[14] The Bar U Ranch played a role in this genetic revolution as it had in so many previous developments in the cattle industry.

The men who owned the Bar U during these rapidly changing times were remarkable people. Each in his own way was linked to the past history of the ranch. Allen Baker was an impetuous, hard-driving maverick of a man. He bought the ranch partly because it was a challenge. Like George Lane, Baker was never content to maintain a comfortable and successful status quo, but rather he constantly pursued new and risky entrepreneurial horizons. The Wambeke family, already well established in the agricultural community along the Highwood, bought the Bar U headquarters site and part of the old ranch with a view to expanding their integrated production system, which combined a cow-calf operation with a feedlot and a farm. They mirrored on a local scale the vision of Patrick Burns. Melvin Nelson was an accomplished cowboy like George Lane; in fact he was lucky enough to be able to enjoy cowboying and horse training into his old age. He was seventy-three when he bought the Bar U, and gained great satisfaction from owning for a brief time two of the most famous ranches in Canada, the Gang Ranch and the Bar U.[15] Sometime, the stories of these cattlemen and their triumphs and trials deserve to be told in full.[16] All I can hope to do here is to outline briefly some of the main events.

Nobody in the long history of the ranch relished his ownership of the Bar U more than did Allen Baker. He enjoyed his nickname "Bar U Baker," and claimed quite rightly that he owned the ranch for longer than anybody else.[17] Allen was only thirty years old when he bought the ranch and had already earned a reputation as one of the leading cattle auctioneers in Alberta. When the corporate ranch was up for sale, Baker pushed his credit to the limit in order to acquire as much land as possible. His Bar U continued to be one of a very few large foothills ranches and covered thirty sections of deeded and leased land.[18] For a decade or more, Baker ran a traditional cow-calf operation with dedication and distinction. At first he was very short of working capital and had to take short cuts to increase his calf crop.[19] As financial pressures began to ease so Baker began to "cull heavy." He rid the herd of troublesome cows and selected his bulls with an eye to trouble-free calving. Soon his exemplary herd of commercial Herefords was flourishing under a system of benign neglect. Baker boasted that he did not have to feed his herd for twenty years.[20] One cowboy remarked after a day working on the Baker range: "It was like gathering elk out there!" Yet Allen consistently achieved an 85 per cent calf crop and had a phenomenal memory for the pedigree of all his animals. In the early 1960s Baker sold his calves at the Lethbridge market and confided to a friend, with a grin of satisfaction, that he was finally free of the bank.

Baker was knowledgeable and shrewd, but at the same time he was impetuous and attracted by new ideas. As an aggressive innovator he stood out among his more conservative peers. He was one of the first to adopt pregnancy testing for his cows and to experiment with artificial insemination. Having read somewhere that cattle were short of vitamin A, he peppered his lease land with mineral blocks. He saw the potential of 'cattle-liners' and helped a local man get started in the business of hauling cattle. It was a forgone conclusion that Baker would embrace the move toward exotic cross-breds with enthusiasm. He agreed to allow his neighbour, Harry Hays, who had bought the Bar U farmlands, to run some big fleshy Holstein bulls with his Hereford cow herd. The aim was to get a larger calf and more milk production which would in turn lead to faster weight gain. In this way Baker contributed to the first synthetic breed developed in Canada, "the Hays Converter," which was registered as a distinct breed by Canada Agriculture in 1978.[21] Baker's initial connection with Hays, who was now Minister of Agriculture, led to other tasks. Nobody knew more about Canadian beef breeds than Baker, and soon he was touring the stock farms of France, Italy, and Switzerland on behalf of the Department of Agriculture. At his home place, the tough, durable herd of commercial Herefords which Baker had carefully tailored to fit his style of ranching were replaced by a succession of experimental cross-bred herds. From where we stand today, it is clear that these innovations had much to offer Canadian ranchers. However, Baker was one of the first in the field, and in the first flush of a new fashion, values for imported bulls and semen were highly inflated. Large investments in fancy bulls did not always pay off, and the larger calves they sired led to problems. During this time of innovation and experimentation every facet of stock rearing demanded meticulous attention to detail. This was not Allen Baker's strong suit. He still spent a lot of time auctioneering and travelling. Things started to go wrong, and as losses mounted Baker was forced to sell land to appease the bank. Years later, Allen admitted to an old friend that he had sold off half the ranch because he did not need the work it entailed. Unfortunately, both men knew that he would never have sold an acre of the ranch he loved if he could have avoided it.

"To commemorate and celebrate ..."

The purchase of the Bar U Ranch headquarters site by Parks Canada in 1991 was the culmination of a search and evaluation process which stretched back over more than twenty years.[22] In 1968 the Historic Sites and Monuments Board of Canada came to the conclusion that "the ranching industry is of national historic significance." They suggested that a historic ranch be acquired at which the industry could be commemorated. During the 1970s a survey of seventeen ranches was completed and criteria for evaluating their comparative attributes were developed.[23] This flurry of activity led to an attempt to buy the Bar U in 1977 which was unsuccessful. After that it was more than a decade before the search for a suitable ranch was reactivated and another shortlist of possibilities was drawn up. This exercise reached the following conclusion:

> If the Board wishes to commemorate the ranching industry through the acquisition of a historic ranch and the preservation of significant in situ resources, then the Bar U Ranch is the best candidate. It was arguably the most important single ranch in western Canada and it has much the best surviving in situ resources of any of the very large ranches of the golden age of ranching.[24]

The negotiations to purchase the headquarters site of the Bar U and some ninety-nine hectares of adjacent land was facilitated by Macleod Member of Parliament Ken Hughes, and by a group of local ranchers who worked with Parks Canada.[25] A deal was struck with Melvin Nelson that included a clause by which the Nelsons would continue to use the ranch facilities until they developed a new ranch headquarters. The establishment of the Bar U Ranch National Historic Site was one of the first projects to be undertaken under the federal government's "Green Plan." From the first, Parks Canada aimed to develop the new site in a co-operative manner with the local ranching community playing an important part in decision making. The Friends of the Bar U Historic Ranch Association was established in 1992, and a joint management agreement between the Association and Parks Canada was worked out in 1993.[26] Together the partners worked on a management plan. Its objective was to

> commemorate the history and importance of ranching in Canada and the Bar U's role in that industry. The Bar U was selected for its long and significant history and for the important cultural resources and landscape features which remain at the site.[27]

The Bar U Ranch headquarters does indeed have a number of advantages as a historic site. First, it possesses a remarkable set of some thirty-five historic buildings. These range in date from eight log buildings of the 1890s to a feed mill and horse barn built in the 1920s. Most of the buildings have been in regular use until very recently, and it has been their continuing utility that has ensured their longevity. The only obvious omission is the original frame and log ranch house, which was completely destroyed by fire in 1927. The site is easily accessible from Highway 22 South, but it is tucked away from casual view along the flood plain of Pekisko Creek. Sight lines are framed by the banks of the coulee, and the eye is led westward to the grassed foothills and the Rocky Mountains.

During the 1990s, the new Historic Site became the focus of intense research activity. The detailed history of the ranch was explored in the archives, an ongoing archaeological program was initiated, oral history interviews were undertaken, and the overwhelming interest and generosity of those who had lived and worked at the ranch led to the rapid growth of a remarkable library of photographs. On the site itself, buildings which were deemed at risk were stabilized and the long-term restoration of key buildings has begun. A visitor centre was built and the Friends produced an award-winning video to help orientate visitors to the Bar U.[28] The site was officially opened on 30 July 1995.[29] During the season casual visitors and school groups can enjoy guided interpretative tours of the site. Special events, held at weekends throughout the summer, like horse-shoeing competitions, ranch rodeos and demonstrations of polo, have proved very popular. The dream of recreating a "living working ranch" has been partially realized with the advent of a small herd of Shorthorn cattle, several working Percherons, and omnipresent riders on horseback.

What does the history of the Bar U tell us about the origins of ranching in the Canadian prairie west? Obviously, one cannot "test a hypothesis" using a sample of one! On the other hand, a detailed case study cannot help but be suggestive and illuminating. The first and overwhelming impression one gets as one reviews the story is one of complexity, of interwoven elements involving people and ideas from eastern Canada and the United States being worked out under the umbrella provided by the British Empire at its zenith.

The Canadian government played an important role in the establishment and development of ranching in western Canada. Prime Minister Macdonald saw the vital need to occupy the western borderlands of the Dominion to foil American dreams of "manifest destiny." More pragmatically, it was not acceptable to have to rely on Montana merchants to feed Canadian Indians, supply the police, and even to relay mail and telegraph messages. But the fledgling country had few resources to spare. Framing advantageous grazing regulations by which investors might be lured to the western grasslands was an inspired (and cheap) way of achieving geopolitical ends. The Department of the Interior encouraged, cajoled, and regulated cattlemen during their first decades on Canadian range, while the Department of Agriculture supervised both the import of purebred stock and the export of cattle across the Atlantic.

The North West Cattle Company was above all a Canadian enterprise. Fred Stimson, and those who accompanied him on his exploratory journey west in 1881, were stockmen from the Eastern Townships of Quebec. They saw the opportunity to replicate in Canada the international investment in ranching which had apparently been so successful on the Great Plains of the United States. The Allans, impressed with the rapid growth of the cattle trade, were persuaded to underwrite the enterprise. The North West Cattle Company was among the first to run a large commercial herd and benefited enormously from lucrative government contracts. In 1882, the Dominion was only fifteen years old and citizenship and nationality were somewhat moot concepts on the western plains. Many of those who occupied the first bunkhouse at the Bar U were frontiersmen like Tom Lynch, George Emerson, and Henry Minesinger.[30] By 1891, three quarters of those working on the range were born in either Canada or Great Britain, and the Bar U community reflected this generality. However, certain key roles were occupied by men born in the United States. By 1901 the Bar U roundup wagon was overwhelmingly Canadian, although many of the young riders like Neils Olsen were new Canadians. George Lane and Herb Millar, who played such important parts in the story, were both born in the United States but thought of themselves as Canadians.

The techniques and equipment used to manage large herds under open range conditions had been developed in the southern Great Plains. As they spread northward, a variety of regional adaptations occurred. Canadians learned ranching from those who drove the first herds in from Montana, and American cowboys played a vital part in this diffusion process. They were especially sought after by large corporate outfits which had so much to lose from inept management. While many of the skills of cowboying were quickly learned, "cow sense" could only be acquired through years of experience. Both George Lane, the first foreman, and Ebb Johnson, his successor, had served intense apprenticeships on the US range. However, it should be noted that such men were not numerous, nor was the employment of an American foreman a necessary condition for success.

The Bar U had extraordinarily close ties to the "mythic west" created by journalists, novelists, impresarios, and artists. This artificial vision constituted the reality for millions of easterners and Europeans who had never seen a range steer or a cow pony. Ebb Johnson was the model for Owen Wister's *Virginian*, the first and greatest Western novel, which sold fifty thousand copies during its first two months in print. George Lane had ridden with the cowboy artist Charlie Russell in Montana, and Elizabeth Lane corresponded frequently with Nancy Russell after the Lanes hosted the Russells during the 1912 Calgary Stampede. Russell painted a picture of Lane and the Wolves, and George bought another picture by Charlie to present to the Prince of Wales. Not only is the Bar U linked to the pre-eminent artist of the range, but through the presentation to Prince Edward, Russell was regarded in Britain as "Canada's Cowboy Artist too."[31] Finally, the presence of the outlaw Sundance Kid in the Bar U bunkhouse created another connection between the Canadian ranch and the wild west depicted in dime novels and later in western films. It is ironic that the Bar U, which epitomized in many ways the distinctly Canadian character of ranching in the Alberta foothills, should have these unequivocal links with the symbolic west of popular culture.

Great Britain made a vital but indirect contribution to the development of ranching in western Canada. It was the growing demand for imported meat for industrializing Britain that spurred the "beef bonanza." For about twenty years the best grass-fed steers from western ranges were sold in Liverpool, Glasgow, and London. Moreover, the United Kingdom was the centre of breeding excellence, and men like Cochrane, Stimson, and McEachran were profoundly influenced in their choice of breed and in their knowledge of breeding technology by current thought in "the Old Country." Immigrants from Britain trod the broad walks and frequented the bars of Calgary. While some "young gentlemen" became notorious as shiftless remittance men, the majority learned the survival skills of their new home quickly and made a major contribution to its development. At the Bar U one thinks of Mrs. Bedingfeld and her son Frank, of Joseph (7U) Brown, and Samuel Leighton the bookkeeper, to name but a few. Although I have always been impressed with the speed and degree to which many British immigrants assimilated, at a deeper level their worldview remained unchanged. Douglas Francis has demonstrated how the Canadian west was seen as both the embodiment of Canadian national unity and also a flowering of imperial grandeur.[32] The strength of these imperial ties was demonstrated in the response of westerners of British descent to the call to the colours in 1914, and the enthusiastic reception given to the Prince of Wales in 1919.

Enough has been said to demonstrate that the Bar U has a complex genealogy. But if we regard ranching first and foremost as an economic activity, and I think we must, then it is clear that the ranch was above all a Canadian enterprise. Imperial ties to Britain and links to American popular culture notwithstanding, the land, the capital, and the entrepreneurship which nurtured the big ranch were distinctly and unambiguously Canadian.

Nature Versus Nurture on the Range?

In the preface I said that I hoped that I would be in a position to comment on the relative significance of the physical environment, economic, and institutional factors, in the growth of ranching. The story of the Bar U pushes me toward two conclusions: first, any attempt to separate these three variables is doomed to failure, as they are too closely linked; secondly, any explanation which does not include an analysis of all three factors is likely to be inadequate. For example, the devastating winter of 1906–7 had obvious and direct effects on the cattle industry. But its impacts were by no means uniform. In some locations herds escaped relatively unscathed while others were wiped out. We need to base our evaluations on good science, not folklore. Moreover, cattle herds were being reduced severely from 1903 onward as prices plunged and the industry experienced a period of herd liquidation in the North American cattle cycle. In the medium and longer term, it was the spread of farm settlement and the crowning of "King Wheat" that had the most significant impact on ranching. Another example: what caused Lane's difficulties in the years following the First World War? The bitter winter of 1919 led to skyrocketing feed costs. Cattle losses were not great, but financial losses were staggering. At the same time the impact of the winter was compounded by a slump in prices and by a protective tariff introduced by the United States. This cut Lane and his peers off from their preferred market. But Lane's Achilles' heel was his huge investment in wheat production at Namaka Farms. He was poised to reap a fortune, but the rains failed to come. Surely in this instance there is a direct link between environment and bankruptcy? This logic is compelling, but, with the benefit of hindsight, one might well question whether labour-intensive mega-farms would have survived even with adequate precipitation, given the sudden rise in wages during the war.

Over the course of more than a century, it seems to me that the stroke of a pen in Ottawa, London or Washington did more to alter the direction of the cattle industry than did the vagaries of the climate. From the scheduling of Canadian cattle by Britain in 1892; through the imposition of the Young Emergency Tariff in 1919 and the Hawley-Smoot tariff in 1929; to the Canadian government's controls on exports during the Second World War; to the present Free Trade era, the fortunes of western cattlemen have been decided by the roll of the dice in faraway political antechambers. This view of the past means that we must be concerned about the present drive in the United States toward "country of origin labelling."

A major reason for the long history of the Bar U Ranch as a corporate enterprise was its unique links to markets. In the first decade Stimson snatched up government contracts, and later the influence of the Allans smoothed the flow of Bar U cattle to Britain. Lane's close association with the exporting firm of Gordon, Ironside and Fares maintained these beneficial links. Pat Burns was a past master at knowing where and when to sell cattle, and the company which inherited the management of the ranch after his death benefited from his connections. The Bar U was a cattle buyer's dream: it produced large quantities of a uniform and consistent product. The ranch built up relationships of mutual respect first with purchasers in Britain and then with commission agents in Chicago. These links in the production chain played a key role in insuring the continuing viability of the ranch, and the economic variable held the trump.

An Honourable Legacy

A number of looming threats cast dark shadows over the foothills ranges in the first decade of the new millennium. The narrow band of rough fescue grassland bordering Highway 22 south from Priddis to Lundbreck, which includes the old Bar U leases, is under enormous pressure. Population growth channelled by increasing affluence and mobility into the rural/urban fringe represents the greatest threat to the maintenance of healthy and sustainable grassland ecosystems. Not only are western cities growing outward as planned developments swallow agricultural land around their margins; small towns and villages within ever-widening commuter sheds are also experiencing rapid expansion. Around each of these nuclei occurs concomitant spread of acreages and "ranchettes." Urbanites in search of recreation add to the pressure. Francis Gardner of the Mount Sentinel Ranch refers to this movement as "an invasion into an intact bioregion that allows for long-term fracturing and nibbling away at the ecosystem."[33] The danger is fragmentation. Even quite modest residential developments can destroy the integrity of whole watersheds. More people, vehicles, fences, dogs, and horses occupy the green zones and make them less accessible to wildlife while disrupting movement from one patch of habitat to another. Environmentalists from Alberta to Arizona recognize that prosperous and well-run ranches may be the most effective barrier to insidious urban-related growth. "Cows not Condos" is a popular bumper sticker on the Great Plains.[34]

Another important threat to the integrity of the last fragile areas of natural grassland is the oil and natural gas industry. Gas production will likely peak in 2003, and companies are turning their attention to exploring areas in the foothills which were bypassed earlier because of their steep topography and inaccessibility. New techniques and higher prices make production in these areas increasingly attractive. Of course, the roads and drilling sites would occupy a tiny fraction of the land, but ranchers point out the dangers of doing irreversible damage. Stephen Hughes, a local rancher and member of the Pekisko Land Owners Association, explained:

> This is an intact system that cannot be restored. It's a functioning, sustainable grassland ecosystem. People should be aware that if we care for this, it's here forever.... But when it's gone, it's gone.[35]

Others have emphasized the importance of undisturbed grasslands for watershed management, as carbon sinks, wildlife refuges and as areas of unparalleled natural beauty. They argue that preservation of this valuable sustainable resource will make more sense in the long term than destroying it for short-term gains. The ranchers won a skirmish when Vermilion Resources withdrew a natural gas application on rangeland located inside the Foothills Parkland Natural Area, but they would like to see a moratorium on drilling throughout the area.[36]

As encouraging as any small victory is the increasing sophistication of foothills ranchers in their grasp of grassland ecology and their growing acceptance of a role as stewards of the ranges which they occupy. Ranchers from Cochrane to Waterton are coming together in a number of associations like the Southern Alberta Land Trust Society and Action for Agriculture. These organizations help individuals realize that they do have options and support as they face seemingly inexorable pressures. For some, conservation easements arranged through the Nature Conservancy of Canada may be an answer.[37] There are workable ways of preserving undisturbed rangeland, but they won't happen without the support of the urban majority who hold the political power. The public have to be educated to appreciate how precious the surviving native grasslands are. The adoption of rough fescue as a provincial emblem is a small step in the right direction.[38] Francis Gardner again stresses the crucial importance of these areas:

> The land of the eastern slopes is a relic, a portion of what was once the great native grasslands of North America. We live on a fringe, a non-ploughed ancient island of biodiversity that has resulted from the interaction of topography, wind, buffalo and fires.[39]

It will be a challenge to ensure that our children's children will be able to enjoy the intoxicating vistas of rangeland and aspen woodland along Highway 22 South.

The legacy of the Bar U Ranch will play a part in the struggle to preserve unfragmented areas of native grassland. Geography and historical accident have meant that most of the leased land obtained by George Lane in 1905, within the great curve of the Highwood River, remains relatively unchanged to the present. Across Pekisko Creek this range merges with that of the D Ranch and the Spruce Grazing Cooperative on the old 7U Brown range, and these in turn lead southward toward the huge Walrond Grazing Cooperative pasture. Physically and tangibly, the old Bar U leases form the northern bastion of a large corridor of precious natural rangeland. At the same time, it is part of the mandate of the Bar U Ranch National Historic Site to explain the nature and importance of rough fescue grass to both past and present ranchers. This is a task which they have tackled with enthusiasm spurred on by the Friends. I am convinced that Stimson, Lane, and Burns would have heartily approved the ongoing role of the Bar U Ranch.

Appendix

The appendix provides the statistical background to several of the topics dealt with in this book. Five groups of figures and tables present data on: the killing winter of 1906–7, the effects of the Great Depression, the impact of the Second World War, the rising labour costs after the war and the contribution of the First Nations people, and, finally, the sale of the ranch.

An Overview

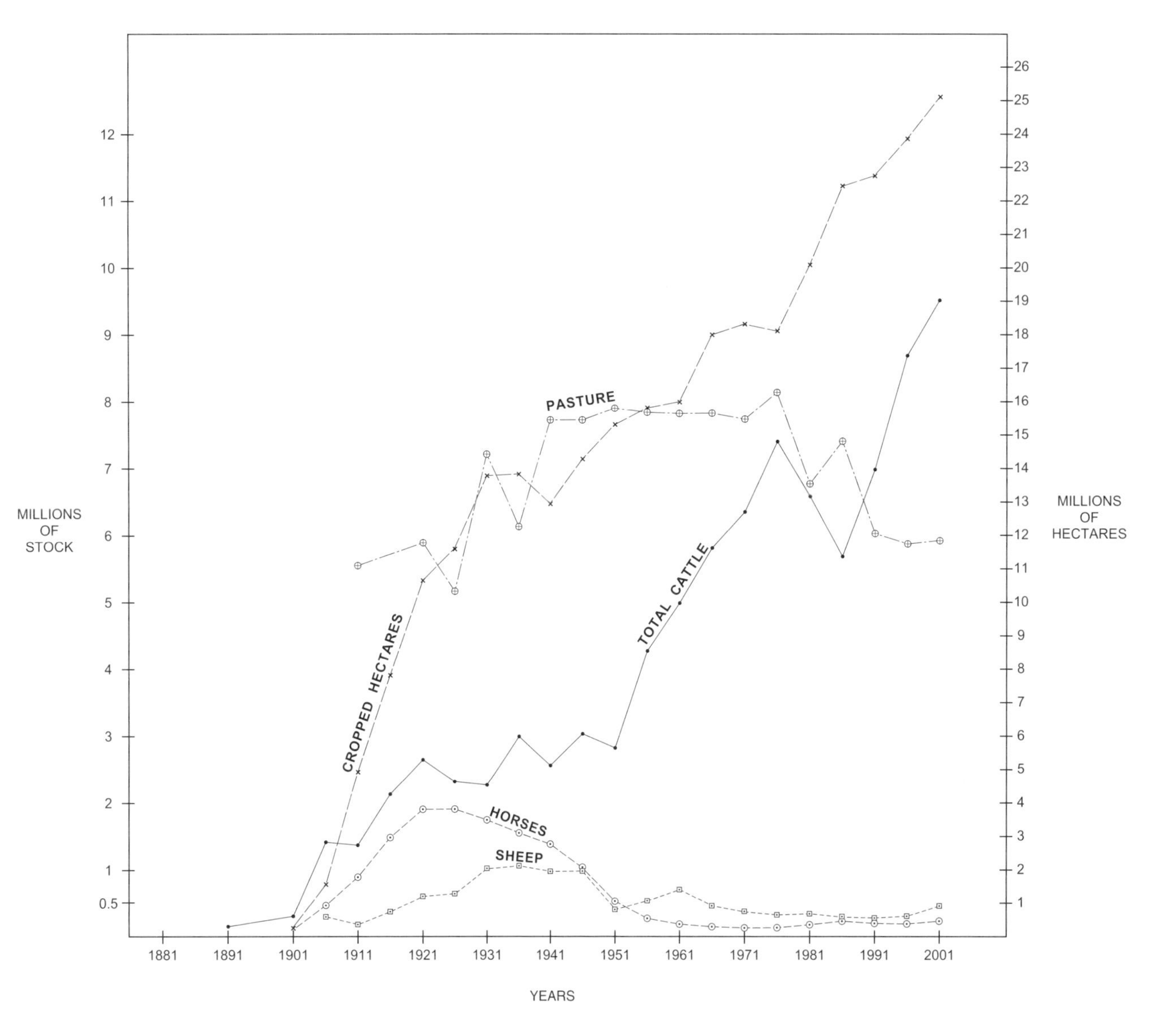

Figure 2.1 An overview: stock crops and pasture, Alberta and Saskatchewan, 1891–2001. Drawn from Census of Canada figures.

Just as the maps used as the endpieces provide a locational context for many of the ranches mentioned in the text, so Figure 2.1 gives an overview of the number of stock in Alberta and Saskatchewan and the expansion of the cropped acreage, as shown by the Census of Canada. Notice that the "buildup" phase of cattle numbers between 1901 and 1921 was interrupted between 1906 and 1911. Between 1921 and 1951 there was little overall growth, and a number of "cattle cycles" are evident. The most outstanding feature of the graph is the prodigious and sustained growth in cattle numbers between 1951 and 1976, and again between 1986 and 2001. I find this graph a useful image to bear in mind when discussing the possibility of overgrazing during the last decade of the nineteenth century. If it did exist, it must have been very local in extent and have resulted from small stockmen not wanting to allow their stock to wander far from their home places. Note that the number of horses on farms in the west grew at much the same rate as did cattle, and remained comparable in numbers until about 1931. It was not until after 1941 that the numbers of horses started to decline rapidly. The definition of "pasture on farms" has changed marginally over the years, but it continues to define unimproved land used for pasture.

The Winter of 1906–7

Tables 4.1, 4.2, and 4.3 provide some background to the discussion of the effects of the severe winter (pp. 129–42). Table 4.1 looks at the figures for "other horned cattle" for four census periods both at the provincial scale and in selected Census Districts chosen to cover the Grazing District. Table 4.2 complements the figures for cattle by illustrating the rapid growth in the area of field crops sown during the same period. Table 4.3 provides an estimate of Lane's losses at his various ranch locations.

Table 4.1a Beef Cattle in Western Canada.

	Alberta		Saskatchewan	
1901	329,391		160,613	
1906	849,387	+158%	360,236	+124%
1911	592,076	–30%	452,470	+25%
1916	893,886	+51%	697,237	+54%

Source: Census of Canada.

Table 4.1b Beef Cattle by Selected Census Districts.

Year	District	Cattle	
1906			
	Alberta[a]	373,482	
	Calgary	262,293	
	Qu'Appelle	37,099	
		Total:	672, 874
1911			
	Calgary	41,156	
	Macleod	85,490	
	Medicine Hat	144,471	
	Moose Jaw	97,484	
		Total:	368,601[b]
1916			
	Bow River	78,030	
	Calgary East	41,431	
	Calgary West	24,982	
	Lethbridge	35,704	
	Macleod	99,667	
	Medicine Hat	43,206	
		Total:	323,020
	Assiniboia	30,310	
	Maple Creek	46,116	
	Moose Jaw	19,443	
	Swift Current	20,579	
	Weyburn	14,986	
		Total:	131,434
	Combined 1916 Total		454,454[c]

The Districts chosen cover the area from Calgary southward to the International boundary, and comprise "the grazing district." With each succeeding census the divisions were divided up into more subdivisions, but the total area remained very much the same.
a Alberta census district, not the whole province.
b A loss of 304,273 or – 45.2%
c A gain of 85,853, or + 23%

Table 4.2 Areas of Field Crops, 1906, 1911, and 1916 (by Census District).

		Acres	
1906	Alberta (district)	270,767	
	Calgary	170,026	
	Qu'Appelle	1,012,092	
		Total	1,452,885
1911	Calgary	227,129	
	Macleod	772,490	
	Medicine Hat	800,654	
	Moose Jaw	1,438,061	
		Total	3,238,334
1916	Bow River	882,602	
	Calgary East	264,684	
	Calgary West	93,120	
	Lethbridge	441,051	
	Macleod	579,892	
	Medicine Hat	780,199	
		Total	3,011,548
	Assiniboia	1,048,656	
	Maple Creek	1,356,833	
	Moose Jaw	849,315	
	Swift Current	1,520,015	
	Weyburn	1,134,443	
		Total	5,909,262
	Combined 1916 Total		8,920,810[a]

a Census districts selected correspond to "the grazing district."

Horse Power

Figure 5.1 provides some background for George Lane's decision to develop a Percheron stud during the first decade of the twentieth century. It shows the number of horses reported on farms in both Canada and Alberta. See pp. 149–53 in the text.

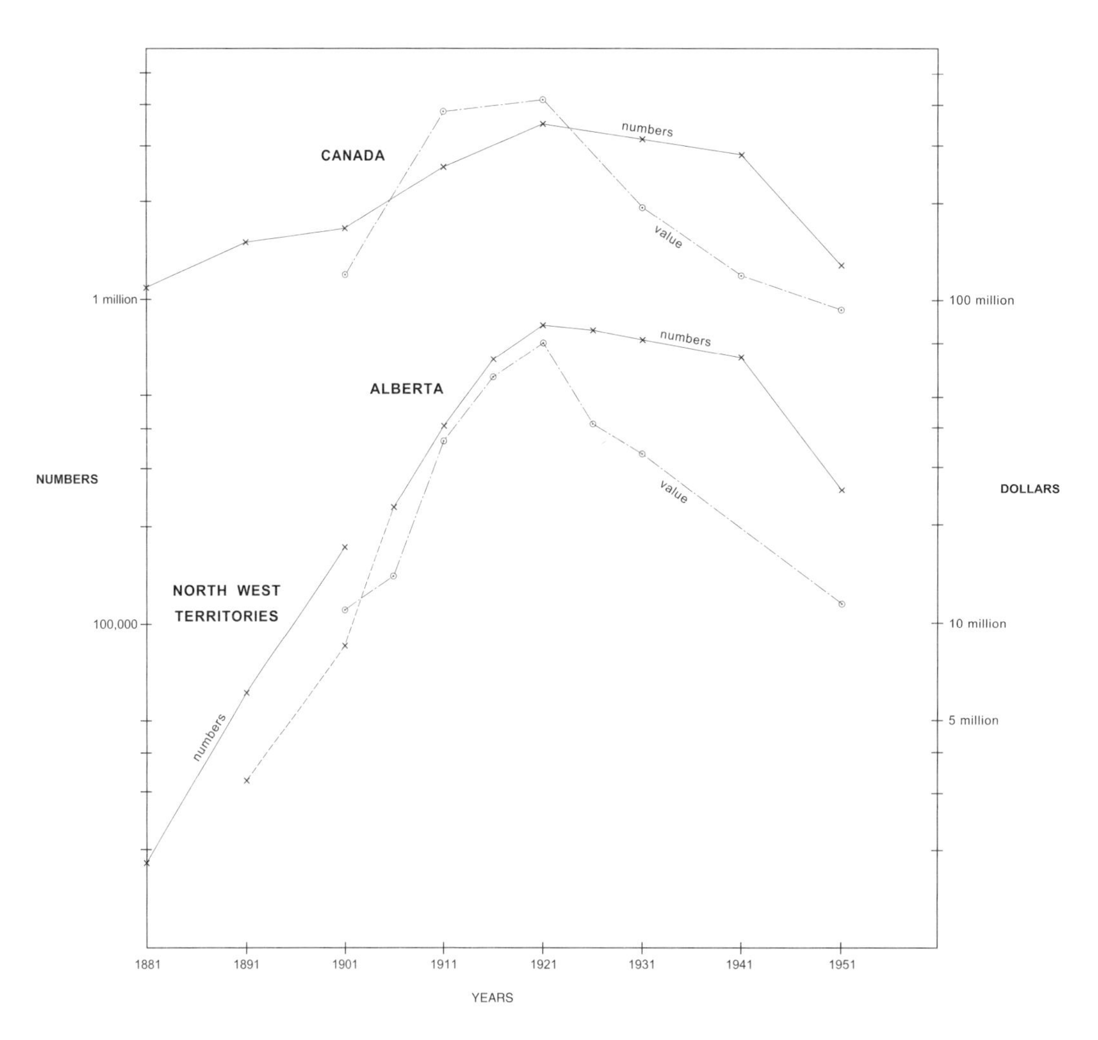

Figure 5.1 Number and value of horses on farms, Canada and Alberta, 1881–1951. Source: *Census of Canada*, 1881–1951

Note: Semi-logarithmic graph paper. No attempt was made to standardize the buying power of the dollar.

Table 4.3 Estimate of George Lane's Losses during the Winter of 1906–7.

Bar U and Willow Creek Ranches (10,000 head)

	Losses	$ Per head	Total $
Calves	350	12.00	4,200
Cows	350	30,00	10,500
Yearlings	200	20.00	4,000
Dry cattle 2years +	100	40.00	4,000
Total	1,000 [10%]	Total	22,700

Big Bow Bar U, near Bassano (10,000 head)

	Losses	$ Per head	Total $
First Year stockers (25% of 3000)	750	30.00	22,500
2nd Year stockers (20% of 3000)	600	50.00	30,000
Ranch bred cattle (15% of 4000)	600	50.00	30,000
Total loss	1,950 [20%]		82,500

Notes: Losses amounted to almost 3,000 head out of 20,000. Their value was in excess of $105,000. Assumptions: Lane owned about 20,000 head before the winter. He estimated his overall losses at 15 per cent, but losses would have been different in different areas of the range, and among different classes of stock.

Table 7.1 Benchmark Evaluations for Land and Stock.

	1927[a]	1937[b]	1947[c]
Land Deeded (acres)	26,842	27,679	29,675
Value	$297,000	$187,636	$593,500
Value per acre	$11.00	$6.78*	$20.00**
Leased acres	45,619	41,490	42,685
Stock			
Cattle total	4,996	5,151	6,637
Steers 3 y.o.	830	206	
2 y.o.	1,422	4	302
1 y.o.	152	1066	
Cows 3 y.o.	1,159	1675	3,891
2 y.o.	534	800	1066
1 y.o.	886		
Bulls	80	72	106
Calves	971	1562	n/a
Horses	273	283	334
Total value	$509,000	$370,000	$860,000

Note: This table compares land, stock, and total values of the Bar U Ranch at three crucial times: when Burns bought the ranch in 1927, at Burns' death in 1937, and in 1947 two years before the sale of the corporate ranch.
* Land value was appraised by the government at $6.78 per acre and by the company at $3.00 per acre.
** Estimate based on figures for 1949.

Sources: 1927: GA M160 f. 93 Estimate of Bar U share.
1937: GA M160 f. 234 "Abstract of valuations."
1947: GA M160 f. 103 "P. Burns Ranches Ltd. Information on ranches."

The five tables and two figures in this section of the appendix provide statistical information on the impact of the Great Depression on agriculture in general and the cattle industry in particular. A variety of scales are employed, ranging from the whole nation to individual ranches. Table 8.1 deals with cattle values and exports, and shows the major changes which took place in both indicators. Table 8.2 illustrates the efforts made at the Bar U to control expenditures through the 1930s. It also shows the rapid escalation in wages during and after the Second World War. Table 8.3 shows the net income of the Bar U from 1937 onward; unfortunately data is not available on a ranch-by-ranch basis for the earlier years of the Depression. Figure 8.1 demonstrates that, in spite of fluctuations in herd size, there was no consistent effort to reduce herd size radically during the Depression, but there was a substantial buildup toward the end and just after the war. Tables 8.4 and 8.5 present data from specific Burns ranches in different ecological zones. Finally, Figure 8.2 shows how the whole group of Burns ranches fared from 1928 through 1946. It suggests that the consistent losses suffered until 1938 were more than recouped during the war years and just after.

Table 8.1 Cattle Values and Exports through the Depression Years.

	A	B	C	D
	Total Value of cattle & calves, Alberta in $000s	Av. Yearly farm price, cattle Calgary per cwt.	Exports of live cattle to US from Canada	Exports of live cattle to UK from Canada
1926	17,255	5.95	92,962	76,985
1927	16,334	7.06	204,336	8,263
1928	19,441	9.09	166,469	405
1929	18,171	8.71	160,103	nil
1930	9,006	7.68	19,483	5,400
1931	6,560	4.71	9,159	27,149
1932	4,084	4.02	9,010	16,568
1933	3,842	3.27	5,686	50,317
1934	5,892	3.68	6,341	53,852
1935	11,067	4.89	102,934	6,704
1936	12,039	4.30	191,149	38,495
1937	17,088	6.28	208,552	9,610
1938	13,565	5.18	147,000*	27,300*
1939	15,344	6.08	284,000*	4,000*
1940	18,018	6.72	229,000*	nil*

Sources: A & B, J.K. Woychuk (ed.), *Alberta Agriculture: A History in Graphs* (Edmonton: Queen's Printer, 1972), pp. 90, 108. C & D, *Canadian Cattlemen*, Vol. 1, No. 1 (June 1938), "Canadian Cattle Exports." The figures marked * were added from G.E. Britnell and V.C. Fowke, *Canadian Agriculture in War and Peace, 1939–50* (Stanford: Stanford University Press, 1962).

Table 8.2 Costs of Board, Wages, and Expenses, Bar U Ranch, 1934–50.

	Board	General Expenses[a]	Wages	Total Expenses
1934	1880	4266	5409	11,556
1935	1610	4366	7027	13,004
1936	1885	3202	6507	11,594
1937	1646	3452	7820	12,919
1938	1632	6918	9326	17,877
1939	1974	5586	8978	16,539
1940	2657	6704	11,059	20,421
1941	2394	8245	13,811	24,453
1942	2796	17,435[b]	19,427	39,661
1943	3686	11,855	19,538	35,080
1944	3423	14,819	22,218	40,460
1945	3705	20,223	23,022	46,951
1946	4105	19,366	27,751	51,224
1947	5026	16,471	30,171	51,669
1948	5825	25,566	31,606	62,997
1949	6716	22,519	31,454	60,689
1950	5586	23,610	30,404	59,600

a. Includes fencing, trucking, light, heat and power, stable, repairs, gas and oil, and miscellaneous.
b. This figure probably reflects the costs of modernizing the cookhouse.
Source: These annual figures were worked out from Clifford/Alwood Collection, Monthly Statements.

Table 8.3 Bar U Ranch Net Income after Expenses and Depreciation.

1937	–5,052
1938	17,528
1939	32,555
1940	27,754
1941	55,009
1942	43,243
1943	49,547
1944	67,287
1945	10,097
1946	103,105
1947	133,105
1948	239,230

Source: GA M160, f. 329, P. Burns Ranches, Income, Profit and Loss Accounts.

Table 8.4 Circle Ranch Expenses, January 1929 to December 1939.

	Feed	Total Expenses	Feed as %
1929	378	7,382	5
1930	1,171	12,253*	10
1931	2,119	7,430	28
1932	1,934	6,371	30
1933	5,745	9,948	58
1934	8,269	12,984	64
1935	7,707	13,038	59
1936	13,443	19,084	70
1937	15,300	21,958	70
1938	5,865	11,598	50
1939	7,842	13,123	60

* $5,000 of fencing, a one time capital expenditure.
Source: GA M160 f. 270, "Circle Ranch Expenses."

Table 8.5. "76" Ranch, Profit and Loss and Cumulative Deficit, 1929–39.

	Profit or Loss ($)	Deficit ($)
1929	–31,273	71,348
1930	–11,808	83,197
1931	–11,257	109,499
1932	–7,155	116,694
1933	+17,822	98,911
1934	+5,282	93,658
1935	–2,794	96,482
1936	–25,832	122,345
1937	–26,298	148,673*
1938	–1,678	150,381
1939	–1,183	125,171**

* Most of the herd was disposed of during this year.
**Deficit after sale of ranch.
Source: GA M7771, f. 25, Audited accounts of "76" Ranch.

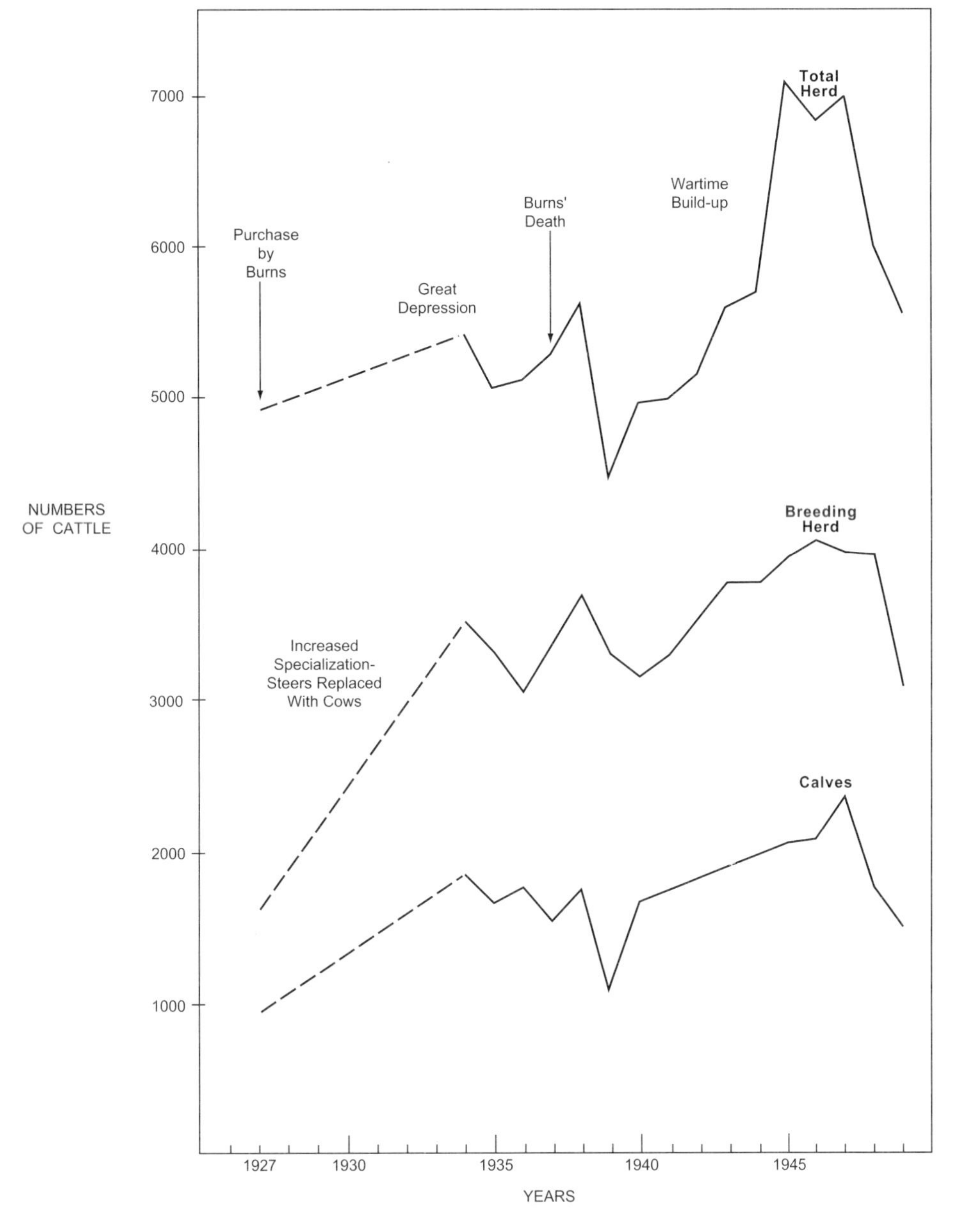

Figure 8.1 Stock on the Bar U Ranch, 1927 – 49

Source: GA M160,f. 92, *P. Burns Ranches, Profit and Loss Statement*, 1934–40 and inventories to 1950. Figures for herd make up provided in these statements differ somewhat from the Stock Books which only provided data on cows and calves, not on 1 and 2 year olds held over to replace cows culled from the herd. Figures refer to December 31st of the year in question.

Note: numbers for the breeding herd includes heifers held over to join the breeding herd when they matured.

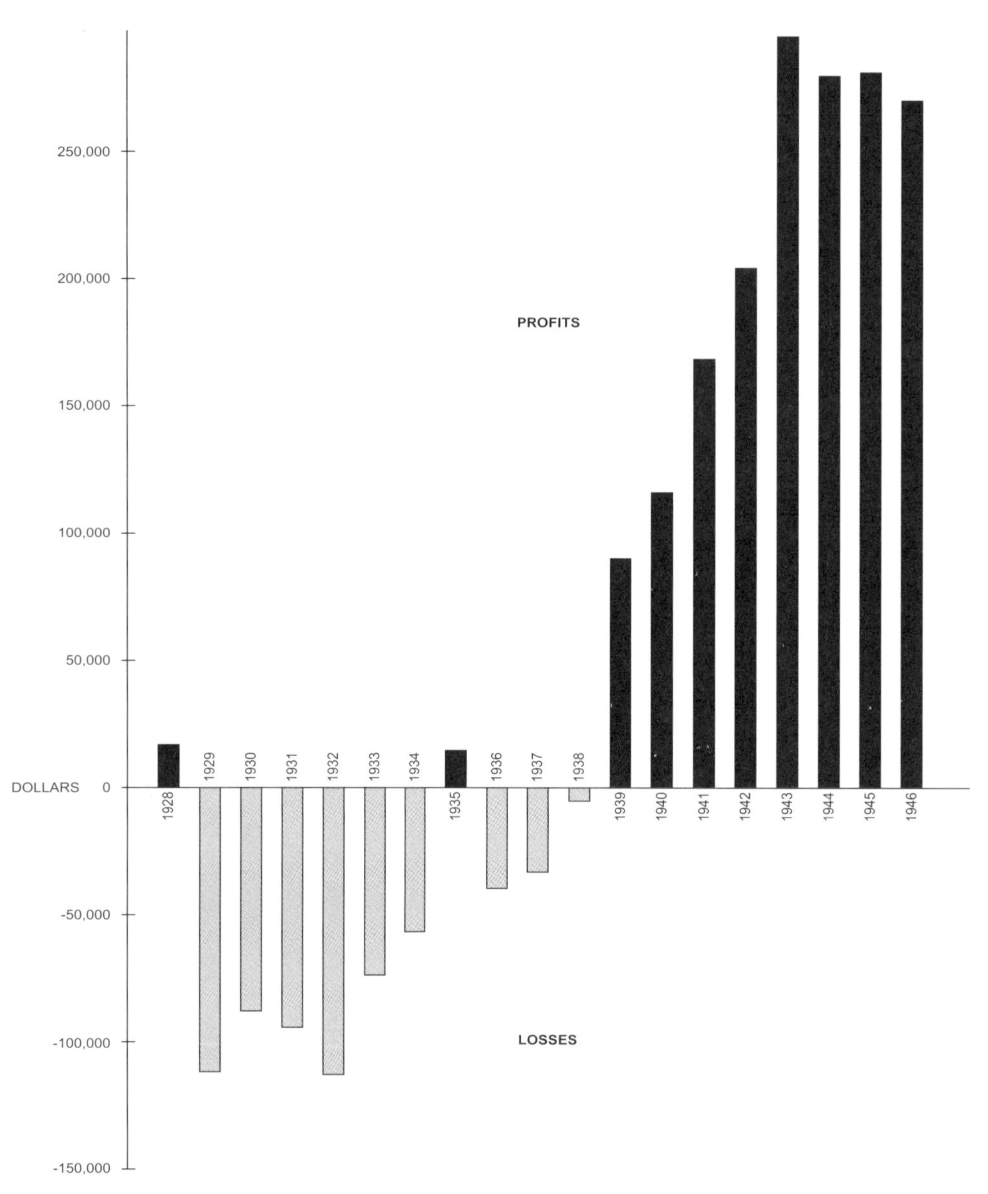

Figure 8.2 P. Burns Ranches Limited net ranch profit or loss, 1928–46. This graph shows both the losses of the 1930s and the wartime profits.

Note: These figures refer to all 12 Burns Ranches
Source: GA M7771, f. 48 " Brief to Excess Profit Tax Act."

The Great Depression gave way to the Second World War, and demand for cattle rose dramatically. It was, however, controlled by the Canadian government in the interests of the war effort. Table 9.1 shows how the direction of Canadian exports changed from the United States to the United Kingdom in 1943 and 1944. Table 9.2 deals with much the same time Period in Alberta. Tables 9.3, 9.4, and 9.5 provide data on all Burns Ranches and the Bar U in particular. There is some overlap with material previously presented, but the sources are different and the figures provide confirmation from a slightly different point of view. Tables 9.6 and 9.7 provide some data on First Nations people employed at the ranch.

Table 9.1 Cattle on Farms in Canada and Exports through the Depression and the Second World War.

	Livestock on Farms, "Other Cattle" (1,000s)	Exports, live cattle U.K. ('000s)	Exports, live cattle U.S.A. ('000s)	Value of Exports Millions ($)	
1930	4,453	5.3	54.7	3.4	
1931	4,601	27.1	24.6	3.6	
1932	4,956	16.6	12.9	2.1	
1933	5,264	50.3	5.7	3.7	
1934	5,209	53.8	6.6	4.0	
1935	5,132	6.7	123.5	6.9	U.S. Quota system
1936	5,024	38.5	241.7	12.4	
1937	5,071	9.6	309.0	15.7	
1938	4,761	27.3	147.4	9.1	
1939	4,693	4.0	284.5	15.3	U.S. reduces duty
1940	4,730	Nil	229.5	12.4	
1941	4,891	Nil	250.6	16.8	
1942	5,099	Nil	212.4	17.6	
1943	5,499	Nil	58.6	9.5	Canadian embargo
1944	5,876	320.0*	53.3	9.0	
1945	6,001	573.0*	70.7	12.3	
1946	5,689	478.0*	96.4	18.0	
1947	5,674	102.0*	74.5	15.0	
1948	5,627	110.0*	450.5	73.0	Canadian embargo lifted
1949	5,404	Nil	417.6	61.4	
1950	5,224	Nil	456.7	79.1	

* Frozen beef export agreements, cattle equivalents.
Sources: Britnell and Fowke, *Canadian Agriculture*, pp. 454, 461. M.C. Urquhart and K.A.H. Buckley (eds.), *Historical Statistics of Canada* (Toronto: MacMillan, 1965), p. 377.

Table 9.2 Value and Price of Cattle through the Depression and the Second World War in Alberta.

	Cattle on Farms (1,000s)	Value of Cattle ($000s)	Price of Good Steers ($ per cwt)
1930	1,037	9,006	7.68
1931	1,124	6,560	4.71
1932	1,244	4,084	4.02
1933	1,519	3,842	3.27
1934	1,642	5,892	3.68
1935	1,697	11,067	4.89
1936	1,554	12,039	4.30
1937	1,515	17,088	6.28
1938	1,376	13,565	5.18
1939	1,311	15,344	6.08
1940	1,299	18,018	6.72
1941	1,342	19,064	8.29[a]
1942	1,429	23,566	9.90
1943	1,548	28,525	11.30
1944	1,640	35,611	11.13
1945	1,708	49,136	11.37
1946	1,601	48,613	12.08
1947	1,630	45,974	13.85[b]
1948	1,663	79,612	18.75
1949	1,595	84,403	20.05
1950	1,563	93,771	26.13

a. Price controls imposed.
b. Price controls removed.
Source: John K. Woychuk (ed.), *Alberta Agriculture: A History in Graphs* (Edmonton: Queen's Printer, 1972), tables 47, 48, and 55.

Table 9.3 P. Burns Ranches: Profits and Losses, 1930s and 1940s.

	Profit/Loss from stock sales ($)	Net Income or Loss ($)	Net profits P. Burns Ranches & Western Ranches ($)
1928	16,670	4,628	
1929	–111,408	–134,875	
1930	–89,328	–118,279	
1931	–94,563	–129,397	
1932	–11,763	–138,650	
1933	–73,414	–95,676	
1934	–59,179	–82,567	
1935	15,518	1,152	
1936	–40,570	–60,191	
1937	–34,013	–53,906	–53,938
1938	–399	38,811	–23,895
1939	89,889	83,978	88,065
1940	116,422	86,398	117,610
1941	169,531	155,814	171,433
1942	202,634	192,714	228,232
1943			295,610
1944			280,659
1945			281,916
1946			270,187

Sources: For the first and second columns: GA M7771, f. 48, "Brief to Excess Profits Tax Act."
Third column: GA M160, f. 329.

Table 9.4 Bar U Ranch, Income and Profits, 1934–50.

	No. of Cattle	No. of Calves	Gross Income $	Profits $
1934	5,416	1,826		
1935	5,036	1,650		
1936	5,080	1,770		
1937	5,151	1,562	24,491	–5,052
1938	5,603	1,741	47,969	17,528
1939	4,431	1,032	63,297	32,555
1940	4,904	1,660	62,639	27,754
1941	4,982	n.a.	86,826	55,009
1942	5,180	n.a.	104,700	43,243
1943	5,581	1,824	108,462	49,547
1944	5,659	1,881	130,464	67,287
1945	7,102	2,015	97,212	10,097
1946	6,812	2,090	216,290*	103,105*
1947	7,006	2,368	331,425*	133,105*
1948	6,061	1,769	452,028*	239,230*
1949	5,512	1,502		
1950	2,169*	1,111*		

Sources: GA M160, f. 329. Figures marked * were added from GA M160, f. 92.

Table 9.5 Bar U Ranch: Labour Costs, 1934–50.

	A Monthly Wages ($)[a]	B Total Wage Bill ($)	C Board ($)	D Total Labour ($)	Labour as % of Expenses
1934	25	5,409	1,880	7,289	
1935	25	7,027	1,610	8,637	
1936	25	6,507	1,885	8,392	
1937	25	7,820	1,646	9,466	34%
1938	25	9,326	1,632	10,958	
1939	25	8,978	1,975	10,953	
1940	30	11,058	2,658	13,716	
1941	40	13,811	2,396	16,207	
1942	55	19,427	2,798	22,225	39%
1943	70	19,538	3,686	23,224	
1944	70	22,218	3,423	25,641	
1945	70	23,022	3,705	26,727	
1946	70	27,751	4,106	31,857	47%
1947	70	30,171	5,026	35,197	
1948	70	31,606	5,825	37,431	
1949	70	31,454	6,716	38,170	
1950	70	30,404	5,585	35,989	

a. Riders' wages; these men received median pay, i.e., not as much as the foreman or the cook, but more than the choreboys.

Sources: Month wages are from oral reports, except for 1939, 1942, 1943, and 1946. Information in columns, B, C, and D from Clifford/Alwood Collection, "Monthly Statements Burns' Ranches Limited." Column E, GA M160, f. 329, P. Burns Ranches, "Income, Profit and Loss Accounts for ten years, December 1946."

Table 9.6 Payments to Stoneys in 1942 and 1946–47.

	1942 ($)	1946/47 ($)
November	n.a.	213.00
December	n.a.	10.00
January	2.50	n.a.
February	22.77	n.a.
March	n.a.	47.32
April	104.00	252.41
May	34.15	439.98
June	619.37	649.00
July	8.00	111.70
August	27.00	57.00
September	n.a.	n.a.
October	186.98	516.60
November	84.98	n.a.
December	322.50	n.a.

Source: Clifford/Alwood Collection, Receipt Books.
Note: Wages paid to Indians in 1942 were 7 per cent of the total ranch wage bill; in 1946–47 the percentage was 7.6 percent.

Table 9.7 Family Names of Stoneys Employed at the Bar U Ranch, 1941–49.

	1941/42	1946/47	June 1946	June 1949
Jonas Rider	X	x	x	x
Matthew Rider	X	x	x	
Neil Dixon	X	x	x	
Peter Dixon	X	x		
Judea Dixon		x	x	x
Stewart Dixon				x
Eddie Lefthand	X	x		x
Webster Lefthand	X	x	x	x
Frank Lefthand		x		
Carl Lefthand		x		
Mark Lefthand		x		
Wing Lefthand		x		
Wayne Lefthand		x		
Billy Bearspaw	X			
Albert Bearspaw		x	x	
Melvin Bearspaw				x
Frank Powderface	X	x		
Sam Powderface		x		
Iza Powderface		x		
Johnson Powderface		x		
Peter Pucette		x	x	x
Peter Pucette (jr.)				x
Paul Wesley		x		
Fred Satler		x		x
George Smalleyes		x		
Hanson Two Youngman		x		
Howard Rollingmud		x		
Dimmit Adams		x		
Wallace Ear		x		x
R. Ear		x		
Smoky Adams			x	
Albert Amos			x	
Pete Labelle		x		

Source: Clifford/Alwood Collection, Receipt Books and Monthly Time Books.

Tables 9.8–9.11 provide some data and estimates concerning the sale of the ranch in 1949. The land was sold to neighbours at prices that varied according to the quality and location of the particular acreage concerned (see Table 9.8). Returns from the sale of this small portion of the ranch are used to estimate the total returns from land sales (see Table 9.9). Table 9.10 shows Tom Farrell's guide to stock prices. It is used to develop two estimates of the actual sales returns generated by stock (see Table 9.11).

Table 9.8 Fifth and Sixth Tenders, 1949.

	Leased	Deeded	Price ($)	Price/Acre ($)
Fifth tender				
Hogg brothers	960	600	12,539	20.89
H.W. Rowland	960	640	18,555	28.99
Doreen & Derek Runciman	240	480	10,013	20.86
G.K.R. Nelson	320	320	7,730	24.16
R.E. Nicholls	720	1,760	33,661	19.12
J.A. Hughes	1,120	1,760	33,661	23.35
Totals	4,320	5,560	123,603	22.89 Av.
Sixth tender				
R. Clifford		380	12,449	32.76
Roy Henderson		2,410	97,676	40.53
Rex R. McLean		1,020	37,792	37.05
Totals		3,810	147,917	36.78 Av.

Note: Acreages estimated from land descriptions.
Source: GA M7771, f.53, "Real Estate: Sale of Land."

Table 9.9 Estimated Returns from Land Sales: Bar U Ranch, 1949–50.

		$30	$36.78
Twp. 17, Rge 2			
J. Allen Baker	3,840 acres farmland	115,200	141,235
	3,810 acres farmland	114,300	140,132
Remaining land	6,400 acres farmland	192,000	235,392
		421,500	516,759
Twp. 16, Rge. 2			
9 Sections	5,760 acres farmland	172,800	211,853
7 Sections	4,480 acres leased	11,200	11,200
		184,000	223,053
		$20	$22.89
Twp. 16, Rge. 3			
1 Section	640 acres rangeland	12,800	14,650
4 Sections	2,560 acres leased	6,400	6,400
		19,200	21,050
Twp. 17, Rge. 3			
J. Allen Baker	1,280 acres rangeland	25,600	29,299
Remainder	640 acres rangeland	12,800	14,649
J. Allen Baker	13,760 acres leased	34,400	34,400
Remainder	1,600 acres leased	4,000	4,000
		76,800	82,348
Twp. 17, Rge. 4			
1 Section	640 acres rangeland	12,800	14,650
15 Sections	9,600 acres leased	24,000	24,000
		36,800	38,650
Twp.18, Rge. 3			
Fifth Tender	5,560 acres rangeland	111,200	127,268
Fifth Tender	4,320 acres leased	10,800	10,800
		122,000	138,068
	Totals	860,300	1,019,928

Note: The two columns of figures show the results of two different assumptions. In column 1, very conservative values of $30 and $20 have been used. In column 2, the actual averages derived from the tender figures available have been used.

Table 9.10 Prices of Stock.

Cattle:	$ per head
Wet Cows	135
Two-year-old heifers	165
Yearling heifers	125
Bulls	300
Calves	75
Work horse and saddle horses	50
Hogs:	
Brood sows	50
Boars	75
Weaner pigs	Market value

Source: GA M160 f. 92, "Memo re Sale of Bar U Ranch," November 8, 1949.

Table 9.11 Estimated Revenue from Sales of Stock.

	Conservative		Less Conservative	
Steers	1,500 at $150	225,000	1,500 at $150	225,000
Cows and calves	4,957 at $140	693,980	4,957 at $180	892,260
Bulls	100 at $300	30,000	100 at $300	30,000
Horses	300 at $50	15,000	300 at $80	24,000
Totals:		963,980		1,171,260

Source: GA M160, f. 103, Information re Ranches and Farms.

Selected Bibliography of Books

For Government Publications and Journal Articles please see the chapter notes.

Abbott, Edward Charles 'Teddy Blue,' and Helena Huntington Smith. *We Pointed them North: Recollections of a Cowpuncher*. Norman: University of Oklahoma Press, 1939.

Adams, Andy. *The Log of a Cowboy*. 1903, Lincoln: University of Nebraska Press, 1964.

Alsager, Dale. *The Incredible Gang Ranch*. Surrey, B.C.: Hancock House, 1990.

Armitage, Susan, and Elizabeth Jameson (eds.). *The Women's West*. Norman: University of Oklahoma Press, 1987.

Atherton, Lewis. *The Cattle Kings*. Lincoln: University of Nebraska Press, 1961.

Baillargeon, Morgan, and Leslie Tripper. *Legends of Our Times: Native Cowboy Life*. Vancouver: University of British Columbia Press, 1998.

Baker, William M. (ed.). *The Mounted Police and Prairie Society, 1873–1919*. Regina: Canadian Plains Research Center, 1998.

Bassano History Book Club. *The Best in the West by a Dam Site*. Calgary: Friesen, 1974.

Beahen, William, and Stan Horrall. *Red Coats on the Prairies: The North-West Mounted Police, 1886–1900*. Regina: Printwest, 1998.

Bellavance, Marc. *A Village in Transition: Compton, Quebec, 1880–1920*. Ottawa: National Historic Parks and Sites, 1982.

Bennett, John W. *Northern Plainsmen: Adaptive Strategy and Agrarian Life*. Arlington Heights, Ill.: AHM, 1976.

——— and Seena B. Kohl. *Settling the Canadian-American West, 1890–1915: Pioneering Adaption and Community Building*. Lincoln: University of Nebraska Press, 1995.

Bennett, Russell. *The Compleat Rancher*. New York: Rhinehart and Co., 1946.

Berry, Gerald L. *The Whoop-Up Trail*. Edmonton: Applied Art Productions, 1953.

Berton, Pierre. *Marching As To War: Canada's Turbulent Years, 1899–1953*. Toronto: Doubleday Canada, 2001.

———. *The Great Depression, 1929–1939*. Toronto: McClelland and Stewart, 1990.

———. *The Promised Land*. Toronto: McClelland and Stewart, 1984.

Bicha, Karel Denis. *The American Farmer and the Canadian West, 1896–1914*. Lawrence: Coronado Press, 1968.

Blasingame, Ike. *Dakota Cowboy: My Life in the Old Days*. Lincoln: University of Nebraska Press, 1958.

Brado, Edward. *Cattle Kingdom: Early Ranching in Alberta*. Vancouver: Douglas and McIntyre, 1984.

Broadfoot, Barry. *Ten Lost Years, 1929–1939: Memories of Canadians who Survived the Depression*. Toronto: McClelland and Stewart, 1997.

Breen, David H. *The Canadian Prairie West and the Ranching Frontier, 1874–1924*. University of Toronto Press, 1983.

Brisbin, General James A. *The Beef Bonanza: or, How to Get Rich on the Plains*. Norman: University of Oklahoma Press, 1959.

Britnell, G.E., and V.C. Fowke. *Canadian Agriculture in War and Peace, 1935–1950*. Stanford: Stanford University Press, 1962.

Bronson, Edgar Beecher. *Reminiscences of a Ranchman*. Lincoln: University of Nebraska Press, 1962.

Brown, Craig (ed.). *The Illustrated History of Canada*. Toronto: Lester and Orpen Dennys, 1987.

Brown, Mark H., and W.R. Felton, *Before Barbed Wire*. New York: Bramhall House, 1956.

Butala, Sharon. *The Garden of Eden*. Toronto: Harper Collins, 1998.

———. *The Perfection of the Morning: An Apprenticeship in Nature*. Toronto: Harper Collins, 1994.

Byfield, Ted (ed.). *Alberta in the Twentieth Century*, Vols. 1–3. Edmonton: United Western Communications, 1991–94.

Cavanaugh, Catherine, and Jeremy Mouat (eds.). *Making Western Canada: Essays on European Colonization and Settlement*. Toronto: Garamond Press, 1996.

Christianson, Chris J. *Early Rangemen*. Lethbridge: Southern Printing, 1973.

———. *My Life on the Range*. Lethbridge: Southern Printing, 1968.

Clawson, Marion. *The Western Range Livestock Industry*. New York: McGraw Hill, 1950.

Clay, John. *My Life on the Range*. Norman: University of Oklahoma Press, 1962.

Clifford, Ann. *Ann's Story*. Calgary: JR's Printing Services, 1995.

Cochrane and Area Historical Society. *Big Hill Country*. Calgary: Friesen, 1977.

Collins, Ellsworth, and Alma M. England. *The 101 Ranch*. Norman: University of Oklahoma, 1938.

Connell, Evan S. *Son of the Morning Star*. New York: Harper and Row, 1984.

Cotton, E.J. (Bud), with Ethel Mitchell. *Buffalo Bud: Adventures of a Cowboy*. Vancouver: Hancock House, 1981.

Craig, John R. *Ranching with Lords and Commons*. Toronto: William Briggs, 1903.

Daggett, Dan. *Beyond the Rangeland Conflict: Towards a West that Works*. Layton, Utah: Gibbs Smith, 1995.

Dale, Edward Everett. *Cow Country*. Norman: University of Oklahoma Press, 1942.

———. *The Range Cattle Industry: Ranching on the Great Plains, 1865–1925*. Norman: University of Oklahoma Press, 1960.

Dary, David. *Cowboy Culture: A Saga of Five Centuries*. New York: Knopf, 1981.

Dempsey, Hugh A. *The Golden Age of the Canadian Cowboy: An Illustrated History*. Saskatoon: Fifth House, 1995.

Dempsey, Hugh A. (ed.). *Men in Scarlet*. Calgary: McClelland and Stewart West, 1974.

Dobie, J. Frank. *Cow People*. Boston: Little, Brown, 1964.

———. *The Longhorns*. New York: Grosset and Dunlap, 1941.

Doig, Ivan. *Dancing at the Rascal Fair*. New York: Penguin, 1987.

———. *English Creek*. New York: Penguin, 1984.

———. *Ride With Me, Mariah Montana*. New York: Penguin, 1990.
———. *This House of Sky: Landscapes of the Western Mind*. New York: Harcourt Brace, 1978.
Donahue, Debra L. *The Western Range Revisited: Removing Livestock from Public Lands to Conserve Native Biodiversity*. Norman: University of Oklahoma Press, 1999.
Dunae, Patrick A. *Gentlemen Emigrants: From the British Public Schools to the Canadian Frontier*. Vancouver: Douglas and McIntyre, 1981.
Dunae, Patrick A. (ed.). *Rancher's Legacy: Alberta Essays by Lewis G. Thomas*. Edmonton: University of Alberta Press, 1986.
East Longview Historical Society. *Tales and Trails: A History of Longview and Surrounding Area*. Calgary: Northwest Publishing, 1973.
Elofson, Warren M. *Cowboys, Gentlemen and Cattle Thieves*. Montreal: McGill-Queen's University Press, 2000.
Epp, Henry. *Three Hundred Prairie Years*. Regina: Canadian Plains Research Center, 1993.
Ernst, Donna B. *Sundance, My Uncle*. College Station, Texas: Creative Publishing, 1992.
Evans, Simon M. *Prince Charming Goes West: The Story of the E.P. Ranch*. Calgary: University of Calgary Press, 1993.
Evans, Simon, Sarah Carter, and Bill Yeo (eds.). *Cowboys, Ranchers and the Cattle Business*. Calgary: University of Calgary Press, 1997.
Ewers, J.C. *The Blackfeet: Raiders of the Northern Plains*. Norman: University of Oklahoma Press, 1958.
Ewing, Sherm. *The Ranch: A Modern History of the North American Cattle Industry*. Missoula: Mountain Press, 1995.
———. *The Range*. Missoula: Mountain Press, 1990.
Fletcher, Robert H. *Free Grass to Fences*. New York: University Publishers, 1960.
Foran, Maxwell L. *Trails and Trials: Markets and Land Use in the Alberta Beef Industry, 1881–1948*. Calgary: University of Calgary Press, 2003.
Francis, R. Douglas. *Images of the West: Responses to the Canadian Prairies*. Saskatoon: Western Producer, 1989.
Francis, R.D., and H. Ganzevoort (eds.). *The Dirty Thirties in Prairie Canada*. Vancouver: Tantalus, 1973.
Frewen, Moreton. *Melton Mowbray and Other Memories*. London: Herbert Jenkins, 1924.
Friesen, Gerald. *The Canadian Prairies: A History*. Toronto: University of Toronto Press, 1984.
Frink, Maurice W., W. Turrentine Jackson, and Agnes Wright Spring. *When Grass Was King: Contibutions to the Western Range Cattle Industry Study*. Boulder: University of Colorado Press, 1956.
Galaty, J.G., and D.L. Johnson. *The World of Pastoralism: Herding Systems in Comparative Perspective*. New York: Guilford Press, 1990.
Gardiner, Claude. *Letters from an English Rancher*. Calgary: Glenbow-Alberta Institute, 1988.
Gayton, Don. *Landscapes of the Interior*. Gabriola Island, B.C.: New Society Publishers, 1996.
———. *The Wheatgrass Mechanism*. Saskatoon: Fifth House, 1990.
Gipson, Fred. *Cowhand: The Story of a Working Cowboy*. College Station: Texas A&M University Press, 1953.
Graber, Stan. *The Last Roundup: Memories of a Canadian Cowboy*. Saskatoon: Fifth House, 1995.
Gray, James H. *The Winter Years: The Depression on the Prairies*. Toronto: Macmillan, 1966.
———. *A Brand of its Own: The 100 Year History of the Calgary Exhibition and Stampede*. Saskatoon: Western Producer, 1985.
Gressley, Gene M. *Bankers and Cattlemen*. New York: Alfred A. Knopf, 1966.
Hage, Wayne. *Storm Over Rangelands: Private Rights in Federal Lands*, 3rd ed. Bellevue, Wash.: Free Enterprise Press, 1994.
Haley, J. Evetts. *The XIT Ranch of Texas and the Early Days of the Llano Estacado*. Norman: University of Oklahoma Press, 1953.
Hall, D.H. *Clifford Sifton*, 2 Vols. Vancouver: University of British Columbia Press, 1985.
Hargreaves, Mary Wilma M. *Dry Farming in the Northern Great Plains, 1900–1925*. Cambridge, Mass.: Harvard University Press, 1957.
Harris, R. Cole. *The Resettlement of British Columbia: Essays on Colonialism and Geographical Change*. Vancouver: University of British Columbia Press, 1997.
———, and Elizabeth Phillips. *Letters from Windermere, 1912–1914*. Vancouver: University of British Columbia Press, 1984.
High River Pioneers' and Old Timers' Association. *Leaves from the Medicine Tree*. Lethbridge: Lethbridge Herald Press, 1960.
Higinbotham, John David. *When the West Was Young*. Toronto: Ryerson, 1933.
Hildebrandt, Walter, and Brian Hubner. *The Cypress Hills: The Land and its People*. Saskatoon: Purich Publishing, 1994.
Holliday, Barbara. *To Be a Cowboy: Oliver Christensen's Story*. Calgary: University of Calgary Press, 2003.
Hopkins, Monica. *Letters from a Lady Rancher*. Halifax: Formac Publishing, 1983.
Houle, Marcy. *The Prairie Keepers: Secrets of the Grassland*. Reading, Mass.: Addison-Wesley, 1995.
Ings, Fred W. *Before the Fences: Tales from the Midway Ranch*. Calgary: McAra Press, 1980.
Iverson, Peter. *When Indians Became Cowboys: Native Peoples and Cattle Ranching in the American West*. Norman: University of Oklahoma Press, 1992.
Jacobs, Frank. *Cattle and Us, Frankly Speaking*. Calgary: Detselig, 1993.
Jameson, Sheilagh S. *Ranches, Cowboys and Characters: The Birth of Alberta's Western Culture*. Calgary: Glenbow-Alberta Institute, 1987.
Joern, Antony, and Kathleen H. Keeler (eds.). *The Changing Prairie: North American Grassland*. New York: Oxford University Press, 1995.
Johnston, Alex. *Cowboy Politics: The Western Stock Growers' Association and its Predecessors*. Calgary: Western Stockgrowers, 1972.
Jones, David C. *Empire of Dust: Settling and Abandoning the Prairie Dry Belt*. Edmonton: University of Alberta Press, 1987.
Jones-Hole, Jo Ann. *Calgary Bull Sale, 1901–2000*. Calgary: Friesen, 2000.
Jordan, Teresa. *Cowgirls: Women of the American West*. Lincoln: University of Nebraska Press, 1992.
Jordan, Terry G. *North American Cattle Ranching Frontiers: Origins, Diffusion, and*

Differentiation. Albuquerque: University of New Mexico Press, 1993.
———. *Trails to Texas: The Southern Roots of Western Cattle Ranching*. Lincoln: University of Nebraska Press, 1981.
———, Jon T. Kilpinen, and Charles F. Gritzner, *The Mountain West: Interpreting the Folk Landscape*. Baltimore: Johns Hopkins University Press, 1997.
Kelly, Leroy V. *The Rangemen*. 75th Anniversary ed. High River: Willow Creek Publishing, 1988.
Kerr, Donald, and Deryk Holdsworth (eds.). *Historical Atlas of Canada*, Vol. 3. Toronto: University of Toronto Press, 1990.
Klassen, Henry C. *A Business History of Alberta*. Calgary: University of Calgary Press, 1999.
Klinkenborg, Verlyn. *Making Hay*. New York: Nick Lyons Books, 1986.
Limerick, Patricia N. *The Legacy of Conquest: The Unbroken Past of the American West*. New York: W.W. Norton, 1987.
Livingstone, Donna. *Cowboy Spirit: Guy Weadick and the Calgary Stampede*. Vancouver: Douglas and McIntyre, 1996.
Long, Philip S. *Seventy Years a Cowboy*. Billings, Mont.: Cypress Books, 1976.
———. *The Great Canadian Range*. Toronto: Ryerson Press, 1963.
Loveridge, D.M., and Barry Potyondi. *From Wood Mountain to the Whitemud: A Historical Survey of the Grasslands National Park Area*. Ottawa: Environment Canada, 1983.
MacEwan, Grant. *Blazing the Old Cattle Trail*. Saskatoon: Western Producer, 1975.
———. *John Ware's Cattle Country*. Saskatoon: Western Producer, 1974.
———. *Pat Burns: Cattle King*. Saskatoon: Western Producer, 1979.
———. *Charles Noble: Guardian of the Soil*. Saskatoon: Western Producer, 1983.
———. *Heavy Horses: Highlights of their History*. Saskatoon: Western Producer, 1986.
———. *Hoofprints and Hitching Posts*. Saskatoon: Western Producer, 1968.
———. *The Breeds of Farm Live-Stock in Canada*. Toronto: Thomas Nelson, 1941.
MacInnes, C.M. *In the Shadow of the Rockies*. London: Rivington's Press, 1930.
Mackintosh, W.A., and W.L.G. Joerg (eds.). *Canadian Frontiers of Settlement*. Toronto: MacMillan, 1934.
MacLachlan, Ian. *Kill and Chill: Restructuring Canada's Beef Commodity Chain*. Toronto: University of Toronto Press, 2001.
Macoun, John. *Manitoba and the Great North-West*. Guelph: World Publishing Co., 1882.
MacRae, Archibald Oswald. *History of the Province of Alberta*. Calgary: Western Canada History Co., 1912.
Malin, James C. *The Grasslands of North America: Prolegomena to its History*. Gloucester, Mass.: Peter Smith, 1967.
Malone, Michael P., Richard B. Roeder, and William L. Lang. *Montana: A History of Two Centuries*. Seattle: University of Washington Press, 1991.
Manning, Richard. *Grasslands: The History, Biology,Politics and Promise of the American Prairie*. New York: Viking Penguin, 1995.
Martin, Chester. *"Dominion Lands" Policy*. Toronto: McClelland and Stewart, 1973.
McCarthy, Cormac. *All the Pretty Horses*. New York: Alfred A. Knopf, 1993.
McCowan, Don C. *Grassland Settlers: The Swift Current Pegion During the Era of the Ranching Frontier*. Saskatoon: Canadian Plains Resource Center, 1975.
McCumber, David, *The Cowboy Way*. New York: Bard, 1999.
McHugh, Tom. *The Time of the Buffalo*. Lincoln: University of Nebraska Press, 1972.
McKinnon, J. Angus. *The Bow River Range*. Calgary: McAra Press, 1974.
McKinnon, Lachlin. *Lachlin McKinnon, Pioneer, 1865–1948*. Calgary: McAra Press, 1956.
McMurtry, Larry. *Lonesome Dove*. New York: Simon and Schuster, 1985.
McQuarrie, John. *Cowboyin': A Legend Lives On*. Toronto: Macmillan, 1994.
Milk River Historical Society. *Under Eight Flags: Milk River and District*. Lethbridge: Graphcom Printers, 1989.
Mothershead, Harmon S. *The Swan Land and Cattle Company Limited*. Norman: University of Oklahoma Press, 1971.
Naftel, William. *The Cochrane Ranch*, Canadian Historic Sites, no. 16. Ottawa, 1974.
Namaka Community History Committee, *Trails to Little Corner: A Story of Namaka and Surrounding Districts*. Strathmore, Alberta, 1983.
Nanton and District Historical Society. *Mosquito Creek Roundup*. Nanton: Nanton and District Historical Society, 1975.
Nash, Gerald D., and Richard W. Etulain (eds.). *The Twentieth Century West: Historical Interpretations*. Albuquerque: University of New Mexico Press, 1989.
Nelson, J. Gordon. *The Last Refuge*. Montreal: Harvest House, 1973.
Oliphant, J. Orin. *On the Cattle Ranges of the Oregon Country*. Seattle: University of Washington Press, 1968.
Osgood, Ernest Staples. *The Day of the Cattleman*. Chicago:University of Chicago Press, 1929.
Patterson, R.M. *The Buffalo Head*. New York: William Sloane Associates, 1961.
Pearce, William M. *The Matador Land and Cattle Company*. Norman: University of Oklahoma Press, 1964.
Perren, Richard. *The Meat Trade in Britain, 1840–1914*. London: Routledge and Kegan Paul, 1978.
Phillips, Paul C. *Granville Stuart, Pioneering in Montana: The Making of a State, 1864–1887*. Lincoln: University of Nebraska Press, 1977.
Potyondi, Barry. *In Palliser's Triangle: Living in the Grasslands, 1850–1930*. Sakatoon: Purich Publishing, 1995.
Potyondi, Barry. *Where the Rivers Meet: A History of the Upper Oldman Basin to 1939*. Pincher Creek: Southern Alberta Water Science Society, 1992.
Powell, John Wesley. *Report on the Arid Region of the United States*. Wallace Stegner (ed.). Cambridge: Belknap Press of Harvard University, 1962.
Raban, Jonathan. *Bad Land*. London: Macmillan, 1996.
Rainbolt, Jo. *The Last Cowboy: Twilight Era of the Horseback Cowhand, 1900–1940*. Helena: America and World Graphic Publications, 1992.
Rees, Ronald. *New and Naked Land: Making the Prairies Home*. Saskatoon: Western Producer Books, 1988.
Rifkin, Jeremy. *Beyond Beef: The Rise and fall of the Cattle Culture*. New York: Dutton, 1992.
Rollins, Philip Ashton. *The Cowboy: His Character, His Equipment and His Part in the Development of the West*. New York: Scribner, 1922.

Rosenvall, L.A., and Simon M. Evans (eds.). *Essays on the Historical Geography of the Canadian West*. Calgary: University of Calgary, Department of Geography, 1987.

Russell, Andy. *The Canadian Cowboy: Stories of Cows, Cowboys and Cayuses*. Toronto: McClelland and Stewart, 1993.

Rutherford, J.G. *The Cattle Trade of Western Canada*. Ottawa: King's Printer, 1909.

Sanders, Alvin Howard. *A History of the Percheron Horse*. Chicago: Breeder's Gazette, 1917.

Sandoz, Mari. *The Cattlemen from Rio Grande Across the Far Marias*. 4th ed. New York: Hastings House, 1975.

Savage, Candice. *Cowgirls*. Vancouver: Greystone Books, 1996.

Savage, William W. *Cowboy Life: Deconstructing an American Myth*. Denver: University Press of Colorado, 1993.

Schlebecker, John T. *Cattle Raising on the Plains, 1900–1961*. Lincoln: University of Nebraska Press, 1963.

Schmidt, John. *Western Stock Growers' Association: An Experiment that Worked*. Calgary: Western Stock Growers' Association, 1994.

Sharp, Paul F. *Whoop-Up Country: The Canadian American West, 1865–1885*. Minneapolis: University of Minnesota Press, 1955.

Shepard, G. *Brave Heritage*. Saskatoon: Modern Press, 1967.

———. *The West of Yesterday*. Toronto: McClelland and Stewart, 1965.

Sheppard, Bert. *Just About Nothing*. Calgary: McAra Press, 1977.

———. *Spitzee Days*. Calgary: McAra Press, 1971.

Sherow, James E. *A Sense of the American West*. Albuquerque: University of New Mexico Press, 1998.

Skaggs, Jimmy M. *Prime Cut: Livestock Raising and Meat Packing in the United States, 1607–1983*. College Station: Texas A&M University Press, 1986.

———. *The Cattle-Trailing Industry: Between Supply and Demand, 1886–1890*. Lawrence: University Press of Kansas, 1973.

Skelton, Oscar D. *General Economic History of the Dominion, 1867–1912*, Vol. 9 of *Canada and Its Provinces*, Adam Short and Arthur G. Doughty (eds.). Edinburgh: T. and A. Constable for the Publisher's Association of Canada, 1913.

Slatta, Richard W. *Cowboys of the Americas*. New Haven: Yale University Press, 1990.

———. *Comparing Cowboys and Frontiers*. Norman: University of Oklahoma Press, 1997.

———. *The Cowboy Encyclopedia*. Santa Barbara: ABC-CLIO, 1994.

Spector, David. *Agriculture on the Prairies, 1890–1914*. Parks Canada: Environment Canada, 1983.

Starrs, Paul F. *Let the Cowboy Ride: Cattle Ranching in the American West*. Baltimore: Johns Hopkins University Press, 1998.

Stegner, Wallace. *The American West as Living Space*. Ann Arbor: University of Michigan Press, 1987.

———. *Wolf Willow*. New York: Viking Press, 1955.

———. *Where the Blue Bird Sings to the Lemonade Springs: Living and Writing in the West*. New York: Random House, 1992.

Strange, Thomas Bland. *Gunner Jingo's Jubilee*. Edmonton: University of Alberta Press, 1988.

Stuart, Granville. *Forty Years on the Frontier*. Cleveland: Arthur H. Clark, 1925.

Sutherland, Rev. A. *A Summer in Prairie-Land: Notes on a Tour through the North-West Territory*. Toronto: Methodist Book and Publishing, 1881.

Symons, R.D. *Where the Wagon Led: One Man's Memories of the Cowboy's Life in the Old West*. Toronto: Doubleday, 1973.

Taylor, Lonn, and Ingrid Marr. *The American Cowboy*. Washington, D.C.: Library of Congress, 1983.

Thomas, Lewis G. *The Prairie West to 1905: A Canadian Source Book*. Toronto: Oxford University Press, 1975.

Thompson, John Herd. *The Harvests of War*. Toronto: McClelland and Stewart, 1978.

Toole, K. Ross. *Montana: An Uncommon Land*. Norman: University of Oklahoma Press, 1984.

Vanderhaeghe, Guy. *The Englishman's Boy*. Toronto: McClelland and Stewart, 1996.

Voisey, Paul. *Vulcan: The Making of a Prairie Community*. Toronto: University of Toronto Press, 1988.

Wald, Johanna, et al. *How Not to be Cowed: Livestock Grazing on Public Lands, an Owner's Manual*. San Francisco: Natural Resources Defence Council, 1991.

Walker, Don D. *Clio's Cowboys: Studies in the Historiography of the Cattle Trade*. Lincoln: University of Nebraska, 1981.

Ward, Faye E. *The Cowboy at Work: All About his Job and How He Does It*. Norman: University of Oklahoma Press, 1987.

Webb, Walter Prescott. *The Great Plains*. New York: Gosset and Dunlap, 1931.

Weir, Thomas R. *Ranching in the Southern Interior Plateau of British Columbia*. Ottawa: Geographical Branch, Mines and Technical Surveys, 1964.

Wheeler, Richard S. *Buffalo Commons*. New York: Tom Doherty, 1998.

Wilson, A.A. *Culture of Nature: North American Landscape from Disney to the Exxon Valdez*. Cambridge, Mass.: Blackwell, 1992.

Woolliams, Nina G. *Cattle Ranch: The Story of the Douglas Lake Cattle Company*. Vancouver: Douglas and McIntyre, 1982.

Wyman, Walker D. *Nothing But Prairie and Sky: Life on the Dakota Range in the Early Days*. Norman: University of Oklahoma Press, 1988.

Zarn, George. *Above the Forks*. Calgary: Zarn, 1978.

NOTES

Notes to Preface

1 Captain Bernal Diaz Del Castillo, *The True History of the Conquest of Mexico* (1568), quoted in Stanley P. Hirshson, *General Patton: A Soldier's Life* (New York: Harper Collins, 2002).

2 See Simon Evans, *Prince Charming Goes West* (Calgary: University of Calgary Press, 1993).

3 Lane died in 1925, but the management did not change radically until the ranch was bought by Burns in 1927. Similarly, Pat Burns died in 1937, but Burns and Company continued to operate the ranch until 1950.

4 The characteristics of this "cattle compact" and its ongoing power to shape policy in the ranching country is a major theme of David Breen's foundation study of ranching in Canada. David H. Breen, *The Canadian Prairie West and the Ranching Frontier, 1874–1924* (Toronto: University of Toronto Press, 1983), 39ff..

5 Wordfest: Banff-Calgary International Writers Festival, October 16–20, 2002, Session 34, "History and Historical Novels."

6 Breen, *Canadian Prairie West*.

7 Terry G. Jordan, *North American Cattle-Ranching Frontiers: Origins, Diffusion, and Differentiation* (Albuquerque: University of New Mexico Press, 1993); Richard W. Slatta, *Cowboys of the Americas* (New Haven, CT: Yale University Press, 1990); and Paul F. Starrs, *Let the Cowboy Ride: Cattle Ranching in the American West* (Baltimore: Johns Hopkins University Press, 1998).

8 Simon Evans, Sarah Carter, and Bill Yeo (eds.), *Cowboys, Ranchers and the Cattle Business: Cross Border Perspectives on Ranching History* (Calgary: University of Calgary Press, 2000).

9 Of course, several of the other authors have published work on a variety of topics, notably Henry Klassen's two books on business history. Henry C. Klassen, *A Business History of Alberta* (Calgary: University of Calgary Press, 1999); and *Eye on the Future: Business People in Calgary and the Bow Valley, 1870–1900* (Calgary: University of Calgary Press, 2002).

10 Warren M. Elofson, *Cowboys, Gentlemen, and Cattle Thieves: Ranching on the Western Frontier* (Montreal: McGill-Queen's University Press, 2000); quote from Alan McCullough's review of the book in *Prairie Forum* 27, no. 2 (Fall 2002): 265.

11 Debra L. Donahue, *The Western Range Revisited: Removing Livestock from Public Lands to Conserve Native Biodiversity* (Norman: University of Oklahoma Press, 1999); Antony Joern and Kathleen H. Keeler (eds.), *The Changing Prairie: North American Grasslands* (New York: Oxford University Press, 1995); Richard Manning, *Grassland: The History, Biology, Politics, and Promise of the American Prairie* (New York: Viking Press, 1995); and for references to articles see Simon M. Evans, "Grazing the Grasslands: Exploring Conflicts, Relationships and Futures," *Prairie Forum* 26, no. 1 (Spring 2001): 67–84.

12 Maxwell L. Foran, *Trails and Trials: Markets and Land Use in the Alberta Beef Cattle Industry, 1881–1948* (Calgary: University of Calgary Press, 2003).

13 Ian MacLachlan, *Kill and Chill: Restructuring Canada's Beef Commodity Chain* (Toronto: University of Toronto Press, 2001).

14 Harmon Ross Mothershead, *The Swan Land and Cattle Company Limited* (Norman: University of Oklahoma Press. 1971); and William M. Pearce, *The Matador Land and Cattle Company* (Norman: University of Oklahoma Press, 1964).

15 Nina Woolliams, *Cattle Ranch: The Story of the Douglas Lake Cattle Company* (Vancouver: Douglas and McIntyre, 1979); and Dale Alsager, *The Incredible Gang Ranch* (Surrey, B.C.: Hancock House, 1990).

16 Edward Brado, *Cattle Kingdom: Early Ranching in Alberta* (Vancouver: Douglas & McIntyre, 1984); C.W. Buchanan, "History of the Walrond Cattle Ranche Ltd.," *Canadian Cattlemen* 8 (March 1946): 171ff.; and William H. McIntyre, "A Brief History of the McIntyre Ranch," *Canadian Cattlemen* 10, no. 2 (September 1947): 86ff..

17 Evans, *Prince Charming Goes West*.

Notes to Chapter 1

1 Jill Ker Conway, *The Road from Coorain* (New York: Random House, 1990), p. 186.

2 G.M. Dawson, *Montreal Gazette*, 17 November 1881. Dawson was a young scientist from Montreal who had been employed as geologist and surveyor on the boundary survey of 1875. His work had a profound effect on the optimistic reports of John Macoun, which did so much to dispel the myth of the Great American Desert. See John Warkentin, "Steppe, Desert and Empire," in Anthony W. Rasporich and Henry C. Klassen (eds.), *Prairie Perspectives 2* (Montreal: Holt, Rhinehart and Winston, 1972), p. 125.

3 See, for example, National Archives of Canada (hereafter NAC), RG 15, f. 142709, M.H. Cochrane to Col. J.S. Dennis, 26 November 1880; RG 15, f. 175296, C. Stimson to Lindsay Russell, 29 May 1882; RG 15, f. 179180, J.E. Chipman to Col. J.S. Dennis, 28 May 1881; and Glenbow Archives (hereafter GA), ms. "Trip of Charles Edward Harris to the Canadian North-West, 1882." Fred Stimson gave power of attorney to Thomas D. Milburne during his absence. A.B. McCullough, "The Formation of the North West Cattle Company," unpublished report, Parks Canada, 1994.

4 *Montreal Gazette*, 17 November 1881.

5 For an overview of his life, see Jaques Ferland, "Matthew Henry Cochrane," in *Dictionary of Canadian Biography*, vol. 13, 1901–1910 (Toronto: University of Toronto Press, 1994), pp. 208–9; and William Naftel, *The Cochrane Ranch*, Canadian Historic Sites, no. 16 (Ottawa, 1974).

6 Breen, *Canadian Prairie West*, p. 27.

7 Alan B. McCullough, "Not an Old Cowhand: Fred Stimson and the Bar U Ranch," in Evans, Carter, and Yeo, *Cowboys, Ranchers and the Cattle Business*.

8 Brado, *Cattle Kingdom*, pp. 140–42.

9 Sir John A. Macdonald remarked that some "officers especially those of a commercial or speculative turn of mind, have been employing themselves looking after herds of cattle, etc., and have thrown off the soldier too much to indulge in such pursuits." Canada, House of Commons Debates, 28 April 1880, p. 1814.

10 There was also a "mammoth stock of business suits, shirts and fine underwear for Gentlemen," according to an advertisement in the *Fort Benton Weekly Record*, 30 June 1881. The following season the Red Cloud hit a snag just below the mouth of the Musselshell and sank in two minutes.

11 Alice Sharples Baldwin, "The Sharples," *Alberta Historical Review* 21, no. 1 (Winter 1973): 12–17.

12 This was the section of the river which had so intrigued Captain Merriweather Lewis on his exploratory voyage seventy-six years before. See Bernard DeVoto (ed.), *The Journals of Lewis and Clark* (Boston: Houghton Mifflin, 1953): 115–32.

13 For a great portrait of the town and the context of the times, see Paul F. Sharp, *Whoop-Up Country: The Canadian American West, 1865–1885* (Minneapolis: University of Minnesota Press, 1955).

14 This situation was, of course, to be short lived, for the railway reached Calgary in 1883. This caused a remarkably rapid economic realignment which left Fort Benton as a sleepy agricultural village by the 1890s. For discussion see Peter Darby, "From Riverboat to Rail Lines," in L.A. Rosenvall and S.M. Evans (eds.), *Essays on the Historical Geography of the Canadian West* (Calgary: University of Calgary, Department of Geography, 1987); and James M. Francis, "Montana Business and Canadian Regionalism in the 1870s and 1880s," *Western Historical Quarterly* 12 (July 1981): 291–304.

15 Sharp, *Whoop-Up Country*, p. 210.

16 *Fort Benton Weekly Record*, 16 June 1881: "Think of it! Four boats tying up here at the same time. Expect a lively town about next Tuesday."

17 Joel F. Overholser, *Centenary History of Fort Benton, Montana, 1846–1946* (pamphlet published by Fort Benton Centennial Association, 1946).

18 Al Noyes, "Story as told by S.C. Ashby," ms., Historical Society of Montana Library, Helena.

19 *Fort Benton Weekly Record*, 30 June 1881.

20 D. McEachran, "A Journey Over the Plains," *Montreal Gazette*, 4, 18, and 29 November, and 2 and 5 December 1881.

21 J.C. Ewers, *The Blackfeet: Raiders of the Northern Plains* (Norman: University of Oklahoma Press, 1958).

22 Hugh A. Dempsey, "Alexander Cuthbertson's Journey to the Bow River," *Alberta Historical Review* 19, no. 4 (1971): 8–20.

23 Canada, Department of the Interior, "Map of the South West Part of the North West Territories Showing Grazing Country Adjacent to the Rocky Mountains" (Ottawa: Queen's Printer, 1881). It is significant that this map extended well to the south of the international border to include the upper Missouri and the town of Fort Benton.

24 Canada, *Sessional Papers*, 1882, vol. 8, no. 18, "Extract from the Report of Mr Montague Aldous, D.T.S.," pp. 36–37.

25 Canada, *Sessional Papers*, 1881, vol. 3, no. 3, "Report of Superintendent W. Winder," p. 24.

26 Correspondents from the *Edinburgh Courant*, *Edinburgh Scotsman*, *London Graphic*, *London Times*, and the *London Telegraph* accompanied the tour, along with a reporter from the *Toronto Globe* who paid his own way.

27 Begg says, "I journeyed through a large portion of Montana and through the Bow River district as far north as Edmonton (about five hundred miles north from Fort Benton) in August, September and October 1881." See John Macoun, *Manitoba and the Great North-West* (Guelph: World Publishing Co., 1882), p. 270. See also Rev. A. Sutherland, *A Summer in Prairie-Land: Notes of a Tour through the North-West Territory* (Toronto: Methodist Book and Publishing, 1881).

28 See G.M. Dawson, *Montreal Gazette*, 17 October 1881.

29 Tom McHugh, *The Time of the Buffalo* (Lincoln: University of Nebraska Press, 1972).

30 In 1874 George McDougall estimated that more than fifty thousand buffalo robes had been traded by the Blackfeet each year for a number of years. The peak of the trade was in 1876 when some sixty thousand robes were shipped. In 1877 this trade fell to half its previous total. Peter Darby, "From River Boat to Rail Lines," in Rosenvall and Evans, *Historical Geography*.

31 See J.G. Nelson, "Some Reflections on Man's Impact on the landscape of the Canadian Prairies and Nearby Areas," in P.J. Smith (ed.), *The Prairie Provinces* (Toronto: University of Toronto Press, 1972), p. 43.

32 Both McEachran and Reverend Sutherland saw something of the remaining northern herd on their passage on the Missouri River both going to and coming from Fort Benton. McEachran, "A Journey," p. 53; and Sutherland, *A Summer in Prairie Land*, p. 23.

33 McEachran's comments were by no means unusual in the writings of the period; however, everything we know about the man and his later attitude to squatters on the Walrond grazing lease suggest that he was of a particularly rigid and unbending disposition. In

describing the Indians forced to congregate near Fort Macleod he wrote, "... a more lazy, filthy race could scarcely be imagined." McEachran, "A Journey," p. 17.

34 Macoun, *Manitoba and the Great North-West*, p. 258.

35 *Montreal Gazette*, 17 November 1881.

36 McEachran, "A Journey," p. 23.

37 This range, on the upper waters of the Oldman watershed between the Livingstone Range and the Porcupine Hills, was where McEachran chose to locate the leases of the Walrond Ranch. Today, much of the area is run by the Walrond Grazing Cooperative.

38 Alexander Begg, "Stock Raising in the Bow River District compared with Montana," in John Macoun, *Manitoba and the Great North-West*, p. 281.

39 They arrived back in Montreal at the end of August having been away almost three months.

40 Barbara Holliday, "Herb Millar of the Bar U," *Alberta History* 44, no. 4 (Autumn 1996): 15–21.

Notes to Chapter 2

1 Paul Baumann in *A Tremor of Bliss: Contemporary Writers on the Saints* (New York: Harcourt, 1994), p. 202.

2 Alan McCullough, for many years a historian with Parks Canada in Ottawa, used a variety of government and church documents to piece together a detailed picture of the Stimson family. His draft paper "Fred Stimson and the North West Cattle Company," research paper, Parks Canada, 24 January 1994, uses material from the National Archives of Quebec at Hull for births, marriages, and deaths, as well as the nominal census of Canada. He was even able to find the will and the inventories of the estate of Fred Stimson's parents. Thus anybody interested in Fred Stimson owes him a considerable debt of gratitude. The following paragraphs draw heavily on his published and unpublished work, although I accept full responsibility for emphasis and interpretation. Factual references are from McCullough's paper cited above and will not be footnoted separately.

3 Alan B. McCullough, "Not an Old Cowhand – Fred Stimson and the Bar U Ranch," in Evans, Carter and Yeo, *Cowboys, Ranchers and the Cattle Business*.

4 Duncan McEachran, "Impressions of Pioneers of Alberta as Ranching Country," Glenbow Archives (hereafter GA), pamphlet 636.08.

5 McCullough urges caution here, for the witnesses' names are not written in full; however, it seems to me unlikely that there were two twenty-one-year-old Allans with the same initials.

6 In 1879, Charles Stimson signed a lease with Andrew Allan, showing that business exchanges took place between the families from time to time.

7 Simon M. Evans, "Canadian Beef for Victorian Britain," *Agricultural History* 53, no. 4 (October 1979).

8 For review of this speculative boom in Canada, see Breen, *Canadian Prairie West*, pp. 25–26; for Britain, W. Turrentine Jackson, "British Interests in the Range Cattle Industry," pt. 2 of *When Grass Was King*, Maurice Frink, W. Turrentine Jackson and Agnes Wright Spring (eds.) (Boulder: University of Colorado Press, 1956); and for the United States, Ernest Staples Osgood, *The Day of the Cattleman* (Chicago: University of Chicago Press, 1929, Phoenix edition, 1966), and Brisbin, *Beef Bonanza*.

9 Oscar D. Skelton, *General Economic History of the Dominion, 1867–1912*, vol. 9 of *Canada and Its Provinces*, Adam Shortt and Arthur G. Doughty (eds.) (Edinburgh: T and A. Constable for the Publishers' Association of Canada, 1913), p. 164.

10 Alexander Begg included a detailed analysis of the costs and returns involved in starting a ranch in his chapter on "Stock Raising." The work had been done by Professor W. Brown of the Ontario Agricultural College. It outlined the truly spectacular profits that might be made given a variety of different assumptions about size of the initial herd, etc. For example, after five years, stock originally worth $5,000 would be worth $55,000! Alexander Begg, "Stock Raising in the Bow River District Compared with Montana," in Macoun, *Manitoba and the Great North-West*.

11 *Canadian Gazette*, 11 June 1881, p. 1757. For an overview see Alan. B. McCullough, "The Formation of the North West Cattle Company," Parks Canada, research paper, 11 February 1994.

12 NAC, RG 95, vol. 2616, North-West Cattle Company Limited.

13 Brian J. Young, and Gerald Tulchinsky, "Sir Hugh Allan," in *Dictionary of Canadian Biography*, vol. 11, 1881–1890 (Toronto: University of Toronto Press, 1982), p. 13.

14 Senator Cochrane was his neighbour, and J.H. Pope, the Minister of Agriculture, was his Member of Parliament. Charles Colby, who organized the Rocky Mountain Cattle Company, was M.P. for the neighbouring county of Stanstead, while E.T. Brooke, the M.P. for Sherbrooke, founded the Eastern Townships Ranch Company.

15 See, for an excellent overview, Marc Bellavance, *A Village in Transition: Compton, Quebec, 1880–1920* (Ottawa: National Historic Parks and Sites, 1982).

16 J.I. Little, "Social and Economic Development of Settlers in Two Quebec Townships, 1851–1870," in Donald H. Akenson (ed.), *Canadian Papers in Rural History*, vol. 1 (Gananoque, Ont.: Langdale Press, 1978), 89–113.

17 Montana Historical Society, MC 55, T.C. Power Papers, box 166, f. 6, Stimson to T.C. Power, 26 April 1882 and 20 May 1882.

18 High River Pioneers' and Old Timers' Association, *Leaves from the Medicine Tree* (High River, 1960), p. 23; for description of the

role of 'frontiersmen' see Simon M. Evans, "Tenderfoot to Rider: Learning 'Cowboying' on the Canadian Ranching Frontier during the 1880s," in Evans, Carter and Yeo, *Cowboys, Ranchers and the Cattle Business*, pp. 61–80.

19 See, for example, "Statement showing the number of Horses, Cattle and Sheep entered into the District of Alberta from the 1st of June 1880." NAC, RG 15, f. 11007, Department of Interior, 9 September, 1885.

20 The 1881 Cochrane herd averaged $18 per head. Prices were rising with demand. The *Chicago Times* noted: "The strong prices paid for cattle the past two months, have affected their value to the furthest limits of the western grazing country.... A large amount of idle capital seeking a safe and profitable investment will be put into stock this year all over the west." *Chicago Times* quoted in *Montreal Gazette*, 9 June 1881, p. 8, "The Western Cattle Trade."

21 Montana Historical Society, MC 55, T.C. Power papers, box 166, f. 6, Stimson to Power, 26 April 1882; and vol. 167, f. 4, 28 July 1882; *Leaves of the Medicine Tree* specifies $19 a head: see p. 252.

22 J. Orin Oliphant, *On the Ranges of the Oregon Country* (Seattle: Washington University Press, 1968), p. 174.

23 Richard W. Slatta, *The Cowboy Encyclopedia* (Denver: ABC-CLIO, 1994), p. 371.

24 Frederick William Ings, *Before the Fences* (Calgary: McAra, 1980), p. 21.

25 Grant MacEwan, *Blazing the Old Cattle Trail* (Saskatoon: Western Producer, 1972), p. 100.

26 Evans, "Tenderfoot to Rider," p. 76.

27 This brand is shown among several others belonging to the Bar U in the Alberta brand book of 1889.

28 Brado, *Cattle Kingdom*, p. 112.

29 Angus McKinnon, "Bob Newboldt, Pioneer of 1884," *Canadian Cattlemen* (February 1960): 26.

30 NAC, RG 15, f. 11007, Department of Interior, 9 September 1885, "Statement showing the number of horse, cattle and sheep entered into the District of Alberta from the 1st of June, 1880."

31 *Macleod Gazette*, 4 September 1882.

32 See L.V. Kelly, *The Rangemen* (Toronto: William Briggs, 1913), p. 75, for an account of this storm, the same one that caused such devastation to the second Cochrane herd, purchased from Poindexter and Orr.

33 *Leaves of the Medicine Tree*, p. 252.

34 See Alan McCullough, "Not an Old Cowhand," pp. 29–42.

35 Kelly, *Rangemen*, p. 75; and MacEwan, *Blazing the Old Cattle Trail*, p. 100.

36 MacEwan, *Blazing the Old Cattle Trail*, p. 102.

37 Oral tradition has usually credited twenty-year-old Millar with accomplishing this complex journey on his own, but in fact he was under the orders of Andrew Bell, a mature stockman from Compton who was trusted by Stimson. See Montana Historical Society, MC 55, T.C. Power Papers, box 166, f. 6, Stimson to Power, 20 May 1882; and Barbara Holliday, "Herb Millar of the Bar U," *Alberta History* 44, no. 4 (Autumn 1996): 15–21.

38 He arrived on the *Black Hills*, which was the third boat to arrive in Benton that year. *Benton Weekly Record*, 18 May 1882.

39 Brado, *Cattle Kingdom*, p. 108.

40 Canada, *Sessional Papers*, 1887, vol. 6, no. 7, "Report of the Dominion Lands Commission," p. 19. See also Appendix, Figure 2.1 for an overview.

41 Canada, *Sessional Papers*, 1904, vol. B, "Census of Canada, 1901," p. 262; and for an overall assessment see H.S. Arkell, "The Cattle Industry," in Henry J. Boam (ed.), *Twentieth Century Impressions of Canada* (Montreal: Sells Ltd., 1914), pp. 247–54. See also Appendix, Figure 2.1.

42 Simon M. Evans, "American Cattle on the Canadian Range, 1874–1914," *Prairie Forum* 4, no. 1 (1979): 121–35.

43 "Milling" is the term used to describe the manner in which cowboys attempted to cope with a herd which had stampeded: the leaders were turned back toward the herd so the stock circled and gradually slowed down and mixed. See Fay E. Ward, *The Cowboy at Work* (Norman: University of Oklahoma Press, 1958), p. 24.

44 NAC, RG 15, f. 11007, , Department of Interior, 9 September 1885, "Statement showing the number of horse, cattle and sheep entered into the District of Alberta from the 1st of June, 1880."

45 NAC, MG 26, Macdonald Papers, A1(d), vol. 420, J.L. Evans to Sir John A. Macdonald, 8 October 1885; and T. White to Sir John A. Macdonald, 20 November 1885.

46 Terry G. Jordan, *Cattle-Ranching Frontiers*, p. 299.

47 Jordan, *Cattle-Ranching Frontiers*, Table 1, p. 214.

48 The classic study is J. Orin Oliphant, *On the Ranges of the Oregon Country* (Seattle: Washington University Press, 1968); see also, for a general review, Simon M. Evans, "Stocking the Canadian Range," *Alberta History* 26, no. 3 (Summer 1978): 1–8.

49 See Jordan, *Cattle-Ranching Frontiers*, p. 300. He recounts a case of direct importation of a Durham bull from Kentucky to the Madison Valley in Montana.

50 William A. Baillie-Groham, "The Cattle Ranches in the Far West," *Fortnightly Review* 38 (October 1880): 447.

51 Quoted by Oliphant, *On the Ranges*, p. 165.

52 Mark H. Brown and W.R. Felton, *Before Barbed Wire* (New York: Bramhall House, 1956), p. 101.

53 Granville Stuart, *Forty Years on the Frontier*, vol. 2 (Cleveland: Arthur H. Clark, 1925), pp. 185–86.

54 For example Henry S. Boice specialized in purchasing cattle at Holbrook, Arizona, and finishing them along the Little Missouri in Dakota. Brown and Felton, *Before Barbed Wire*, p. 104.

55 Andy Adams, *The Log of a Cowboy*, 1903 (Lincoln: University of Nebraska, 1964). Adams recounts an epic five-month trail drive from the Rio Grande to the Blackfoot Agency in northwestern Montana during 1882.

56 See Charles L. Wood, "Upbreeding Western Range Cattle: Notes on Kansas, 1880–1920," *Journal of the West* 16, no. 1 (January 1977): 16–28. A British journalist commented on the earlier animals in the following words: "The quality of their beef is naturally inferior, the grain coarse and the percentage of bone and muscle very high. Their hair is of very coarse quality, and the back too truly of the Gothic style of structure to carry a large quantity of roasting beef." John MacDonald, *Food from the Far West: or American Agriculture* (London: William P. Nimmo, 1878), p. 268; for an overview of these cattle see J. Frank Dobie, *The Longhorns* (New York: Gosset and Dunlap, 1941).

57 General James S. Brisbin, *The Beef Bonanza: or, How to Get Rich on the Plains* (Norman: University of Oklahoma Press, 1959), p. 170.

58 Robert H. Fletcher, *Free Grass to Fences* (New York: University Publishers Inc., 1960), p. 146.

59 Canada, *Sessional Papers*, 1882, vol. 7, no. 11, "Annual Report of the Minister of Agriculture," p. viii.

60 Canada, *Sessional Papers*, 1882, vol. 15, no. 11, "Report of Chief Inspector of Stock," pp. 130–31.

61 In 1886, an outbreak of pleuro-pneumonia occurred among a consignment of purebred stock from Scotland, and the whole herd was slaughtered to contain the disease. John Stewart lost twenty-two pedigree Polled Angus, while Andrew Allan of the NWCC lost twenty-nine bulls. Canada, *Sessional Papers*, 1887, vol. 10, no. 12, "Annual Report on Cattle Quarantines, Quebec and the Maritime Provinces," p. 201.

62 *Calgary Herald*, 5 March 1884.

63 Alfred Pegler, "A Visit to Canada and the United States in Connection with the Meetings of the British Association, Montreal 1884," GA, pamphlet 971. 376v.

64 Canada, *Sessional Papers*, 1885, vol. 5, no. 8, "Annual Report of the Brandon Immigration Agent," p. 143.

65 Professor Sheldon, "A Visit to the Bar U," *The Western World*, June 1891, reprinted in *Alberta History* 48, no. 1 (Winter 2000): 21–26; see also Simon Evans, "More on the Bar U," *Alberta History* 49, no. 1 (Winter 2001): 21–22. Sheldon made two visits to the Bar U, one in 1884 and another in 1887.

66 Alan B. McCullough, "Land, Cattle, Markets and Settlers," Parks Canada, research paper, 1993; and "Book Counts," January 1994.

67 GA, Walrond Ranche Stock Book. The following totals were recorded: 1885 – 1810; 1886 – 1134; 1887 – 1352; and 1888 – 1374. The Little Bow Cattle Company, on the other hand, mentioned losses of 25 per cent in their Herd Account Book. The numbers of calves branded were: 1885 – 133; 1886 – 171; 1887 – 131; and 1888 – 176. GA, M6552, no. 4, "Cochrane, William Edward, notebook and stock records."

68 NAC, RG 15, f. 192192, "Stock Returns from Ranches in the North West Territories," Department of Interior, 15 January 1890.

69 Order-In-Council, 24 May 1886, Department of Interior, vol. 8, p. 425; and Pearce's covering letter which accompanied the stock returns.

70 Leases were to be stocked at the rate of one head to twenty acres. The NWCC had 158,000 acres under lease and therefore required 7,900 cattle to meet the requirements of the regulations.

71 NAC, RG 15, f. 192192, William Pearce to Department of Interior, 26 January 1891. A careful count of the Walrond herd, recorded in the livestock notebook, indicated a total of 6,345 head in July 1890, while a second count the following year recorded 8,599 – not too far off the 7,000 head claimed.

72 Canada, North-West Mounted Police, 1896, "Report of Superintendent Howe," pp. 146–49.

73 *Calgary Herald*, 5 March 1884; and Canada, *Sessional Papers*, 1885, vol. 5, no. 8, "Annual Report of Brandon Immigration Agent," p. 143.

74 Sheldon, "Visit to the Bar U," p. 24.

75 *Leaves from the Medicine Tree*, pp. 357–58.

76 Ings, *Before the Fences*, p. 14.

77 Brado, *Cattle Kingdom*, p. 148; see also GA, M2435, f. 5, Quorn Ranch Day Book, typical entries include: August 1892, "cattle delivered to Gordon at Strathmore with North West Cattle Company"; or 20 November 1892, "Stimson paid for shirt, but 75c lost, C.W. Martin."

78 See, for example, Canada, *Sessional Papers*, 1887, vol. 20, no. 12, "Remarks Respecting Horses and Horse Breeding in Canada," p. 246.

79 Osgood, *Day of the Cattleman*, p. 93.

80 Grant McEwan, *John Ware's Cow Country* (Edmonton: Institute of Applied Arts Ltd., 1960), p. 134.

81 Canada, *Sessional Papers*, 1896, vol. 5, no. 8, "Report of the Minister of Agriculture," p. xxii.

82 Canada, *Sessional Papers*, 1897, vol. 10, no. 13, "Report of the Superintendent of Mines," p. 28.

83 North-West Territories, Department of Agriculture, *Annual Report*, 1901, p. 68.

84 These range cattle were slaughtered at the ports anyway, as they were too wild to be attractive to farmers. Thus the western ranchers were not really affected directly by the British "schedule." Ironically, it was perceived as a major obstacle and much energy was put into trying to remove it!

85 *The Nor'West Farmer*, February 1897, p. 35; and June 1897, p. 183.

86 Canada, *Sessional Papers*, 1894, vol. 6, no. 8, "Report of the Emerson Quarantine," p. 56.

87 Canada, *Sessional Papers*, 1894, vol. 10, no. 13, "Report of the Superintendent of Mines," p. 22.

88 Canada, *Sessional Papers*, 1894, vol. 5, no. 8, "Report of the Commissioner, North-West Mounted Police," p. 93.

89 Evans, "American Cattle on the Canadian Range"; and Canada, *Sessional Papers*, 1897, vol. 5, no. 8, "Report on the Cattle Quarantine," p. 105. Turning back drifting American cattle became one of the most onerous tasks of the overburdened North West Mounted Police force. See the reports of Superintendent R. Burton Deane from Lethbridge, which were later used as the basis of a chapter entitled "Wholesale Cattle Smuggling," in his book *Mounted Police Life in Canada* (London: Cassell, 1916).

90 See, for example, McEachran's report on a visit he made to the Liverpool lairages. Canada, *Sessional Papers*, 1899, vol. 6, no. 8, "Report on Cattle Quarantines," p. 65.

91 Osgood, *Day of the Cattleman* p. 83

92 *Breeder's Gazette*, 27 September 1883.

93 *Breeder's Gazette*, 26 June 1884. The naïveté of this comment is highlighted when it is remembered that most stock purchases at this time were based on 'book counts,' and neither buyer nor seller really knew how many head were actually included in a deal.

94 GA, Pamphlet 636.081, Moreton Frewen, "Free Grazing: A Report to the Shareholders of the Powder River cattle Company Limited, 1883," p. 15.

95 Ibid., p. 16.

96 C.N. Dilke, *Greater Britain*, 2 vols. (London, 1869), quoted by Osgood, *Day of the Cattleman*, p. 83; and Brisbin, *Beef Bonanza*, p. 13.

97 Alexander Begg, "Stock Raising in the Bow River District,", p. 276.

98 "A thousand of these animals are kept nearly as cheaply as a single one, so with a thousand as a starter and an investment of $5000 in the start, in four years the stock raiser has made from $40,000 to $45,000." *Breeder's Gazette*, 27 September 1883.

99 *Statutes of Canada*, 1881, c. 16, s. 8; and for a general review see Chester Martin, *"Dominion Lands" Policy* (Toronto: McClelland and Stewart Ltd., 1973); and Kirk N. Lambrecht, *The Administration of Dominion Lands, 1870–1930* (Regina: Canadian Plains Research Center, 1991).

100 Osgood, *Day of the Cattleman*, p. 199.

101 Ibid., p. 194; the cattleman was long regarded as "a trespasser on the public domain, an obstacle to settlement, and at best but a crude forerunner of civilization of which the farmer was the advance guard and the hoe the symbol." Walter Prescott Webb, *The Great Plains* (New York: Grosset and Dunlap, 1931), p. 424.

102 Donahue, *Western Range Revisited*, p. 12. This book reviews the historical context of cattle ranching in the United States. Donahue demonstrates that there was formidable opposition to cattlemen both from eastern-based congressmen who were imbued with the agrarian ideal and from growing numbers of dry farmers, irrigators, and state builders who saw ranches as a barrier.

103 Annual Report of the Commissioner of the General Land Office, 1875, pp. 6–7; and Annual Report of the Secretary of the Interior, 1877, p. 20; cited by Osgood, *Day of the Cattleman*, p. 194. Few concessions were made to the stockman. The *Timber Culture Act* (1873), and the *Desert Lands Act* (1877) were outgrowths of the 1862 Homestead Act and had the objective of luring the farmer westward, not of providing a realistic framework within which ranching could be pursued legitimately. This argument is developed in Walter M. Kollmorgan, "The Woodsman's Assaults on the Domain of the Cattleman," *Annals of the Association of American Geographers* 59 (March 1969): 215–39.

104 James Daniel Richardson (ed.), *Messages and Papers of the Presidents*, vol. 8, p. 476, quoted by Roy M. Robbins, *Our Landed Heritage: The Public Domain, 1776–1936* (Lincoln: University of Nebraska Press, 1962), p. 220.

105 John Wesley Powell, *Report on the Arid Region of the United States*, ed. Wallace Stegner (Cambridge: Belknap Press of Harvard University, 1962), p. xxiv.

106 Quoted by Osgood, *Day of the Cattleman*, p. 199.

107 Joseph Nimmo, *The Range and Ranch Business of the United States* (Washington, 1885), p. 6.

108 Osgood, *Day of the Cattleman*, p. 114.

109 Gerald Friesen, *The Canadian Prairies* (Toronto: University of Toronto Press, 1984), pp. 133–35; Walter Hildebrandt and Brian Hubner, *The Cypress Hills* (Saskatoon: Purich Publishing, 1994); and for a graphic fictional account see Guy Vanderhaeghe, *The Englishman's Boy* (Toronto: McClelland and Stewart, 1996).

110 Of course, the degree to which this was planned or a pragmatic response to an emergency is difficult to determine: see Simon M. Evans, "The Origin of Ranching in Western Canada: American Diffusion or Victorian Transplant," in Rosenvall and Evans, *Historical Geography*, p. 76.

111 Brian Fitzpatrick, "The Big Man's Frontier and Australian Farming," *Agricultural History* 21 (January 1947): 8–12.

112 Hugh A. Dempsey (ed.), *Men in Scarlet* (Calgary: McClelland and Stewart West, 1974); William Beahen and Stan Horrall, *Red Coats on the Prairies: The North-West Mounted Police, 1886–1900* (Regina: Printwest, 1998); and William M. Baker (ed.), *The Mounted Police and Prairie Society, 1873–1919* (Regina: Canadian Plains Research Center, 1998).

113 See correspondence between Cochrane and Macdonald in NAC, RG15, B2a, 10, 142709, pt. 1. This exchange is described in Breen, *Canadian Prairie West*, p. 17; and by A.B. McCullough, "Eastern Capital, government purchases and the development of Canadian ranching," *Prairie Forum* 22, no. 2 (Fall 1997): 219.

114 *Statutes of Canada*, 1876, c. 19, s. 15.

115 Canada, Department of Interior, vol. 3, p. 617, Order-In-Council, 20 May 1881.

116 Canada, Department of Interior, vol. 3, p. 805, Order-In-Council, 23 December 1881.

117 *Statutes of Canada*, 1881, c. 16, s. 8.

118 Donald G. Godfrey and Brigham Y. Card, *The Diary of Charles Ora Card: The Canadian Years* (Salt Lake City: University of Utah Press, 1993), pp. 51–53; see also A. Wilcox, "Founding of the Mormon Community in Alberta," unpublished MA thesis, University of Alberta, 1950, pp. 23–24.

119 David H. Breen, "The Canadian Prairie West and the 'Harmonious' Settlement Interpretation," *Agricultural History* 47 (January 1973): 63–75.

120 Some closed leases had very long lives. About fifty thousand acres leased first to the North West Cattle Company in 1881 was passed on to George Lane in 1903, and then to Pat Burns in 1928. It continued to make up the greater portion of the Bar U Ranch until 1950. Part of the lease is still held by the Nelson family, from whom Parks Canada acquired the headquarters site.

121 A township contains 36 sections each of 640 acres for a total of 23,040 acres. Thus, a 100,000-acre lease consisted of more than four townships, or 144 square miles.

122 Canada, Department of Interior, vol. 3, p. 805, Order-In-Council, 23 December 1881.

123 Canada, Department of Interior, vol. 7, p. 115, Order-In-Council, 6 April 1885.

124 Canada, Department of Interior, vol. 10, p. 153, Order-In-Council, 19 March 1888; and Simon M. Evans, "The Passing of a Frontier: Ranching in the Canadian West, 1882–1912," unpublished PhD dissertation, University of Calgary, 1976, p. 165 and Figure.

125 The following figures show receipts from grazing leases from 1 November to 31 October:

	$		$
1882	10,100	1885	20, 300
1883	19,300	1886	47,300
1884	10,600	1887	39,600

Canada, *Sessional Papers*, Annual Reports of Timber, Mines and Grazing Lands. Figures rounded to nearest $100; rents were doubled in 1886.

126 Chester Martin, *"Dominion Lands" Policy* (Toronto: McClelland and Stewart, 1973), p. xxi.

127 During 1903, R.H. Campbell reviewed the situation with reference to grazing leases in western Canada. He compared practice in other parts of the world, and observed: "In Australia, from which the lease system which has been followed in Canada is borrowed, the history has been somewhat similar to our own." NAC, RG 15, f. 145330, Canada, Department of Interior, "Grazing Regulations," 6 November 1903.

128 A.J. Christopher, "Government Land Policies in South Africa," in R.G. Ironside (ed.), *Frontier Settlement* (Edmonton: University of Alberta, 1974).

129 Stephen H. Roberts, *History of Australian Land Settlement, 1788–1920* (Melbourne: MacMillan, 1924), p. 177.

130 Ibid., pp. 186–92.

131 For a discussion see Brian Fitzpatrick, "The Big Man's Frontier and Australian Farming," *Agricultural History* 1 (1947): 8–13; and Trevor Langford-Smith, "Murrumbidgee Land Settlement, 1817–1912," in G.H. Dury and M.I. Logan (eds.), *Studies in Australian Geography* (Melbourne: Heinemann Educational, 1968), pp. 99–138.

132 Simon Evans, "Spatial Aspects of the Cattle Kingdom: The First Decade, 1882–1892," in Antony W. Rasporich and Henry Klassen (eds.), *Frontier Calgary* (Calgary: McClelland and Stewart West, 1975), pp. 41–56.

133 Canada, *Sessional Papers*, 1885, vol. 7, no. 13, "Annual Report of the Department of the Interior," p. xii.

134 Canada, *Sessional Papers*, 1884, vol. 8, no. 14, "Annual Report of the Minister of Agriculture," p. vii.

135 Canada, *Sessional Papers*, 1884, vol. 7, no. 12, "Annual Report of the Department of the Interior," p. xv.

136 For details about the Orders-in-Council for each lease and the land patents for deeded land, the reader is referred to the following draft research papers completed for Parks Canada: Simon M. Evans, "Land Acquisition: The Bar U Ranch"; and Alan B. McCullough, "Land Holdings of the Bar U". During 1992 and 1993, the author was working in Calgary on this topic while McCullough was working in Ottawa.

137 Chester Martin, *"Dominion Lands" Policy* (Toronto: McClelland and Stewart, 1973), pp. 13–17; and D.W. Thomson, *Men and Meridians: The History of Surveying and Mapping in Canada* (Ottawa: Supply and Services, 1967).

138 John A. MacDonald, Debates of the House of Commons, 1883, p. 874, quoted in Martin, *Dominion Lands*, p. 17.

139 See Friesen, *Canadian Prairies*, pp. 182–83; and John L. Tyman, *By Section, Township and Range: Studies in Prairie Settlement* (Brandon: Assiniboine Historical Society, 1972).

140 Details concerning quarter sections will be given later in a section dealing with deeded land.

141 Nanton and District Historical Society, *Mosquito Creek Roundup: Nanton and Parkland* (Nanton, Alberta: Nanton and District Historical Society, c. 1975), pp. 34–35.

142 Alan B. McCullough, "The Formation of the North West Cattle Company," unpublished research paper, Parks Canada, 1994.

143 By 1888, the North West Cattle Company controlled three leases: ranch 1, 44,000 acres; ranch 2, 58,960 acres; and ranch 35, 55,000 acres, for a total of about 150,000 acres, or more than seven complete townships. These figures were repeated in the lists of leases published annually by the Department of the Interior

until 1895. See, for example, Canada, *Sessional Papers*, 1887, vol. 6, no. 7, "Timber, Mines and Grazing Lands," p. 36. It seems that the Department of the Interior had interpreted the clause which limited the size of a lease to 100,000 acres to refer to each individual lease. It did not stand in the way of companies holding several leases. In fact the North West Cattle Company's holdings were quite modest when compared with Senator Cochrane's six leases, which amounted to 359,000 acres. Similarly, Sir John Walrond leased 100,000 acres in 1883, and doubled his holding by absorbing Charles Rea's lease. His brother, Arthur Walrond, also held 100,000 acres, while the ranch manager, Duncan McEachran, leased a further 60,000, for a corporate total of 360,000 acres.

144 Canada, Department of Interior, vol. 3, p. 612, Order-In-Council, 20 May 1881.

145 NAC, Macdonald Papers, MG26, A1(e), vol. 525, pp. 165–66, J.A. Macdonald to A. Campbell.

146 Canada, Department of Interior, vol. 7, p. 361, Order-In-Council, 27 August 1885.

147 D.H. Breen, "The Canadian Prairie West and the 'Harmonious' Settlement Interpretation," *Agricultural History* 47 (1973): 68–73.

148 Brado, *Cattle Kingdom*, pp. 146–48.

149 See section map, Calgary 114 (1955). For discussion see Simon M. Evans, "Evolution of a Transport Network in the Pekisko Region: Routes to the Bar U Ranch," unpublished research paper, Parks Canada, 1973.

150 GA, M337, J.L. Douglas, "Journal of a Four Months Holiday to Canada and USA, August to December, 1886." p. 34.

151 NAC, RG 15, B2a, vol. 171, 145330, pt. 3, Robert Findlay et al. to Frank Oliver, February 1897.

152 Breen, *Canadian Prairie West*, pp. 18–19.

153 For a detailed account see McCullough, "Land Holdings of the Bar U," and Evans, "Bar U Ranch: Deeded Land Acquisition."

154 See section map Township 17, Range 2, West of the fifth principal meridian, published 14 May 1884.

155 NAC, RG 15, vol. 698, f. 45364, extract from the minutes of NWCC, Department of Interior to E.W. Riley, 17 August 1899.

156 For example, see Debates of the House of Commons, 1891, XXIII, 25 September, pp. 6466–76; a special committee of the North West Legislative Assembly forwarded a petition to the federal government which included the following clause: "That owing to the growing settlements in the grazing districts, a change is desirable both in the existing leases and in the system of leasing lands." Order-In-Council, 7 March 1892, Department of Interior, vol. 14, p. 93; for the dramatic story of the editor and the 'cattle compact' see Breen, *Canadian Prairie West*, pp. 60–61.

157 Order-In-Council, 11 January 1886, Department of Interior, vol. 8, p. 41.

158 Canada, *Sessional Papers*, 1887, XXI, vol. 12, no. 14, "Report of the Department of the Interior," p. xx.

159 NAC, RG 15, f. 141376, letter from William Pearce to Department of Interior, 11 November 1886.

160 NAC, RG 15, f. 141376, "Petition to Department of Interior from M.H. Cochrane, D. McEachran, H. Montague Allan and George Alexander," 9 March 1886.

161 Order-In-Council, 17 September 1889, Department of Interior, vol. 11, p. 459.

162 See Chapter 8, "Pat Burns Takes Over, 1927–1937".

163 Haystacks were burnt on the Walrond Ranch and McEachran was shot at. The Dunbar family, who had settled on the Walrond lease prior to the lease legislation, were evicted and their house was pulled down by Walrond cowboys. *Macleod Gazette*, 30 July and 3 September 1891; and David H. Breen, "The Canadian Prairie West and the 'Harmonious' Settlement Interpretation," *Agricultural History* 47 (1973): 68–73.

164 Breen, *Canadian Prairie West*, p. 76

165 Ibid., p. 81.

166 John R. Craig, *Ranching with Lords and Commons* (Toronto: William Briggs, 1903), p. 293.

167 Canada, *Sessional Papers*, 1894, vol.10, no.13, "Report of the Superintendent of Mines," p. 19.

168 Canada, *Sessional Papers*, 1894, vol. 10, no. 13, "Report of the Department of Interior," p. xx.

169 Canada, *Sessional Papers*, 1896, vol. 10, no. 13, "Report of the Department of Interior," p. xxi.

170 William Pearce, "Historical narrative, Part I," p. 32, unpublished ms., Pearce Papers, University of Alberta Archives, 19–12; the Cochrane Ranch applied to purchase 48,000 acres. NAC, RG 15, f. 45330, Canada, Department of Interior, T. Mayne Daly to the Governor General in Council, December 1893.

171 Half the amount charged for pre-empted acres in the Fort Macleod region. NAC, RG 15, vol. 698, f. 45364, 22 April 1893.

172 In the United States, too, the deeded land base of cattle companies was recognized as a valuable investment. In 1892, a shareholder of the famous Swan Land and Cattle Company argued against the liquidation of the company on the ground that "the residue of the property is too valuable to be sacrificed." W. Turrentine Jackson, "British Interests in the Range Cattle Industry," in Maurice Frink, W. Turrentine Jackson and Agnes Wright Spring (eds.), *When Grass Was King* (Boulder: University of Colorado Press, 1956), p. 270.

173 NAC, RG 15, vol. 698, f. 345364, pt. I, Stimson to Burgess, 25 October 1893.

174 Ibid., E.W. Riley to Department of Interior, 19 October 1894.

175 Ibid., G.W. Ryley to Burgess, 21 November 1894.

176 Ibid., Riley to Department of Interior, 9 May 1895; and Periera to Riley, 18 May 1895.

177 Ibid., Riley to Department of Interior, 20 September 1895.

178 Canada, Department of Interior, "Plan of Township no. 16, Range 2, West of the Fifth Meridian," Second Edition, 26 May 1899.

179 The company was entitled to additional 1767 acres.

180 NAC, RG 15, vol. 698, f. 45364, Part I, Loughead to Smart, 3 September 1901; Loughead to Department of Interior, 4 November 1901 and 21 November 1901. For details see Evans and McCullough, research papers, Parks Canada, referred to in footnote 1.

181 Brisbin, *Beef Bonanza*.

182 Still the best contextual reference for conditions in the Montana-Alberta borderlands during this period is Sharp, *Whoop-Up Country*; for a specific report see Canada, *Sessional Papers*, 1880, XIII, vol. 3, no. 4, "Report of Superintendent Winder," p. 9.

183 Henry C. Klassen, "The Conrads in the Alberta Cattle Business, 1875–1911," *Agricultural History* 64, no. 3 (1990):54.

184 Alan B. McCullough, "Eastern Capital, Government Purchases and the Development of Canadian Ranching," *Prairie Forum* 22, no. 2 (Fall 1997): 213–35.

185 Canada, *Sessional Papers*, 1882, vol. 7, no. 11, "Report of the Minister of Agriculture," p. viii.

186 GA, M1303, Cochrane Ranche Notebook.

187 Canada, *Sessional Papers*, 1885, vol. 18, no. 8, "Annual Report of the Minister of Agriculture," pp. 136 and 143. Later, the North West Cattle Company held the contract to supply the Sarcee reserve. Grant MacEwan, *Pat Burns: Cattle King* (Saskatoon: Western Producer, 1979), p. 78.

188 GA, M8688, Walrond Ranch Account Books. There are five books covering the period between 1883 and 1912. (McCullough has only analyzed the period 1883–89).

189 Montana Historical Society, MC 55, T.C. Power Papers, box 166, f. 6, Stimson to Power, 1 January 1882.

190 See McCullough, "Eastern Capital," especially Table 5, p. 231.

191 *Macleod Gazette*, 24 January 1885, "The Price of Beef."

192 The best overall treatment of this is Richard Perren, *The Meat Trade in Britain, 1840–1914* (London: Routledge and Kegan Paul, 1978).

193 D. McEachran had the temerity to comment, "The time occupied in transit from the shipping point on the railway to Montreal (over ten days) should and can easily be reduced to six and a half or seven days (the distance is only 2,333 miles). Canada, *Sessional Papers*, 1890, vol. 23, no. 6, "Report on Quarantines," p. 24.

194 The progress of the cattle trade is recorded in the annual reports of the Minister of Agriculture, which include those of the immigration agents and quarantine officials in his ministry. The Department of the Interior was involved in land disposition and development in the west. The number and location of ranches holding grazing leases, reports concerning the number and quality of their herds, their successes and difficulties, are all described in the reports of the Deputy Minister, A.M. Burgess, and his man on the spot, William Pearce, Superintendent of Mines.

195 Oscar D. Skelton, *General Economic History of the Dominion, 1867–1912*, vol. 9 of *Canada and Its Provinces*, ed. Adam Shortt and Arthur G. Doughty (Edinburgh: T. and A. Constable for the Publisher's Association of Canada, 1913), p. 164.

196 Perren, *Meat Trade in Britain*; James Law, *The Lung Plague*, quoted by Duncan McEachran, Canada, *Sessional Papers*, 1880, vol. 7, no. 10, "Report on the Cattle Trade, 1879," p. 108; and J.R. Fisher, "The Economic Effects of Cattle Disease in Britain and its Containment, 1850–1900," *Agricultural History* 54, no. 2 (April 1980): 278–93.

197 Simon M. Evans, "Canadian Beef for Victorian Britain," *Agricultural History* 53, no. 4 (October 1979): 748–62; and John P. Huttman, "British Meat Imports in the Free Trade Era," *Agricultural History* 52, no. 2 (April 1978): 247–362.

198 These were all eastern cattle. Canada, *Sessional Papers*, 1884, vol. 17, no. 14, "Report of Minister of Agriculture," p. vii.

199 Canada, *Sessional Papers*, 1879, vol. 12, no. 9, "Report of the Liverpool Agent," p. 143.

200 See Nele Loring, "Five Thousand Miles with Range-Cattle," *The Nineteenth Century* 39, no. 170 (April 1891): 648–69; and for a partial rebuttal, see the account by Albert Browning of his experiences with a herd from the Cochrane Ranch, *Macleod Gazette*, 13 and 20 August 1891.

201 As has already been mentioned, it seems likely that the NWCC pioneered the journey to the United Kingdom because the Allan family owned both the ranch and the shipping line and had influence with the railway company.

202 *Calgary Tribune*, 7 October 1887.

203 Canada, *Sessional Papers*, 1888, vol. 21, no. 4, "Report on Cattle Quarantines," p. 198.

204 Ibid., "Report of the Superintendent of Mines," p. 10.

205 Ibid., "Report of the Minister of Agriculture," p. ix.

206 Evans, "Canadian Beef," p. 750.

207 *Macleod Gazette*, 20 June 1888.

208 Canada, *Sessional Papers*, 1890, vol. 23, no. 6, "Report on Cattle Quarantines," p. 24

209 The next paragraphs draw extensively on A.B. McCullough, "Winnipeg Ranchers: Gordon, Ironside and Fares," *Manitoba History* 41 (Spring/Summer, 2001): 18–25.

210 During the 1890s the buyer might have been George Lane.

211 Ironside was a Liberal Member of the Manitoba legislature at the same time as Clifford Sifton, who became federal Minister of the Interior in 1896. He also served as president of the Canadian Livestock Association. Gordon was also an MLA, and a director of several banks and securities firms.

212 In 1893, for example, the Allan Line carried 7,377 head and lost 11; while the Beaver Line carried 11,823 and lost only 14 head. Canada, *Sessional Papers*, 1894, vol. 10, no. 5, "Trade and Commerce," p. 36.

213 Canada, *Sessional Papers*, 1903, vol. 15, no. 52, "Cattle Trade with Europe," p. 146.

214 Ibid., p. 151.

215 Ibid., p. 147.

216 Ibid., p. 144.

217 MacEwan, *Pat Burns.*

218 *Macleod Gazette*, 22 January 1897; see also *Calgary Herald*, 1 April 1897.

Notes to Chapter 3

1 Henry Glassie, *Passing Time in Ballymenone: Culture and History of an Ulster Community* (Philadelphia: University of Pennsylvania Press, 1982), p. 621.

2 John Innes, "A Visit to a Round-Up," *The Canadian Magazine* 16, no. 1 (November 1900): 12.

3 The first temporary base for the NWCC was established just outside the cluster of houses which was to grow into High River. Once the location of the Bar U leases had been established, the permanent headquarters was chosen close to the centre of the lease.

4 It is interesting, considering the classic characteristics of the "open range system," that both Fred Stimson and Duncan McEachran of the Walrond, in their letters to Thomas C. Power, the Fort Benton merchant, inquired about purchasing and shipping mowers and horse-drawn hay rakes for their ranches, even before they had bought any cattle. Montana Historical Society Archives (hereafter MHSA), MC55, T.C. Power Papers, box 166, file 6, Stimson to Power, 26 April 1882.

5 GA, M265, Kenneth Coppock, "Letters and reminiscences of Phil Weinard."

6 *Descriptions of the Townships of the North-West Territories* (Ottawa: McClen Roger, 1886), p. 363.

7 GA, M652, Elizabeth Sexsmith Lane, "A Brief Sketch of Memories of my Family," p. 11

8 GA, M337, J.L. Douglas, "Journals of a Four Month Holiday to Canada and the United States, August to December 1886."

9 Canada, *Sessional Papers*, 1886, vol. 10, no. 9, "Annual Report North West Mounted Police," p. 21.

10 GA, 364.971, G226n, Don Gardner, *North West Mounted Police Outposts of Southern Alberta*, vol. 2, Fort Macleod and Calgary.

11 Rent payments went on until at least 1904. Gardner, p. 451.

12 For the relations between the Police and Ranchers, see D.H. Breen, "The Mounted Police and the Ranching Frontier," in Dempsey, *Men in Scarlet*; and Elofson, *Cowboys, Gentlemen and Cattle Thieves* .

13 Terry G. Jordan, Jon T. Kilpinen and Charles F. Gritzner, *The Mountain West: Interpreting the Folk Landscape* (Baltimore: Johns Hopkins University Press, 1997), p. 6.

14 Terry G. Jordan-Bychkov, "Does the Border Matter? Cattle Ranching and the 49th Parallel," in Evans, Carter and Yeo, *Cowboys, Ranchers and the Cattle Business*, p. 10; see also the close-up photograph, on p. 9, of the dovetail joints used in the Ranchers' Hall from Millarville, now found at Heritage Park in Calgary. For explanation and discussion see Jordan, Kilpinen, and Gritzner, *Mountain West*, pp. 61–63.

15 See the last chapter of this book, "Reviewing Origins."

16 Edward Mills, "Historic Bar U Ranch Headquarters Longview, Alberta," Parks Canada, Federal Heritage Buildings Review Office, Building Report 92–17, pp. 14–15.

17 Jordan, "Does the Border Matter?" p. 9.

18 See Parks Canada, Historic Structures Reports, Bar U Ranch, 1993–95.

19 The nominal returns are made available ninety years after they were collected. Census of Canada, 1891, Provincial Archives of Alberta (hereafter PAA), microfilms T-6425, T-6551, and T-6652.

20 J.R. Weston, "Charlie Miller of the Bar U," *Canadian Cattlemen* 10, no. 2 (September 1947): 168.

21 W.B. Yeo, "Pekisko: Heart of a Rural Community," unpublished research paper, Parks Canada, January 1998, p. 5.

22 Letter dated 16 March 1942, at Toronto, sent to Herb Millar on the occasion of his eightieth birthday:

"I often think of the fine rides we had together... My father enjoyed his life in Mexico. I joined him there and was with them for three years... I still own the ranch, we milk 350 cows a day... Tell Ella (Millar's wife) I still have the big blanket dressing gown she and my mother made me... I hope this reaches you; I just wanted you to know that I had not forgotten old friends." *Canadian Cattlemen* 5, no. 1 (June 1942): 42.

23 Nellie returned to live in Mexico where she died in 1935. She is buried in Compton, Quebec. Alan B. McCullough, "Personnel and Personalities," Parks Canada, research paper, 1995, p. 55.

24 Lane had left to recover from a bout of typhoid fever which nearly killed him.

25 Cochrane and Area Historical Society, *Big Hill Country* (Calgary: Friesen, 1977), p. 318.

26 See Slatta, *Cowboys of the Americas* , p. 205, for an appreciation of the novel; and GA, M4018, Jean L. Johnson, "A Virginian Cowboy: His Life and His Friends," for the connection. In 1904, Wister sent Ebb a signed copy of the book, and they corresponded and met from time to time over the years.

27 NAC, RG 15, f. 142083, Frewen to Department of Interior, 18 February 1886 and subsequent letters to 18 May 1886; and Moreton Frewen, *Melton Mowbray and Other Memories* (London: Herbert Jenkins, 1924), p. 222.

28 GA, M2388, "Stair Ranch Letterbox," p. 350; D.H. Andrews had had an exciting life on the range himself; he became embroiled in the "Johnson County War" in Wyoming, and came to Canada to avoid repercussions. See John Clay, *My Life on the Range* (Norman: University of Oklahoma Press, 1962), p. 264; and "Early Days in Wyoming," *Farm and Ranch Review* 19, no. 23 (December 1923): 7.

29 *Big Hill Country*, p. 318.

30 For an account of the bookkeepers and ranch accounts see, W. B. Yeo, "Debit and Credit, Keeping Accounts at the Ranch," unpublished research paper, Parks Canada, March 1998.

31 This conclusion is supported by the fact that there were a large number of teamsters elsewhere in the High River District. Men described their occupation according to what they spent most time doing. Perhaps the title "Cowboy" was one to be earned.

32 During my time as Project Historian for the Bar U, I received more inquiries about "The Sundance Connection" with the ranch than on any other single topic. Leading the charge were Donna B. Ernst, the author of *Sundance, My Uncle* (College Station, Texas: Creative Publishing, 1992), and her husband Paul, who is the grandnephew of the Sundance Kid. Anne Meadows and Daniel Buck of the National Association for Outlaw and Lawman History were not far behind. For the famous outlaw's doings in Alberta, see Donna B. Ernst, "Sundance in Alberta," *Alberta History* 42, no. 4 (Autumn 1994): 10–15; Anne Meadows and Daniel Buck, "Running down a Legend," *Americas* 42, no. 6 (1990–91): 21–28; and Daniel Buck, "Surprising Development: The Sundance Kid's Unusual, and Unknown, Life in Canada," *Journal of the Western Outlaw-Lawman History Association* 3, no. 3 (Winter 1993): 8–11.

33 Ings, *Before the Fences*, p. 52.

34 Barbara Holliday, "Herb Millar of the Bar U," *Alberta History* 44, no. 4 (Autumn 1996): 17; and *Leaves of the Medicine Tree*, p. 259.

35 Meadows and Buck, "Running Down a Legend," p. 21.

36 For discussion of these ideas, see Jordan-Bychkov, "Does the Border Matter?".

37 See, Simon M. Evans, "Some Observations on the Labour Force of the Canadian Ranching Frontier in its Golden Age, 1882–1901," *Great Plains Quarterly* 15, no. 1 (Winter 1995): 3–18.

38 Up to this date all the foremen at the ranch, Tom Lynch, Abe Cotterell, George Lane, and Ebb Johnson, had been born in the United States.

39 MHSA, MC 55, box 448, file 2. In the first letter, 18 December 1882, McEachran explains to Power that he is planning to establish a new cattle company with English backing. He enlists Power's help and goes on, "the first thing I would want would be an experienced cattle man to put in charge – I wish you would turn this over in your mind and keep your eye on a good man that I can put my hands on if I wanted this spring."

40 Ibid., McEachran to Power, 28 February 1883.

41 Ibid., McEachran to Power, 9 April 1883. The file contains several more letters in the same vein.

42 *Macleod Gazette*, 30 November 1886, quoting the *London Times*.

43 The next few paragraphs draw heavily on W.B. Yeo, "Pekisko."

44 Patrick A. Dunae, *Gentlemen Emigrants* (Vancouver: Douglas and McIntyre, 1981), p. 92.

45 Patrick A. Dunae (ed.), *Ranchers' Legacy* (Edmonton: University of Alberta Press, 1986), p. 14.

46 Harry A. Tatro, "A Survey of Historic Ranches," Study prepared for the Historic Site and Monuments Board of Canada, 1973, p. 182.

47 Interestingly, their ranch developed within the lease of the NWCC; it was eventually purchased by the Prince of Wales in 1919. See Evans, *Prince Charming Goes West*, pp. 49–55.

48 GA, M337, J.L. Douglas, "Journal of Four Months Holiday to Canada and the USA, August to December, 1886," p. 35.

49 Norman Rankin, "The Boss of the Bar U," *Canada Monthly* 9, no. 5 (March 1911): 332.

50 Arni Brownstone, "Tradition Embroidered in Glass," *Rotunda* 29, no. 1 (Summer 1996): 11–21.

51 See, for example, his letter concerning the starvation which was occurring on reserves, which he deemed a disgrace, and urged the government to do something. *Calgary Herald*, 12 February 1885. I had always taken this letter as evidence of Stimson's concern for the state of the Native people. It was only while writing about the economics of the range, and the role of government supply contracts to feed the Indians, that I realized that Stimson's indictment of the government was not without self-interest!

52 Harold Riley stated that Stimson had "a penchant for purloining anything that he could lay his hands on." Harold W. Riley, "The Romance of the Ranching Industry and the Pioneer Ranchers and Cowboys," *Canadian Cattlemen* 3, no. 1 (June 1940): 399.

53 Harold W. Riley, "Herbert William (Herb) Millar," *Canadian Cattlemen* 4, no. 4 (March 1942): 168.

54 John Innes, "A Visit to a Round-Up," *The Canadian Magazine* 16, no.1 (November 1900): 12.

55 Brownstone quotes from an interview with Dr. Dempsey. Brownstone, "Tradition," p. 16; and GA, M4562, f.33, Eleanor Luxton collection, W.E.M. Holmes, "Early Ranches," p. 3.

56 Museum of the Highwood, Bar U National Historic Park Oral History Project, 1992–1993. Interview outlines and cumulative index. Mrs. A.A. Leitch interview. (hereafter OHP).

57 Brownstone, "Tradition," p. 18.

58 *Calgary Herald*, 5 February 1902.

59 Innes, "Round-Up," p. 12.

60 Census of Canada, 1901, Central Alberta, Pekisko District, Schedule 1, page 2, PAA, microfilm T-6551.

61 In 1902 the NWCC's board of directors sold the Bar U Ranch to George Lane and Gordon, Ironside and Fares, without consulting

with, or even informing, their ranch manager of their negotiations. Stimson sued them successfully.

62 Yeo, "Debit and Credit," pp. 12–13.

63 Of this crew only Fred Latur was a U.S. citizen, and even he had been born in England.

64 *Leaves from the Medicine Tree*, p. 446.

65 *Big Hill Country*, pp. 317 and 388.

66 For discussion of the 1891 and 1901 nominal census returns and some of the difficulties in comparing them see, Simon M. Evans, "Some Observations on the Labour Force of the Canadian Ranching Frontier During it Golden Age, 1882–1901," *Great Plains Quarterly* 15, no. 1 (Winter 1995): 3–18.

67 It is still in use today to contrast "the west's open-range environments" with the more intensive methods of eastern Canada. MacLachlan, *Kill and Chill*, p. 33.

68 Jordan, *Cattle-Ranching Frontiers*, pp. 208–40.

69 Jordan, *Cattle-Ranching Frontiers*, p. 236.

70 Frank Wilkeson, "Cattle-Raising on the Plains," *Harper's New Monthly Magazine* 72 (April 1886): 789.

71 Ibid.

72 Jordan, *Cattle-Ranching Frontiers*, pp. 267–307.

73 Ibid., p. 307.

74 MHSA, MC55, box 166, f. 6, Stimson to Power, 26 April 1882. Stimson wrote, "I wish you would send my mowers and horse rakes up as soon as you can as I want to put up my hay early...."; see also "Buying the Home place," above.

75 Craig, *Ranching with Lords and Commons*, p. 9.

76 GA, M6552, William Edward Cochrane fonds, f. 4 "Stock Records and notebooks."

77 GA, M1543, Cross Papers, box 59, f. 467, Douglas to Cross, 16 December 1906.

78 Simon M. Evans, "The End of the Open Range Era in Western Canada," *Prairie Forum* 8, no. 1 (Spring 1983): 71–87.

79 Stan Graber, *The Last Roundup* (Saskatoon: Fifth House, 1995).

80 GA, M6878, Margaret Barry-McGechie, "7U Brown, Alberta Cattle Country, 1881–1883," prepared for Alberta Culture, c. 1985.

81 Simon M. Evans, "The Burns Era at the Bar U Ranch," unpublished research paper, Parks Canada, 1997, p. 32.

82 *Lethbridge News*, 2 March 1905.

83 Alan McCullough produced a research guide to "The Seasonal Round" which drew heavily on contemporary accounts written in diaries. This collection proved invaluable. See A.B. MacCullough, "The Seasonal Round in Ranching Country," unpublished research report, Parks Canada, 1997.

84 PAA,, F.S. Stimson Homestead File 173491, Reels 2010 and 2011.

85 GA, M1543, Cross Papers, box 14, f. 104, Stimson to Cross, 25 May 1898.

86 F.W. Ings, "Roundup Days of the Eighties," *Canadian Cattlemen* 1, no. 3 (December 1938): 108.

87 Buchanan, "History of the Walrond," p. 174.

88 GA, M7888, Edward Hills Letters (hereafter Hills Letters), no. 42, 6 May 1885.

89 Ings, "Roundup Days," p. 108.

90 Ibid., p. 133.

91 Hills Letters, no. 46, 1 September 1885.

92 Some accounts mention several different captains for various portions of the roundup. Cross suggests that the role of captain was assumed by the foreman of the major ranch whose range they were sweeping for cattle. Alfred E. Cross, "The Roundup of 1887," *Alberta Historical Review* 13 (Spring 1965): 26.

93 There was another general roundup after the bad winter of 1906–7. The strategy used was very similar: "Up the three coulees, the Chin, Etzikom and Verdigris, the outfits worked their way with military precision...." Katherine Hughes, *Edmonton Bulletin*, 22 June 1907. Reprinted in *Alberta Historical Review* 11, no. 2 (Spring 1963): 5.

94 Cross, "Roundup of 1887," p. 26.

95 Hills Letters, no. 44, 11 June 1885.

96 Ings, "Roundup Days," p. 133.

97 GA, M8788, Arthur William Turner, "Reminiscences of an English Boy in Canada, 1903."

98 Cross, "Roundup of 1887," p. 26. "Made into stags" means that these big old bulls had to be castrated. This operation was, of course, usually carried out while the calves were small and young. These big brutes had obviously escaped earlier attention.

99 Ings, "Roundup Days," p. 108.

100 Hills Letters, no. 44, June 11, 1885.

101 Ings, "Roundup Days," p. 136.

102 Slatta, *Cowboy Encyclopedia*, p. 335; and Buchanan, "History of the Walrond," p. 174.

103 The practice of branding can be traced back to at least 2000 BCE in Egypt. Slatta, *Cowboy Encyclopedia*, p. 331; and Ward, *Cowboy at Work*, p. 63.

104 Jordan, *Cattle-Ranching Frontiers*, p. 94. Some of the early Spanish outfits were enormous, Jordan mentions that Francisco de Ibarra branded forty-two thousand calves annually.

105 Jordan, *Cattle-Ranching Frontiers*, p. 213. Jordan adds, "So complete was the Anglo-Texan rejection of Spanish brand designs that the term 'with a Mexican brand' appeared in some newspapers as an adequate description of strayed animals." Starrs, *Let the Cowboy Ride*, endorses the historic roots of branding but he suggests that changes and adaptations make it both "home grown and foreign," p. 25.

106 The first Canadian brands were used to identify a herd of some two thousand cattle brought into British Columbia in 1840. See L.S. Hester, "Cattle Brands," *R.C.M.P. Quarterly* 9, no. 3 (1941–42): 305. Of course brands had also been used for hundreds of years to identify ownership of bales of furs and spars of timber.

107 Brand inspectors recovered sixteen thousand head in 2000. *Calgary Herald*, 17 July 2001; and 17 March 2000.

108 "Branding in the West: New Age, old irons," *The Economist*, 22 November 1997.

109 For a detailed account see Ward, *Cowboy at Work*, pp. 59–63.

110 Bert Sheppard, *Just About Nothing* (Calgary: McAra Press, 1977), p. 143.

111 Sheppard reported that Alex Fleming regarded two hundred as a "nice sized branding." The Walrond Ranch livestock notebook shows that branding usually took about seven days and that about two hundred calves were branded each day. GA M8688, f. 37, "Walrond Livestock Notebook." This covers the years 1885–96 and give details of seven brandings.

112 NAC, MG 40, Godsal papers, M21 Reel A-1609, p. 116, June 1882.

113 Frank White Diary, 25 October 1882, *Canadian Cattlemen* 8, no. 4 (March 1946): 245.

114 GA, M2365, "Memorandum re Brand Office: Its Organization and Duties"; and for the role of the police, Breen, "The Mounted Police and the Ranching Frontier."

115 GA, Alberta Brand Book, 1883, p. 821; this was officially registered on 11 January 1882 as the thirty-first brand entered in the new registration book. GA, M2190, Cattle Brands, "A."

116 Hank Pallister, "Early Brands," *Home Quarter* 9, nos. 3 and 4 (Fall and Winter 1989): 1.

117 GA, M2365, F.S. Stimson to Territorial Department of Agriculture, 8 May 1898; Fearon to Deputy Commissioner, 13 June 1898; and Deputy Commissioner to Stimson, 28 July 1898.

118 GA, M2365, George Lane to Brand Recorder, 23 February 1902.

119 Alberta and Saskatchewan Brand Book, 1907. Gordon, Ironside and Fares also owned several brands.

120 MacEwan, *Pat Burns: Cattle King* , pp. 137–44.

121 *Alberta Horse, Cattle and Sheep Brands, 1937* (Edmonton: King's Printer, 1937).

122 For example, the NWMP refer to "Stimson's": see Canada, *Sessional Papers*, 1887, 7-7a, vol. 20, no. 6, "North West Mounted Police, Annual Report, 1886"; "Pekisko" is often used to describe the community of which the ranch was the centre: see W.B. Yeo, "Pekisko"; and Aldous' first survey map of 1883 locates the headquarters of the "North West Cattle Company."

123 None of the other great ranches were identified by their brands. The company name was used for the Walrond, the Oxley, the Quorn, and the Cochrane. On the other hand, the Circle outfit, the 76, and the 44, were all associated with brands rather than their corporate identities.

124 Frank Jacobs, *Cattle and Us, Frankly Speaking* (Calgary: Detselig, 1993), p. 191.

125 MacLachlan, *Kill and Chill*, p. 31.

126 Things did not change soon for a study done in 1938–41 found that on more than a third of foothills ranches the calf crop was less than 55 per cent, and only a quarter of the sample had a crop of over 70 per cent. See G.C. Elliott and G.D. Chattaway, *Cattle Ranching in Western Canada* (Ottawa: Department of Agriculture, 1941), p. 38.

127 The section maps produced from the first surveys, completed in the early 1880s, show these fields in red. For example, Section Map for Township 17, Range 2 shows the Bar U, while Township 18, Range 3 shows the Rio Alto Ranch.

128 GA, Microfilm A1610, Godsal Papers, 31 July 1882.

129 Frank White Diary, 7 November 1882, *Canadian Cattlemen* 9, no. 1 (June 1946): 8.

130 Frank White Diary, 7 July 1884, *Canadian Cattlemen* 11, no. 4 (March 1949): 235.

131 This winter had a devastating impact on the overstocked ranges of the western United States, but it had much more limited effects on newly established foothills ranches. Cattlemen in the High River district estimated their losses at 8 per cent, and their optimism was borne out by the July roundup. *Calgary Herald*, 13 May 1887 and 15 July 1887. The agricultural press in the United States, which had reported on the "Beef Bonanza" with such enthusiasm, now advocated a move to "ranch farming." *The Denver Range Journal* talked of fattening herds on irrigated alfalfa. Closer to home the *River Press* expressed the view that "The range industry is bound to die out and a new order of things will exist." These reports were republished in Canadian newspapers like the *Macleod Gazette* and the *Calgary Herald*. For detailed coverage see Evans, "The Passing of a Frontier," pp. 164–68; and Robert H. Fletcher, *Free Grass to Fences* (New York: University Publishers, 1960).

132 One such crew was run by Ralph Nelson's grandfather. Ralph is a longtime local rancher and president of the Friends of the Bar U Ranch.

133 *Calgary Herald*, 5 February 1902, "Big Transaction." The same year the Walrond Ranch put up two thousand tons of hay. *The Nor'West Farmer*, October 6, 1902, "Ranching in Southern Alberta."

134 Weaning the calves was practised on some ranches in the mid-1880s, but did not become general until about 1890. NAC, RG 15, vol. 1220, f. 192192, Pearce to Ryley, 26 January 1891.

135 Sheldon, "Visit to the Bar U," pp. 21–26; see also Evans, "More on the Bar U," pp. 21–22.

136 NAC, RG 15, vol. 1220, f. 192192, Pearce to Ryley, 26 January 1891.

137 Kelly, *Rangemen*, p. 211.

138 *Calgary Herald*, 5 February 1902.

139 OHP, Allen Baker, 12.03.

140 NAC, MG 40, M21, Reel A-1603, 11 January 1885, pp. 394–99.

141 Warren Elofson has argued that large ranches were inherently inefficient and destined to fail because of the extensive methods which they employed. The case of the Bar U demonstrates that business decisions, not problems at the ranch, were the cause of the sale in 1902. A similar case could, I am sure, be made for the sale of the Cochrane Ranch in 1906. Here, the death of Senator Cochrane, the desire of his sons to retire in Montreal, and the huge opportunity costs generated by rapidly rising land costs meant that the sale was a great business success for those involved. See Elofson, *Cowboys, Gentlemen and Cattle Thieves*, p. 15 and following.

Notes to Chapter 4

1 James P. Ronda, "Why Lewis and Clark Matter," *Smithsonian* 34, no. 5 (August 2003): 101.

2 Norman Rankin, "The Boss of the Bar U," *Canada Monthly* 9, no. 5 (March 1911): 332. As far as we know Lane seldom wore jeans. He always appears in dark trousers and a frock coat. Moreover, ranchers and cattle buyers were accustomed to carrying cash and cheques for large amounts of money.

3 Rankin, "Boss of the Bar U," p. 332.

4 He had purchased the Victor Ranch on Willow Creek, and then the YT Ranch on the Little Bow River in 1898.

5 *The Albertan*, 25 September 1925.

6 Brian J. Young and Gerald Tulchinsky, "Sir Hugh Allan," in *Dictionary of Canadian Biography*, vol. 11, 1881–1890 (Toronto: University of Toronto Press, 1982), p. 11.

7 Gordon Burr, "Andrew Allan," in Henry J. Morgan (ed.), *The Canadian Men and Women of the Time: A Handbook of Canadian Biography of Living Characters*, (Toronto: W. Briggs, 1912), p. 14.

8 It is unlikely that Lane knew the details, but everybody in cattle country knew that the initiative of Stimson to purchase lands was first repudiated by 'head office' in Montreal. See, for example, the letter of William Pearce, NAC, RG 15, vol. 698, f. 345364, pt. 1; and G.W. Ryley to Burgess, 21 November 1894.

9 "Sir Hugh Montagu Allan," in W. Stewart Wallace and W.A. McKay (eds.), *The MacMillan Dictionary of Canadian Biography* (Toronto: MacMillan, 1978).

10 Exports of live cattle to Britain rose from 115,000 head valued at $7.6 million in 1900, to 119,000 at $8.0 million in 1901, and 149,000 at $9.7 million in 1902. GIF handled 60 per cent of this volume, or about 90,000 head valued at $5.8 million in 1902. GA, M2383, "Memorandum Showing Exports of Canadian Live Cattle," Papers of the Canadian Cattle Association of Great Britain, March 1921.

11 NAC, RG 95, vol. 1486, Gordon, Ironside and Fares Company Limited.

12 Alan B. McCullough, "Winnipeg Ranchers: Gordon, Ironside and Fares," *Manitoba History* 41 (Spring/Summer 2001): 22.

13 The exact nature of the relationship between Lane and GIF over the Bar U is by no means transparent. We do know that each party put up half of the purchase price, and that Lane retained executive powers over the ranch. Informal partnerships like this were not unusual where the purchase of large lots of cattle were concerned, but I don't know of another example where a longer lasting purchase of a ranch was involved.

14 The bill of sale has not survived, and we are therefore forced to rely on accounts in the press, and on the reports of the principals involved. *Calgary Herald*, 5 February 1902.

15 See, for example, the discussion of the Powder River Cattle Company in Chapter 3, "The Open Range."

16 Chapter 2, "Buying the Home place."

17 For a discussion of the make up of an ideal herd, see A.B. McCullough, "Book Counts," Parks Canada, unpublished research paper, 1994.

18 Prices paid for mixed foundation herds twenty years before varied from $15 to $32 a head. See "Stocking the Canadian Range," above; and North-West Territories, Department of Agriculture, *Annual Report*, 1901, p. 68, "Railway Compensation Schedule."

19 The Cochrane Ranch obtained $6 an acre for their land in 1906.

20 To recap: cattle, 8000 x $30 = $240,000; horses 500 x $40 = $20,000; and land 1800 x $2 = $36,000; for a total of $296,000.

21 *Calgary Herald*, 5 February 1902.

22 This is the sum mentioned by Hugh Robertson, Manager of Royal Trust and liquidator of the NWCC, Cour Superior, Montreal, 7 May 1907. I am once again indebted to Allan McCullough for these details.

23 Brado, *Cattle Kingdom*, p. 122.

24 Stimson's estimate would have been $220,000 + $100,000 = $320,000; our estimate, based on conservative assumptions, was $296,000. Stimson, as well as being an employee of the NWCC, was a shareholder, and, because his salary was in arrears, he was also a creditor of the company. He sued the company over the appointment of Hugh Montague Allan as liquidator. He was vindicated in court and paid $5,772.97 for his claim and costs. Cour Superieure, Montreal, 23 May 1901, no. 46, "NWCC en liquidation & H.M. Allan et F.S. Stimson requerent."

25 The Allans considered not only the selling price but also the timely nature of the deal. If they had had to spend two or three years winding up the NWCC, they would have had to forego the potential return from their capital once it had been redeployed.

26 Elsie Gordon Collection, owned by Willa Gordon, Oxley Ranch, Author Interview, 28 September, 1992.

27 GA, M652, Elizabeth Sexsmith Lane, "A Brief Sketch of Memories of My Family," p. 11

28 GA, M4026, Edna Kelly, Interviews with the Pioneers, Alex Fleming, p. 159.

29 *Lethbridge Daily Herald*, 29 September 1925.

30 Joy Oetelaar has pointed out that the original version of Mrs. Lane's family memoir states that his birthplace was Indiana, while a later typescript mentions Iowa. There are towns named Boonville in both states. Ritchie, probably following Higinbotham, throws in a casual mention of Des Moines, which is of course in Iowa, but my feeling is that he may have assumed the state and then looked up Boonville, Iowa, and found that it was close to Des Moines. All accounts seem to agree that George's parents were both born in Indiana, and it seems likely that he was born there too. Joy Oetelaar, "George Lane: From Cowboy to Cattle King," in Evans, Carter and Yeo, *Cowboys, Ranchers and the Cattle Business*; J.D. Higinbotham, *When the West Was Young*, 2nd ed. (Lethbridge: Herald Printers, 1978); and C.I. Ritchie, "George Lane, One of the Big Four," *Canadian Cattlemen* 3, no. 2 (September 1940): 415ff.

31 Michael P. Malone, Richard B. Roeder, and William L. Lang, *Montana: A History of Two Centuries* (Seattle: University of Washington Press, 1991), pp. 89–90.

32 *High River Times*, 30 March 1916.

33 K. Ross Toole, *Montana: An Uncommon Land* (Norman: University of Oklahoma Press, 1959), pp. 127–29; and Evan S. Connell, *Son of the Morning Star* (New York: Harper and Row, 1984), pp. 216–17.

34 Jordan, *Cattle-Ranching Frontiers*, pp. 298–307.

35 As we shall see, this was the path followed by Pat Burns in Canada a decade or two later, except that Burns was feeding work crews building the CPR Railway rather than miners.

36 The Grant-Kohrs Ranch is now run by the National Parks Service of the US Department of the Interior. The following paragraphs draw on the pamphlets and interpretative information provided at the site. Note that Lane would have worked with Percheron horses on this ranch as well as being involved in "breeding up" the cattle herd.

37 Lane may well have driven a herd of "CK" cattle south and then wintered over in the Salt Lake City area. McIntyre reported that Lane had worked for him on the "IHL" spread in Utah. William H. McIntyre Jr., "A Brief History of the McIntyre Ranch," *Canadian Cattlemen* 10, no. 2 (September 1947): 87.

38 *Rocky Mountain Husbandman*, 6 September 1883, quoted in Osgood, *Day of the Cattlemen*, p. 181.

39 Granville Stuart, *Forty Years on the Frontier*, vol. 2 (Cleveland: 1925), pp. 187–88.

40 GA, M517, f. 51, J.D. Higinbotham papers, "Old Timer Tells," by William Cousins. Strong did move to Canada and worked for the Circle outfit until his death in 1889. Lane may also have heard about conditions in Canada from two friends, John Sailor and Mr. Orton, who had been as far north as the Peace River Country. Archibald Oswald MacRae, *History of the Province of Alberta*, vol. 1 (The Western Canada History Co., 1912), p. 490. Some accounts suggest that Lane also visited Canada with a herd in 1882.

41 Lane told this story, but I have been unable to find the documents in the Historical Society of Montana Archives in Helena. The letter must have been similar to those written by Duncan McEachran to T.C. Power, which were discussed earlier. The T.C. Power papers have survived while those of the Sun River Stock Association and of Ford himself are very limited.

42 "A Pioneer Stockman," *Farm and Ranch Review* 17 (20 September 1922): 9.

43 GA, M6088, William L. Sexsmith, "Pioneer: The Story of a Ranch and Its People," pp. 172–73. The author was the son of Lem Sexsmith, Elizabeth Lane's brother.

44 Ings, *Before the Fences*, p. 14

45 Ritchie, "George Lane," p. 415.

46 Rankin, "Boss of the Bar U," p. 333.

47 A.E. Cross, "The Roundup of 1887," *Alberta Historical Review* 13, no. 2 (Spring 1965): 23.

48 Higinbotham, *When the West Was Young*, p. 262.

49 "A Pioneer Stockman," p. 9.

50 Ritchie, "George Lane," p. 442.

51 Ritchie, "George Lane," p. 442.

52 *High River Times*, 25 July 1912; and GA, M4689, Lane to W. Greig, 18 June 1912.

53 The Northwest Rebellion or Resistance started at Duck Lake in March 1885 and ended at Batoche in May. The Canadian army hired horses and wagons for transport at rates which were very attractive to those in southern Alberta who were trying to establish homesteads.

54 GA, M7988, Edward F.J. Hills Letters, no. 46, 1 September 1885.

55 Ibid., no. 42, 18 May 1885.

56 *Edmonton Journal*, 25 September 1925.

57 Higinbotham, *When the West Was Young*, p. 263.

58 See Steve Sears, "Foothills Portraits," *The Nanton News*, 1 September 1955; and Rankin, "Boss of the Bar U," p. 328.

59 Ings, *Before the Fences*, p. 15.

60 Elizabeth Lane, "Memories," p. 11. Roy Lane's account of the incident differs in a few details. It is found in Bassano History Book Club, *The Best in the West by a Dam Site* (Calgary: Friesen, 1974), p. 294.

61 *Leaves from the Medicine Tree*, p. 68.

62 George Hill, *Calgary Daily Herald*, 24 September 1925: and Higinbotham, *When the West Was Young*, p. 264.

63 This was the name given to Clifford Sifton by the press. Pierre Berton points out that Sifton owned the Winnipeg Free Press and may have coined the name for himself! See Pierre Berton, *The Promised Land* (Toronto: McClelland and Stewart, 1984), p. 22.

64 See Appendix, "Trade in Live Cattle"; and Simon M. Evans, "Canadian Beef for Victorian Britain," *Agricultural History* 53, no. 4 (October 1979): 748–62.

65 William Pearce seized on the idea enthusiastically and worked out many of the practical details. Pearce Papers, University of Alberta Archives, I-B-8, Pearce to Burgess, 16 January 1895.

66 Breen, *Canadian Prairie West*, pp. 111–15; and Berton, *Promised Land*, pp. 11–12.

67 D.H. Hall, *Clifford Sifton*, 2 vols. (Vancouver: University of British Columbia Press, 1985); and Berton, *Promised Land*.

68 D.H. Hall, *Clifford Sifton, vol. 1: The Lonely Eminence, 1901–1929*, p. 62.

69 GA, M3799, box 2, f. 27, Stimson to Oliver, June 23, 1897.

70 NAC, RG 15, B2, vol. 159, no. 141376, pt. 2, W.M. Gunn to Department of Agriculture, 21 April 1895. "The big stockmen claim almost every section has been allotted and as they are the only ones that seem to know anything about it, would you be kind enough to enlighten us about the matter?"

71 GA, M494, H.M. Hatfield, "Letter to the Provincial Librarian, 1908."

72 David H. Breen, "The Canadian Prairie West and the 'Harmonious' Settlement Interpretation," *Agricultural History* 47 (January 1973): 63–75; and Warren M. Elofson, "The Untamed Canadian Ranching Frontier, 1874–1914," in Evans, Carter, and Yeo, *Cowboys, Ranchers and the Cattle Business*, pp. 81–99.

73 NAC, RG 15, B2a, vol. 171, 145330, pt. 3, Petition to F. Oliver, February 1897, reprinted in Lewis G. Thomas, *The Prairie West to 1905* (Toronto: Oxford University Press, 1975), p. 267. Unconsciously, Findlay articulated a series of complaints which had been made by settlers concerning ranchers throughout the length of the Great Plains since the 1870s.

74 Ibid..

75 Breen, *Canadian Prairie West*, p. 120.

76 Up to this point the ranchers had had to face incursions of home-steaders and squatters, most of whom aspired to become ranchers. After the turn of the century, the great influx of settlers was made up of people of a very different ilk. Many of them were experienced cash grain farmers. Some had sold farms in the United States for $15-$20 an acre, and they were purchasing land in Canada for $4-$5 an acre. They had capital to invest in machinery, and hastened to break up their holdings and extend the acreage sown to wheat. See Karel Denis Bicha, *The American Farmer and the Canadian West, 1896–1914* (Lawrence, Kansas: Coronado Press, 1968).

77 Simon M. Evans, "The End of the Open Range Era in Western Canada," *Prairie Forum* 8, no. 1 (Spring 1983): 71–88; Paul F. Sharp, "The American Farmer and the 'Last Best West,'" *Agricultural History*, 47 (1947): 65–75; and Jack C. Stabler, "Factors Affecting the Development of a New Region: The Great Canadian Plains, 1870–1897," *Annals of the Regional Science Association*, 7 (1973): 75–88.

78 Canada, *Sessional Papers*, 1904, vol. 12, no. 28, "Report of Superintendent P.C.H. Primrose," p. 5.

79 This letter represents the first occasion of many in which Lane acted as spokesman for the ranchers' interests. The letter's lack of passion and the author's ability to see both sides of the question must have impressed Sifton.

80 NAC, RG 15, vol. 1210, f. 145330-3, George Lane to J.W. Greenway, 1 June 1904.

81 Ibid., F. Oliver to C. Sifton, 4 June 1904.

82 NAC, RG 15, vol. 1210, f. 145330, pt. 4, "Memorandum by R.H. Campbell on Grazing Regulations," 6 November 1903.

83 Canada, Department of Interior, Order-In-Council, 30 December 1904. This order was not finally approved until 15 February 1905.

84 NAC, RG 15, vol. 1210, f. 145330, pt. 4, Memorandum, 27 February 1905. Sifton's abrupt resignation came as a complete surprise to his friends and colleagues. It may have been provoked by ill health, fatigue, frustration, and possibly by the threat of scandal. See Berton, *Promised Land*, pp. 196–205.

85 NAC, RG 15, B2a, vol. 172, f. 145330, F. Oliver to Governor General, 27 July 1905.

86 Ibid., pt. 4, Memo Ryley to Oliver, 15 June 1905.

87 NAC, RG 2, Z-28, Reel T-5002, no. 1159D, Order-In-Council, 11 April 1905. The lease lands are described as follows: "Available lands in T13, T14, Rn 29 &30 west of the fourth, and T16, Rn2; T 17 Rn 2; T 17, Rn 3; T 18, Rn 3 and T 18, Rn 4, west of the fifth. Totalling about 62,561 acres. Lease will not include lands occupied by settlers and will not be subject to withdrawal of any lands for homestead entry during the term of the lease, 21 years."

88 Breen, *Canadian Prairie West*, p. 140. The Milk River Cattle Company was also registered in Brandon, Sifton's home town. See *Alberta and Saskatchewan Brand Book, 1907* (Medicine Hat: Department of Agriculture, 1907), p. 78.

89 Lane was to run as a Liberal candidate in the 1913 General election. He was elected, but resigned his seat so that cabinet minister C.P. Mitchell could return to Ottawa. See Oetelaar, "George Lane," p. 57.

90 This meant that the ranch had eighteen thousand acres of deeded land and forty-four thousand acres of leased land for a total of sixty-four thousand acres.

91 This was a reversal of the customary movement eastward to summer pasture along the Little Bow River. It presaged the con-temporary movement of cows and calves into the forest reserves and Kananaskis Country.

92 Nominal Census forms, Dominion Bureau of Statistics, Census of Canada, 1901, PAA, microfilm T-6552.

93 For discussion on this point see Simon M. Evans, "Some Observations on the Labour Force of the Canadian Ranching

Frontier, 1882–1901," *Great Plains Quarterly* 15, no. 1 (Winter 1995): 16–31.

94 In many cases lands assigned to the CPR were not patented until 1910 or later. In this way the company avoided paying tax on the land. Lane did not pay rent either. *Leaves of the Medicine Tree*, p. 21.

95 *High River Times*, 16 April 1908.

96 During the 1890s, Lane had often travelled to Ontario to establish relations with farmers there and to purchase stockers. See GA, M289, box 1, f. 4, W.E. Cochrane to A.E. Cross, 14 August 1895. The Deputy Minister of Agriculture alluded to the movement: "Thousands of yearlings and two year old steers have been brought into Alberta, and should this experiment prove successful, thousands more will be shipped there next summer." Canada, *Sessional Papers*, 1896, vol. 5, no. 8, "Report of the Minister of Agriculture," p. xxxii. Availability of young stockers at competitive prices depended very much on the US market. When Ontario farmers sold to the US midwest, men like Lane had to look further afield for cattle. Gordon, Ironside and Fares had a big ranch in Mexico.

97 Bassano History Book Club, *The Best in the West by a Dam Site* (Calgary: Friesen, 1974), p. 294.

98 Note, for example, the experience of the Cayford family. Alice Carey Cayford, "Homesteading on the Bow," *Alberta History* 41, no. 3 (Summer 1993): 12.

99 GA, RCT 35-1, Bill Henry interviewed by George H. Gooderham, 1966.

100 Bassano, *Dam Site*, p. 295.

101 Sheppard, *Just about Nothing*, pp. 34–41.

102 This term was derived from the Spanish 'caballada,' meaning a band of saddle horses. Richard Slatta, *Cowboy Encyclopedia*, p. 64.

103 Bassano, *Dam Site*, p. 319. Ernest Lane was another of George Lane's sons.

104 See Evans, "The End of the Open Range Era," p. 76. In retrospect it might have been more interesting to try to analyze not THE end, but rather the many and various ends of the open range.

105 Lane had pioneered a small-scale irrigation scheme on his Willow Creek Ranch. Elizabeth Lane, "Memories," p. 13; and Bow River Basin Irrigation, Pioneer Development, 1895–1920, p. 2.

106 *High River Times*, 16 April 1908.

107 *High River Times*, 26 September 1912. Stoney Indians had been contracted to stook the grain.

108 Interview by the *Spokane Review*, 9 January 1912, reprinted in the *Albertan*, 12 September 1912.

109 The man Lane was on his way to see was Dr. Carlyle, whom he later persuaded to come to the Bar U to run the Percheron operation. After a few years Carlyle became manager of the E.P. Ranch.

110 David Spector, *Agriculture on the Prairies, 1890–1914* (Parks Canada/Environment Canada, 1983), p. 139, quoting the *Grain Growers Guide*.

111 Alberta, Department of Agriculture, *Annual Report*, 1912, p. 202.

112 See comments of Mr. Anderson, *High River Times*, 9 July 1908.

113 Interview with *The Shorthorn World*, reprinted in *High River Times*, 24 August 1916.

114 Cochrane raised Galloways, and Lane used these and Hereford bulls to cross with his Shorthorns.

115 *High River Times*, 1 October 1925.

116 NAC, RG 15, B2a, vol. 172, f. 145330, pt. 4, Lane to Greenway, 1 June 1904.

117 In the fall of 1906 Lane was reported to be running thirty thousand head. But, if there really had been an increase of ten thousand head, it would have occurred in herds along the Bow and not along Pekisko Creek.

118 David Breen refers to a display at the Canadian National Exhibition in Toronto in 1908, which showed a field of standing grain with a cowboy in the background. The title of the scene was "Another trail cut off." See Breen, *Canadian Prairie West*, p. 150.

119 Wallace Stegner, *Wolf Willow* (New York: Viking Press, 1955), p. 137.

120 Hugh A. Dempsey, *The Golden Age of the Canadian Cowboy* (Saskatoon: Fifth House, 1995).

121 Joy Oetelaar, "Climatic Risk or Social progress: The Historiography of Ranching in Southern Alberta," unpublished MA thesis, University of Calgary, 2000.

122 This caveat makes a lot of sense to me. I have always placed considerable confidence in the conclusions reached by the Livestock Commissioner, J.G. Rutherford. After all, he was an expert on stock raising, and had lived through the events which he was discussing. Surely his estimate that 50 per cent of the cattle on the range were lost must carry some weight? Oetelaar reminds me that Rutherford had a vision of the prairies settled with mixed farms; stable and protected from both economic and ecological hazards by the variety of their outputs. There was no room in this picture for ranchers who used the natural grassland extensively. Naturally he would exaggerate their plight and stress that they had brought their losses on themselves through their irresponsible behaviour. See J.G. Rutherford, *The Cattle Trade of Western Canada* (Ottawa: King's Printer, 1909), p. 18.

123 So often it is the extremes which make the critical difference between life and death. I worked with monthly mean temperatures for four stations for the period from October 1906 to May 1907. Even the averages showed important spatial contrasts in the severity of the winter. Gleichen recording a January mean of minus14.7 degrees Fahrenheit, which was 23 degrees below the thirty-year average. However, Calgary, Lethbridge, and Medicine Hat all

recorded mean average temperatures ABOVE the thirty-year average for February. See Evans, "Passing of a Frontier, p. 270.

124 A.K. Chakravati, "Precipitation Deficiency Patterns in the Canadian Prairies, 1921–1970," *Prairie Forum* 1, no. 2 (November 1976).

125 *High River Times*, 21 February 1907. For a description of the location of Hull's ranch see Brado, *Cattle Kingdom*, pp. 52–53; reports to A.E. Cross from his manager at the a7 Ranch were equally hopeful. GA, M1543 Cross Papers, box 59, f. 472, McCallum to Cross, 6 February 1907.

126 GA, M1543, Cross Papers, box 59, f. 467, Douglas to Cross, 16 December 1906. Charlie McKinnon, with the Bar U herd at Bassano, was doing the same thing: "We are trying to keep them up where they can get some feed but the cold winds is hard on them and the snow is deep and crusted." McKinnon to Cross, 31 January 1907.

127 GA, M1543, Cross papers, box 59, f. 467 Cross to Douglas, 5 January 1907.

128 Alberta, Department of Agriculture, *Annual Report*, 1907, p. 6. Trying to feed stock in these conditions must have been like trying to launch the lifeboats during a hurricane.

129 Katherine Hughes, "The Last Great Round-Up," *Alberta Historical Review* 2 (Spring 1963): 2

130 GA, Harry Otterson, *Thirty Years Ago on the Whitemud River* (Maple Creek: n.p., 1937), p. 17.

131 T.B. Long, *70 Years a Cowboy* (Regina: Western Printers Association, 1959), p. 22.

132 W.J. Redmond, "The Texas Longhorn on Canadian range," *Canadian Cattlemen* 3 (December 1938): 112.

133 J.W.G. MacEwan, "The Matador Ranch," *Canadian Cattlemen* 2 (March 1940): 358.

134 Simon M. Evans, "The End of the Open Range Era in Western Canada," *Prairie Forum* 8, no. 1 (Spring 1983): 71–87.

135 See Appendix, Tables 4.1 and 4.2. The downturn is even evident on the graph showing cattle numbers in Alberta and Saskatchewan from 1891 to 1996, which is at a very generalized scale: Figure 2.1.

136 In Saskatchewan, the number of cattle in Crop Area #3, which covered the southwestern portion of the province including Maple Creek and Swift Current, dropped from 122,026 in 1908 to 62,308 in 1911. Saskatchewan, Department of Agriculture, *Annual Report*, 1911.

137 Friesen, *Canadian Prairies*, chapter 13, "The Farm, the Village and King Wheat," pp. 301–38; and Ted Byfield (ed.), *Alberta in the Twentieth Century*, vol. 2: *The Birth of the Province* (Edmonton: United Western Communications, 1992).

138 Donald Kerr and Deryk Holdsworth (eds.), *Historical Atlas of Canada*, vol. 3: *Addressing the Twentieth Century, 1891–1961* (Toronto: University of Toronto Press, 1990), plates 17 and 18.

139 Bicha, *American Farmer*..

140 Mary Wilma M. Hargreaves, *Dry Farming in the Northern Great Plains, 1900–1925* (Cambridge, Mass.: Harvard University Press, 1957), p. 87.

141 Canada, *Sessional Papers*, 1909, vol. 16, no. 28, "Report of Superintendent J.O. Wilson, Lethbridge," p. 89.

142 *Lethbridge Herald*, 11 December 1947, p. 13.

143 Canada, *Sessional Papers*, 1909, vol. 21, no. 28, "Report of Superintendent J.V. Begin, Maple Creek," p. 98.

144 Otterson, *Thirty Years Ago*, p. 22; and Don C. McGowan, *Grassland Settlers* (Regina: Canadian Plains Research Center, 1975), pp. 104–6.

145 Margaret V. Watt, "McCord's Ranch: Chronicle of Sounding Lake," *Canadian Cattlemen* 15 (November 1952): 20.

146 Cecil H. Stockdale, "Another Hard Winter Story," *Canadian Cattlemen* 13 (March 1950): 58.

147 Christianson, *Early Rangemen* (Lethbridge: Southern Printing, 1973), p. 91.

148 *Medicine Hat News*, 11 April and 23 May 1907.

149 Alberta, Department of Agriculture, *Annual Report*, 1907, p. 121.

150 M.F. Dunham, "Agriculture in Alberta," *Edmonton Daily Bulletin*, Christmas edition, 1908.

151 Evans, "End of the Open Range," p. 80.

152 *Calgary Herald*, 10 August 1909.

153 *Medicine Hat News*, 3 January 1907, reported that fifteen hundred head had been refused shipment in that district. Moreover, many smaller cattlemen had held back their stock in the hope that prices might improve. Canada, *Sessional Papers*, 1905, vol. 12, no. 28, "Report of Superintendent G.E. Sanders," p. 28.

154 Alberta, Department of Agriculture, *Annual Report*, 1907, p. 6; and 1911, appendix B. Shipments of cattle from Alberta eastward:

1905	83,000	1908	100,000
1906	115,000	1909	128,000
1907	80,000	1910	154,000

155 GA, M651, Lane, George, "Address Given to National Livestock Convention, Ottawa, 1912."

156 "Both the Turkey Track and ourselves had been careless of the winter range and had dumped big herds on the river at different times, and consequently the winter grazing had been reduced, for which we paid dearly." Otterson, *Thirty Years Ago*, p. 16.

157 In general the 'carrying capacity' of winter range reflected the precipitation received during the summer and fall. Both drought and prairie fires could reduce available forage.

158 A comment on the progress of herds during the early part of the winter emphasizes the importance of mange: "The loss amongst cattlemen this year is by no means general and those *who have followed dipping regularly* report that they are practically immune from loss so far." *High River Times*, 10 January 1907 (emphasis added).

159 GA, M1543, Cross papers, box 59, f. 466, Cochrane to Cross, 12 June 1906.

160 Rancher A.E. Cross in an address to the national Livestock Association in 1908. See Oetelaar, "Climatic Risk," p. 80.

161 Canada, *Sessional Papers*, 1909, vol. 7, no. 15, "Report of the Veterinary Director General," pp. 23–24; and McIntyre, "History of the McIntyre Ranch," p. 107.

162 Elizabeth Lane, "Memories," p. 22.

163 *High River Times*, 16 May 1907.

164 For a breakdown of estimated losses and values per head see Appendix, Table 4.3.

165 This is largely supposition. A photograph of the "Big Bow Bar U" does show extensive feeding shelters and some haystacks, but I think it dates from about 1909.

166 See Appendix for breakdown.

167 Other expenditures by Lane help to provide a context for the losses he experienced: the Bar U Ranch cost him $220,000 in 1902; Namaka Farms, purchased in 1913, involved an outlay of $250,000; a single wheat crop on his combined acreages was valued at $250,000 in 1917.

168 Henry C. Klassen, "The Conrads in the Alberta Cattle Business, 1875–1911," *Agricultural History* 64, no. 3 (1990): 56.

169 Pearce, *Matador Land and Cattle Company*, p. 16.

170 See Osgood, *Day of the Cattleman* 216–27; and Robert S. Fletcher, "The Hard Winter in Montana, 1886–1887," *Agricultural History* 4 (1930): 123–30.

171 Oetelaar, "Climatic Risk," p. 82.

172 For one example see Kelly, *Rangemen*, p. 379.

173 Rutherford, *Cattle Trade of Western Canada*, p. 8.

174 Stegner, *Wolf Willow*, p. 137, quoted above.

175 Patricia Nelson Limerick, *The Legacy of Conquest: The Unbroken Past of the American West* (New York: W.W. Norton, 1987), pp. 289–91.

176 Chester Martin, *Dominion Lands Policy* (Toronto: McClelland and Stewart, 1973), p. 98.

177 L.S. Curtis, "The Pre-emption Policy of 1908 Ruined Many Small Ranchers," *Calgary Herald*, 18 August 1949.

178 W.A. Mackintosh, *Prairie Settlement: The Geographical Setting*, vol. 1, *Canadian Frontiers of Settlement*, ed. W.A. Mackintosh and W.L.G. Joerg (Toronto: MacMillan, 1934), pp. 86–104; and Canada, *Sessional Papers*, 1909, XLIII, vol. 14, no. 25, "Report of the Red Deer Immigration Agent," p. 25.

179 *Historical Atlas of Canada*, plate 18.

Notes to Chapter 5

1 John H. Wallace, *The Horse of America in his Derivation, History and Development* (Wilmington, Delaware: Scholarly Resources, 1973. Reprint of 1897 ed.), p. 409.

2 Of course, many of Lane's sales were to other breeders, but the demand for draft animals, and the high prices which ensued, were derived from the influx of farm settlers in the decade before the First World War.

3 Canada, Dominion Bureau of Statistics, Census of Canada, 1921, p. 21; See Appendix, Figure 5.1.

4 Alberta, Department of Agriculture, *Annual Report*, 1906, p. 70.

5 Alberta, Department of Agriculture, *Annual Report*, 1907, p. 7.

6 Alberta, Department of Agriculture, *Annual Report of Alberta Horse Breeders' Association*, 1911, p. 260.

7 J. Angus McKinnon, *The Bow River Range* (Calgary: McAra, 1974), p. 37.

8 Ibid., p. 34.

9 Alberta, Department of Agriculture, *Annual Report of the Alberta Horse Breeders' Association*, 1914, p. 258.

10 Alberta, Department of Agriculture, *Annual Report*, 1916, p. 95.

11 Ibid., 1918, p. 66.

12 Ibid., 1921, p. 12.

13 Alberta, Department of Agriculture, *Report of the Livestock Commissioner*, 1924, p. 23.

14 Ibid., 1926, p. 15.

15 Ibid., 1928, p. 13.

16 Ibid., 1932, p. 14.

17 Brado, *Cattle Kingdom*, pp. 146–48.

18 *Macleod Gazette*, 9 November 1894.

19 *High River Times*, 16 June 1910.

20 Alberta, Department of Agriculture, *Annual Report*, 1916, "The Future of the Horse and the Horse of the Future," p. 93; and Saskatchewan, Department of Agriculture, *Report of the Livestock Commissioner*, 1919, p. 221.

21 David Spector, *Agriculture on the Prairies, 1870–1940* (Parks Canada/ Environment Canada, 1983).

22 Roberta E. Ankli, H. Dan Halsberg, and John Herd Thompson, "The Adoption of the Gasoline Tractor in Western Canada," *Canadian Papers in Rural History*, vol. 2 (Gananoque, Quebec: Langdale Press, 1980).

23 Alberta, Department of Agriculture, 1931, *Report of the Livestock Commissioner*, p. 15.

24 Grant MacEwan, *Heavy Horses: Highlights of their History* (Saskatoon: Western Producer, 1986), p. 6; and see also Appendix, Figure 2.1.

25 MacEwan, *Heavy Horses*, p. 41.

26 *Canadian Percheron News*, June 1938, p. 2.

27 *Live Stock Journal*, 17 November 1916, quoted in Wayne Dinsmore, "Development of the Percheron in Canada," *Nor'West Farmer*, 20 February 1917.

28 GA, M2382, f. 9, Papers of the Canadian Percheron Association, Ross Butler, "The Percheron," p. 1.

29 Parks Canada, Edward Harris Photograph Album. This beautifully illuminated book contains text as well as photographs.

30 Jean Pelatan, *The Percheron Horse Past and Present*, revised and translated by John P. Harris (Montagne-au-Perche: Association des Amis du Perche, 1985), p. 26.

31 Alvin Howard Sanders, *A History of the Percheron Horse* (Chicago: Breeder's Gazette, 1917), pp. 216–18; MacEwan, *Heavy Horses*, p. 48–51; see also GA, M2382, f. 25, "The Oaklawn Stud of Percheron Horses."

32 Ardouin-Dumazet, *Voyage en France* (Paris: Berger-Levrault, 1894), quoted in Pelatan, *Percheron Horse*, p. 32

33 This is probably a very conservative estimate. M.W. Dunham surmised that he had paid some $350,000 to Ernest Perriott over a number of years. See MacEwan, *Heavy Horses*, p. 43.

34 Sanders, *Percheron Horse*.

35 Pelatan, *Percheron Horse*, p. 57.

36 Wayne Dinsmore, "Percherons of the Past and Present," *The Alberta Farmer and Calgary Weekly Herald*, 28 March 1918. To properly differentiate between the two horses, their registration numbers should be included because there are so many horses named Brilliant in the Percheron line. The older horse was numbered (1271) and the younger (11116) in the French stud book.

37 MacEwan, *Heavy Horses*, p. 43.

38 *Nor'West Farmer*, 20 February 1917.

39 MacEwan, *Heavy Horses*, p. 6.

40 Ibid., p. 54.

41 Saunders, *Percheron Horse*, p. 471.

42 This purchase was made in 1898. MacEwan, *Heavy Horses*, p. 34.

43 Elizabeth Lane, "Memories," p. 18

44 In the possession of Mr. Bruce Roy of Cremona, Alberta.

45 MacEwan, *Heavy Horses*, p. 51.

46 *High River Times*, 18 November 1909.

47 *High River Times*, 1 September 1910.

48 *High River Times*, 19 May 1910. Williams was lucky to be alive. Four years before he had been kicked in the head by a horse, and remained unconscious for twenty-one days.

49 This passage was translated and quoted in the *High River Times*, 1 September 1910.

50 *High River Times*, 28 December 1911. Ironically, "Imprecation," highly fitted too long, died shortly afterward, leaving very few American-bred foals and causing a grievous loss to his owners, J. Crouch and Sons, of Lafayette, Indiana. See Bruce Roy, "The Bar U Percherons," *Horses All* 3, no. 4 (January 1980). This piece was originally published in *The Draft Horse Journal* (November 1971). Mr. Roy is editor of *Fetlock and Feather* and has been involved with heavy horses all his life. The next paragraphs draw heavily on his work.

51 *High River Times*, 11 January 1912.

52 Grant MacEwan, *Hoofprints and Hitching Posts* (Saskatoon: Western Producer, 1968), p. 98; and Simon M. Evans, "George Lane: Purebred Horse Breeder," research paper, Parks Canada, 1994, preface.

53 Bruce Roy reports this conversation. Roy, "The Bar U Percherons," p. 13.

54 *Alberta Farmer and Weekly Herald*, 28 March 1918, p. 10.

55 Roy, "The Bar U Percherons," p. 14.

56 *Farm and Ranch Review*, 20 February 1910.

57 Roy, "The Bar U Percherons," p. 23.

58 Sheppard, *Just About Nothing*, p. 69.

59 Ritchie, "George Lane," p. 442.

60 *High River Times*, 7 October 1909; and 19 October 1911.

61 *High River Times*, 8 April 1909.

62 *High River Times*, 7 October 1909.

63 *High River Times*, 7 December 1911.

64 Rankin, "Boss of the Bar U," 333.

65 *Farm and Ranch Review*, 10 July 1910, p. 252.

66 Sheppard, *Just About Nothing*, p. 71.

67 *High River Times*, 9 October 1919 and 15 January 1920; and Roy, *Horses All*, p. 27. John Thornton and Company of Eaton, near Norwich, England, paid $46,300 for thirty-four Percherons.

68 *High River Times*, 9 October 1919.

69 Roy, "The Bar U Percherons," p. 14.

70 Evans, *Prince Charming Goes West*, p. 119.

71 Clifford/Alwood Collection, Studbook for 1908. This hardbound ledger was especially designed for use by stud owners. Each page is devoted to a different mare and includes its number and descriptive characteristics, like colour and markings. Spaces are provided to record the name of the stallion 'bred to,' the dates of service, foaling, and the name of the foal.

72 Clifford/Alwood Collection, Pike's Notebook. F.R. Pike was bookkeeper at the Bar U and secretary of the Percheron Society.

73 Henderson's Directory, 1915, refers to "Alberta Government Telephones, local and long distance service, Lane, Gordon Ironside and Fares agents." Posts and wires show that there were outlets at the post office building and at the main house.

74 A Delco generator and batteries was located in one of the storage sheds opposite the cookhouse. See Norman Mackenzie interview and photograph. After 1917, the system was housed in the basement of the new bunkhouse.

75 While the plans for this scheme were outlined in 1907, it took a decade for them to come to full fruition. GA, M3734, *Bow River Basin Irrigation*, 1895–1920, vol. 7, f. 401.

76 Bert Sheppard, *Spitzee Days* (Calgary: McAra, 1971), p. 121.

77 This transformation went on building by building and area by area, and culminated in a complex and fully developed ranch site by the end of the First World War. The Prince of Wales, who visited in 1919, saw the ranch at the zenith of its development. The objective here is to provide an overview of site changes. Anyone interested in the history of particular extant buildings is referred to the Historic Structures Reports or Stabilization Reports which detail the investigations carried out by Parks Canada, and which are available at the Western Regional Service Center of Parks Canada in Calgary.

78 *High River Times*, 3 February 1910.

79 Ibid.

80 After 1917 it was used primarily by younger workers and temporary hay crews. See, for example, Museum of the Highwood and Bar U National Historic Park, Oral History Project, 1992–93 (hereafter OHP), interview with Gordon Davis, 11.11.

81 Initially, I could not understand where the growing labour force had been housed between 1910, when the first bunkhouse burnt down, and 1917, when the large new frame bunkhouse was constructed. However, during this period four smaller houses were built on the site for individuals or families, and a separate bunkhouse was built at the Bar U farm.

82 See Historic Structures Report, Building no. 14, Blacksmith's Shop, March 1994.

83 W.B. Yeo, "Debit and Credit," p. 11; and "Pekisko."

84 "Alberta's Big Percheron Ranch," *Farm and Ranch Review*, 20 February 1910.

85 Ruth Olsen McGregor, "Memories of the Bar U," unpublished paper, Parks Canada, 1996, p. 1. Professor Carlyle and his wife and daughter lived in 'The Big House' from 1917 when they came to Alberta until 1922 when he moved to the E.P. Ranch .

86 This apparently man-made area has always posed problems to those interested in interpreting the site. At one time we thought it might have been used for polo, but it was too small. In the last years of the Lane era it was sometimes used by Lane for golf practice.

87 McGregor, "Memories," p. 1.

88 Pekisko House was home to Mary Stimson, Ella Bowen, and Mary Bigland, who married the foreman. The Stimson's son Bryce lived at the ranch during the school holidays.

89 The family photograph albums collected by Barbara Holliday for Parks Canada illustrate the continuous presence of children at the Bar U; see, for example, the following collections: R.W. Gardner, Norman Mackenzie, Hugh Paulin, and Alwood/Rousseau collections.

90 Initially, the horses which he had bought in Montana were kept at the YT Ranch. After the purchase of Namaka farms in 1913, many Percherons were moved there, so the period of concentration at the Bar U was 1907–13.

91 Evans, "Purebred Horse Breeder," p. 14.

92 Barbara Holliday, "Bar U Stud Horse Barn Exhibit Text," Parks Canada, 1998.

93 Parks Canada, Historic Structures Report, Building no. 1, Stud Horse Barn, June 1994.

94 *Alberta Farmer and Calgary Weekly Herald*, 28 March 1918.

95 Roy, "The Bar U Percherons," p. 27

96 Parks Canada, Historic Structures Report, Building no. 6, Harness Shop, December 1995.

97 Initially this building was thought to have been a later addition, but the archaeologists found sandstone blocks used in the foundations, and they associate this with a date prior to 1910. Moreover, Lane purchased 100,000 board-feet of lumber in 1909, "Because he was in a hurry to build his horse barns." *Mosquito Creek Roundup*, p. 115.

98 These facilities are shown in a picture which was part of an advertisement for sale of young horses in *Farm and Ranch Review*, December 1909.

99 Parks Canada, Stabilization Report, Building no. 25, Dairy Barn, February 1997.

Notes to Chapter 6

1 Felipe Fernandez-Armesto, *Civilizations* (Toronto: Key Porter Books, 2000), p. xi.

2 *Albertan*, 16 June 1913.

3 *Calgary Daily Herald*, 22 June 1929.

4 Lane was at pains to make it clear that this purchase was a personal investment. The Percherons at Namaka had a separate brand from those at the Bar U.

5 For an overview, see Namaka Community History Committee, *Trails to Little Corner: A Story of Namaka and Surrounding Districts* (Strathmore, Alberta, 1983), pp. 43–44.

6 See Edward Harris Photo Album, which shows pictures of teams of geldings and underneath the following text: "Grade Percheron geldings, work horses on the Namaka farm, bred up from very small stock to as high as 1,800 lbs. It is easily within the power of any farmer with small but sound mares and the exercise of common sense to produce these valuable types from such mares." Lane was aiming to sell horses from Namaka directly to the best and most progressive farmers.

7 *High River Times*, 8 April 1915. Later, in August, fifteen binders were engaged to harvest the crop: *High River Times*, 19 August 1915.

8 *High River Times*, 20 September 1917.

9 *High River Times*, 9 August 1917.

10 *Lethbridge Herald*, 18 April 1918; and John Herd Thompson, *The Harvests of War* (Toronto: McClelland and Stewart, 1978).

11 Grant MacEwan, *Charles Noble: Guardian of the Soil* (Saskatoon: Western Producer, 1983).

12 For example, in 1916 Charles Noble was breaking and seeding the Cameron Ranch. The local reporter wrote: "40 binders were in use simultaneously; 250 horses and mules eating $125 worth of food a day; 150 hired men receiving an average of $4 a day; and total operating costs of $1,200 a day." *Lethbridge Herald*, 23 August 1917.

13 The past few decades have seen tremendous pressure on the family farm and the rise of huge "super farms" owned by agribusiness. It is ironic that commentators in the 1920s were pointing out the advantages of the smaller family-owned enterprise, because the cost of labour was much reduced and the intensity of cultivation and care was greater. See *Calgary Daily Herald*, 22 June 1929.

14 Alex Johnston, *Cowboy Politics: the Western Stock Growers' Association and its Predecessors* (Calgary: Western Stock Growers' Association, c. 1973).

15 Johnston, *Cowboy Politics*, p. 11.

16 See p. 165 above.

17 Alberta, Department of Agriculture, *Annual Reports*, 1911–1915, Appendix A, Alberta Horse Breeders' Association.

18 *High River Times*, 30 May and 6 June 1912.

19 Donna Livingstone, *Cowboy Spirit: Guy Weadick and the Calgary Stampede* (Vancouver: Douglas and McIntyre, 1996), pp. 38–39. Archie McLean was the fourth member of the "Big Four." According to Livingstone he was co-opted on another occasion.

20 James H. Gray, *A Brand of its Own: The 100 Year History of the Calgary Exhibition and Stampede* (Saskatoon: Western Producer, 1985), pp. 36–37; and Guy Weadick, "Origin of the Calgary Stampede," *Alberta Historical Review* 14, no. 4 (Autumn 1966): 20–21.

21 *High River Times*, 4 November 1907; and 10 February 1910.

22 *High River Times*, 27 August 1908.

23 *High River Times*, 27 July, 3 August, and 10 February 1911.

24 See *Calgary Herald*, 14 June and 15 July 1951.

25 *Calgary Herald*, 8 and 15 April 1913.

26 *Calgary Herald*, 18 April 1913.

27 L.G. Thomas, *The Liberal Party in Alberta: A History of Politics in the Province of Alberta, 1905–1921* (Toronto: University of Toronto Press, 1959), p. 144; and *High River Times*, 29 May 1913. Mitchell remained a friend, and was one of Lane's pallbearers in 1925. Lane was also involved in the 1917 federal election. For details see Oetelaar, "George Lane," p. 58.

28 Things changed somewhat when Lane acquired a car in 1910, but the advantages of living in the city remained considerable.

29 Elizabeth Lane, "Memories," pp. 15–16.

30 Yeo, "Debit and Credit," p. 9.

31 Alberta, Municipal Affairs, Corporate Registry, microfiche file on George Lane and Company, Certificate of Incorporation, 3 January 1916; and *High River Times*, 27 January 1916.

32 For general appreciations see: Barbara Holliday, "Herb Millar of the Bar U," *Alberta History* 44, no. 4 (Autumn 1996): 15–21; and Harold W. Riley, "Herbert William (Herb) Millar; Pioneer Ranchman," *Canadian Cattlemen* 4, no. 4 (March 1942): 165ff.

33 *Leaves from the Medicine Tree*, p. 34.

34 Nanton and District Historical Society, *Mosquito Creek Round Up* (Nanton: Nanton and District Historical Society,1975), pp. 114–15.

35 J. Frank Dobie, *Cow People* (Boston: Little, Brown, 1964), p. 32.

36 As with Harry Longabaugh, the Sundance Kid. See Holliday, "Herb Millar," p. 17.

37 Museum of the Highwood, Guest Register of the High River Club.

38 Riley, "Herbert, William Millar," p. 165.

39 For an overview see Barbara Holliday, "Alex Fleming: A Biographical Sketch," Parks Canada, research paper, 1996.

40 *Calgary Herald*, 29 September 1943.

41 *Leaves from the Medicine Tree*, p. 151.

42 Bruce Roy, "The Bar U Percherons," *Draft Horse Journal* 3, no. 4 (January 1980).

43 *High River Times*, 30 September 1943.

44 Sheppard, *Just About Nothing*, p. 139.

45 Florence and Jean Fleming, interview with B. Holliday, 25 and 26 September 1995.

46 Sheppard, *Just About Nothing*, p. 139; and *Mosquito Creek Roundup*, p. 99.

47 *Leaves from the Medicine Tree*, p. 148.

48 Yeo, "Debit and Credit," p. 13.

49 Book recording the birth of Percheron foals and registered mares, 1911–14. See Yeo, "Debit and Credit," p. 9; see also Elsie Gordon Papers, Record Book, Percheron Society of America, 1910.

50 GA, M2180, William Edward Meredith Holmes, "Records of the Gee-Bung Polo Club."

51 Yeo, "Pekisko," p. 13.

52 *Leaves from the Medicine Tree*, p. 148.

53 Most of our knowledge about the Olsen family comes from Ruth Olsen McGregor, who was born at the Bar U Ranch in 1917 and lived there until 1925. She was kind enough to write down her "memories" for the benefit of historians of the ranch. See Ruth Olsen McGregor, "Memories of the Bar U," Parks Canada, research paper, 1996. Mrs. Olsen McGregor was also interviewed by Barbara Holliday of Historical Services, Parks Canada during March and June 1996. See "Oral History Interview with Ruth Olsen McGregor re Neils Olsen and family," Calgary, Parks Canada, 1996. In addition Mrs. McGregor provided more than eighty

photographs which were copied and provide a valuable visual record of the Bar U during this period.

54 *Leaves from the Medicine Tree*, p. 130.

55 Compare the experience of Ted Hills with the Bar U wagon in 1885. GA, M7988, Edward F.J. Hills letters, nos. 43–45.

56 Holliday, Ruth Olsen McGregor interview, p. 2.

57 The Olsen family, Albert, Emil, Ruth, and Irene; the Andrews, Cecil, Joan, and another baby; The Paulins, Hugh, William, George, and David; and the Piersons, with son Manville.

58 Losing governesses and maids was a common "problem" in ranching country. See Sarah Carter, "'He Country in Pants' No Longer – Diversifying Ranching History," in Evans, Carter, and Yeo, *Cowboys, Ranchers and the Cattle Business*, p. 158.

59 Holliday, Ruth Olsen McGregor interview, p. 5.

60 Ibid.

61 For a detailed account of the royal tour and the subsequent purchase of the E.P. Ranch, see Evans, *Prince Charming Goes West*. The following paragraphs draw heavily on this book.

62 Frances Donaldson, *Edward VIII* (London: Weidenfeld and Nicolson, 1986), p. 63; and Philip Ziegler, *King Edward VIII: The Official Biography* (London: Collins, 1990), p. 114.

63 Christopher Warwick, *Abdication* (London: Sidewick and Jackson, 1986), p. 31.

64 Ziegler, *Edward VIII*, p. 117.

65 H.R.H. Edward Duke of Windsor, *A King's Story: Memoirs of the Duke of Windsor* (Toronto: Thomas Allen, 1951), p. 140.

66 Nor had the peace come without its problems. It took too long to absorb returning soldiers back into the economy. Unemployment was high, homelessness and rootlessness were widespread. Only two months before H.R.H. arrived in Winnipeg, the city had been the site of a bitter general strike which had ended on "Bloody Sunday" with clashes between police and strikers which left thirty casualties and one man dead.

67 *Times*, 4 September 1919.

68 *High River Times*, 18 September 1919.

69 Olsen McGregor, "Memories of the Bar U," p. 6.

70 NAC, RG 7 G23(2), HRH to Duke of Devonshire, 9 June 1919.

71 Ziegler, *Edward VIII*, p. 123.

72 *High River Times*, 18 September 1919.

73 *Calgary Herald*, 17 September 1919.

74 *High River Times*, 18 September 1919.

75 Jack Peach, "Prince of Wales set hearts aflutter during visit," *Calgary Herald*, GA clipping file, n.d.

76 *High River Times*, 25 September 1919.

77 NAC, RG 7 G23, vol. 17 (4), no. 29, Henderson to Duke of Devonshire, 24 September 1919.

78 Rupert Godfrey (ed.), *Letters from a Prince* (London: Little, Brown, 1998), pp. 189–90. See the introduction for an overview of the relationship between Freda and the Prince. I have followed Godfrey's lead and left spelling and punctuation as the prince wrote it down. There is, of course, a lot of overlap between the "official version" of events and that recorded by H.R.H.; however, the point of view and language used is so different that it is worth quoting the letters in full.

79 Godfrey, *Letters*, p. 189.

80 Godfrey, *Letters*, p. 190.

81 In his autobiography, Edward wrote, "In the midst of that majestic countryside I had suddenly been overwhelmed by an irresistible longing to immerse myself, if only momentarily, in the simple life of the western prairies. There, I was sure, I could find occasional escape from the sometimes too confining too well ordered island life of Great Britain." *King's Story*, p. 152.

82 *High River Times*, October 16, 1919.

83 Godfrey, *Letters*, p. 209.

84 Godfrey, *Letters*, p. 202.

85 Evans, *Prince Charming*, pp. 115–43.

86 The next paragraphs draw heavily on A.B. McCullough, "Winnipeg Ranchers: Gordon, Ironside and Fares," *Manitoba History* 41 (Spring/Summer 2001): 18–25.

87 Alberta, Department of Agriculture, *Annual Report*, 1925.

88 Foran, *Trails and Trials*, p. 71.

89 *Calgary Herald*, 3 October 1925.

90 This temporary measure was extended indefinitely in November 1921. It was not reduced until 1935.

91 *High River Times*, 27 October 1921.

92 *High River Times*, 2 November 1922. A headline read, "Beef for Chicago, George Lane made a large shipment from Alberta to Chicago Despite Fordney Tariff." Rancher A.E. Cross estimated that it cost him $114.74 to raise a four-year-old steer, and the tariff meant that he lost $30.34 per head. GA, M289, A.E. Cross, box 115, f. 926, 8 September 1922.

93 Elizabeth Lane, "Memories," p. 19

94 The Little Bow River dried up in 1918 and remained dry through 1919. My appreciation of the effects of the winter of 1919 has been much influenced by a thoughtful and innovative thesis written by Joy Oetelaar. See Joy Oetelaar, "Climatic Risk or Social Progress: The Historiography of Ranching in Southern Alberta," unpublished MA thesis, University of Calgary, 2000.

95 *High River Times*, 15 April 1920.

96 Ibid..

97 Ings, *Before the Fences*, p. 72.

98 T.L. Shepherd, "Winters, Mild and Not So Mild," *Canadian Cattlemen* 7, no. 3 (December 1944): 126.

99 One of the best presentations of the contrast between the war years and the post-war depression is the series of graphs showing yields, prices, and returns, in Paul Voisey, *Vulcan: The Making of a Prairie Community* (Toronto: University of Toronto Press, 1988), pp. 109–14.

100 MacEwan, *Charles Noble*, Chapter 4.

101 David C. Jones, *Empire of Dust: Settling and Abandoning the Prairie Dry Belt* (Calgary: University of Calgary Press, 2002).

102 John Herd Thompson, *The Harvests of War* (Toronto: McClelland and Stewart, 1978), p. 175.

103 *High River Times*, 11 October 1923.

104 Alberta, Municipal Affairs, Corporate registry, microfiche file on George Lane and Company, p. 1 (hereafter Registry).

105 Registry, p. 1.

106 Probate Records, Surrogate Court, Calgary. File copied by Parks Canada, Calgary.

107 Charles Noble, George Lane's rival for the title "Largest Wheat Farmer in Calgary," went bankrupt in 1923 and lost his house and everything except his furniture. MacEwan, *Charles Noble*, chapter 4, and *Calgary Herald*, 6 March 1923.

108 Elizabeth Lane, "Memories," p. 20.

109 *High River Times*, 13 September 1923.

110 Elizabeth Lane, "Memories," p. 20.

111 *Calgary Herald*, 28 September 1925.

112 See for example, *Calgary Herald*, 24 and 28 September 1925; and *Farm and Ranch Review* 21 (October 1925).

113 *Calgary Herald*, 24 September 1925.

Notes to Chapter 7

1 Any student of Pat Burns and his times must owe an overwhelming debt of gratitude to Albert Frederick Sproule, who carried out research on Burns during the 1950s. He interviewed many of the key employees in Burns' enterprises as well as members of his family. His findings were presented in his Master's thesis at the University of Alberta in 1962. Specific quotes from his work will, of course, be acknowledged in the usual way, but it is also important to point out that my overall view of Burns has been shaped and informed by Sproule's work. Albert Frederick Sproule, "The Role of Patrick Burns in the Development of Western Canada," unpublished MA thesis, University of Alberta, 1962 (hereafter cited as Sproule).

2 James E. Sherow (ed.), *A Sense of the American West* (Albuquerque: University of New Mexico Press, 1998), p. 21.

3 For an analysis of this point, see MacLachlan, *Kill and Chill*, pp. 4–10; and "California's Agribusiness," *The Economist*, 13 December 1980, p. 68.

4 MacEwan, *Pat Burns*, p. 161.

5 The Burns "empire" also displayed horizontal integration, with diversification into other meat products like pork and lamb, and later into cream, butter, cheese, fresh milk, and fruit products. Burns also held interests in mines and urban real estate.

6 Author Interview, Ann Clifford, May 1994.

7 GA, M160, f. 92, Newton to Wilson, 13 June 1941.

8 Pat Burns' parents, Michael Byrne and Bridget Gibson, were from County Mayo, Ireland. They came to Canada in 1847. Patrick, the fourth of eleven children, was born in 1854. Sproule, p. 1.

9 Terry Jordan explains that a class of professional drovers arose in Highland Britain during the 1500s to handle the growing flow of cattle to London. In Jordan's analysis of the origins of cattle ranching in North America, Highland Britain was one important source region. Jordan, *Cattle-Ranching Frontiers*, pp. 42–55.

10 Sproule, p. 21.

11 MacEwan, *Pat Burns*, p. 60.

12 On this first contract Burns was able to persuade twenty-year-old Walter Wake to join him; Wake was to spend most of the rest of his life working for Burns. Another employee was George Webster, who handled all the paperwork. He later became Mayor of Calgary.

13 MacEwan, *Pat Burns*, p. 1.

14 Sproule, pp. 21–25. This paragraph draws heavily on Sproule, who was able to interview many family members and employees of Burns to help him piece together a picture of the man.

15 Sproule, p. 55.

16 *The Albertan*, 25 February 1937.

17 C.W. Vrooman, "Cattle Price Fluctuations," *Canadian Cattlemen* 4, no. 4 (March 1942): 149.

18 MacEwan, *Pat Burns*, p. 129.

19 For figures on this flow, see Canada, *Sessional Papers*, 1897, vol. 10, no. 13, "Report of the Superintendent of Mines," p. 28; and Alberta, Department of Agriculture, *Annual Reports*, 1898–1902.

20 *Calgary Herald*, 6 May 1897.

21 *Calgary Herald*, 10 January 1901.

22 *Calgary Herald*, 30 November 1904.

23 MacEwan, *Pat Burns*, p. 135.

24 See Maps 5.2 and 5.3. For discussion see "The Agricultural Frontier, 1900–1913," chapter 2 of Foran, *Trails and Trials*.

25 Alberta, Department of Agriculture, *Annual Report*, 1907, p. 6: see also an address by Duncan Anderson entitled "Stay with Cattle," p. 235 in the same report.

26 Alberta, Department of Agriculture, *Annual Report*, 1907, p. 23; and Saskatchewan, Department of Agriculture, *Annual Report*, 1907, p. 25.

27 The deal with Hull was concluded in London just prior to Burns' wedding to Eileen Ellis. MacEwan, *Pat Burns*, pp. 108–110. The Bow Valley Ranch was to remain part of the Burns estate through the 1950s.

28 For details, see Ann Clifford, *Ann's Story* (Calgary: JR's Printing Services, 1995); and GA, M160, Maps of Burns properties.

29 Burns also purchased the C.K. (Charlie Knight) Ranch, the Neilsen place, and the Ricardo Ranch, all of which were close to the growing city of Calgary and were used to produce perishable milk products for urban consumption or to support the packing plant and its feedlots with hay and grain.

30 Sproule, p. 158.

31 Milk River Historical Society, *Under Eight Flags: Milk River and District* (Lethbridge: Graphcom Printers, 1989), p. 21.

32 GA, M160, box 13, f. 94, Balance Sheets.

33 Henry C. Klassen, "The Conrads in the Alberta Cattle Business, 1875–1911," *Agricultural History* 64, no. 3 (Summer 1990): 31–59.

34 See Guy Weadick, "Ballie Buck, Range Cowman," *Canadian Cattlemen* 5, no. 4 (March 1943): 164ff.

35 Brado, *Cattle Kingdom*, pp. 142–46; and Buchanan, "History of the Walrond," pp. 171ff.

36 Breen, *Canadian Prairie West*, p. 149; and Brado, *Cattle Kingdom*, p. 145.

37 GA, M160, f. 224, "Stock and Balance Sheets," Walrond Ranch; and "Statements of Profit and Loss."

38 The 1891 return credits the major companies with the following herds: North West Cattle Company, 10,000; Cochrane, 10,000; Walrond, 13,000; and Oxley 6,500, for a total of 39,500.

39 Foran, *Trails and Trials*, pp. 80–86.

40 Francis Mollison Black, "Patrick Burns as I Knew Him," quoted in Sproule, p. 189.

41 Foran, *Trails and Trials*, pp. 39–46.

42 *Calgary Albertan*, 8 November 1927; *Calgary Herald*, 10 November 1927; Sproule, p. 218; and MacEwan, *Pat Burns*, pp. 155–57.

43 GA, M160, f. 93, Burns to George Lane and Company Limited, 1 November 1927; see also Appendix, Table 7.1.

44 GA, M160, f. 93, Burns to Munson et al., 23 December 1927. Emphasis by the author.

45 Ibid., 16 March 1928.

46 Ibid., Munson to Burns, 20 December 1927.

47 Ibid., Corlet to Munson et al., 27 December 1927.

48 Ibid., Johnson to Burns, 9 December 1927.

49 Ibid., Burns to Recorder of Brands, 30 November 1927.

50 Ibid., Burns to Sexsmith, 8 December 1927.

51 Ibid., Burns to McLean, 7 December 1927.

52 Ibid., Burns to McLaughlin and Sexsmith, 8 December 1927.

53 MacEwan, *Pat Burns*, p. 156.

54 *High River Times*, 1 October 1925.

55 Sproule, p. 222.

56 Sproule, p. 124.

57 Evans, *Prince Charming Goes West*, pp. 117–18.

58 For accounting purposes, Burns "bought back" the ranches for $5 million. For some details of the complex reorganization see Evans, "Burns Era," 25–27

59 The two other big corporate ranches, the Oxley and the Cochrane, had been subdivided for farming and were therefore not available.

60 S.E. Warren, "Some Memories of the Old "76", *Canadian Cattlemen* 12, no. 2 (September 1949): 20; Moreton Frewen, *Melton Mowbray and Other Memories* (London: Herbert Jenkins, 1924), p. 222; and NAC, RG 15, f. 142083, Canada, Department of Interior, series of letters from Frewen to department, 18 February to 18 May 1886.

61 Barry Potyondi, *In Palliser's Triangle: Living in the Grasslands, 1850–1930* (Saskatoon: Purich Publishing, 1995), p. 52.

62 T.B. Long, *Seventy Years a Cowboy* (Regina: Western Printers' Association, 1959), pp. 97–99.

63 Stegner, *Wolf Willow*; and for a recent evocation of the area, see Sharon Butala, *The Perfection of the Morning: An Apprenticeship in Nature* (Toronto: Harper and Collins, 1994).

64 GA, M7771, f. 221, Ledger Book.

Notes to Chapter 8

1 Paul Kennedy, commenting on the work of A.J.P. Taylor, *Atlantic Monthly*, April 2001.

2 Comparison of insurance maps, GA, M160, f. 104 and f. 106, dated 1927 and 1945.

3 Author interview with Alfie Baines, October 1993. The fire broke out on Sunday, 1 April 1928. For an account see *High River Times*, 5 April 1928.

4 OHP, Hugh McLaughlin, 20:03. Hugh was the son of Frank McLaughlin who had been brought in as foreman at the request of Archie McLean who was running the ranch for the bank. The cups and trophies won by the Percheron horses during the Lane era were rescued.

5 *High River Times*, 5 April 1928.

6 University of Calgary Collection, 82J, 1931, A3507, #52.

7 Clifford, *Ann's Story*, p. 13.

8 Millar resented Clifford's arrival and his relationship with Pat Burns and did little to make things easy for him.

9 Yeo, "Pekisko,", p. 11.

10 Clifford, *Ann's Story*, p. 26

11 Archaeological investigations showed charred wood and rocks discoloured by heat. Rod Heitzmann, Preliminary Archaeological

Inventory and Assessment, Parks Canada, 1993. The fire took place between 1929, when Clifford started to visit the ranch regularly, and 1931 when he took up residence. Local legend talks of a party in the barn, and a lantern getting kicked over.

12 Parks Canada, Historic Structures Report, Building no. 3, Chop House, draft report, March 1995.

13 Ironically, it is less easy to be precise about the composition of the community for this time period than for the 1890s or 1900s. No employment books have survived, although they are valuable for the 1940s, and all Pat Burns' papers disappeared at his death. There was a constant turnover of men; many worked for a season or two and then left, only to return again later.

14 There were three Gibson girls; one married a Burns, the second a Clifford, and the third a man named Farrell who was Tommy Farrell's father. Tommy, like Raymond, became one of Pat Burns' most trusted lieutenants.

15 Clifford, *Ann's Story*, pp. 17–18.

16 Stockmen's Foundation, "Heritage Voices," videotaped interview with Raymond Clifford, January 1986.

17 Clifford, *Ann's Story*, p. 20.

18 Museum of the Highwood, videotaped interview with Raymond and Ann Clifford, 1989.

19 Ibid.

20 Author interview with Alfred Baines, 20 October 1992; OHP, 20-1, Hugh McLaughlin; and for context, Barbara Holliday, "Herb Millar of the Bar U," *Alberta History* (Autumn 1996): 15–21.

21 Museum of the Highwood, videotaped interview with Raymond and Ann Clifford, 1989.

22 OHP, 18-6, George Lawrence; and 11-7, Gordon Davis.

23 Author interview with Ann Clifford, 20 April 1994.

24 Museum of the Highwood, videotaped interview with Raymond and Ann Clifford, 1989; and pictures of Ann on her honeymoon: see Clifford, *Ann's Story*, pp. 9 and 10.

25 Clifford, *Ann's Story*, p. 66.

26 "I had to put on the monthly time sheets what the men's jobs were: riding, hauling hay, repairs, fencing, engine repair, driving tractor, plowing, discing harrowing, packing, seeding, cutting crop, stooking, etc." Clifford, *Ann's Story*, p. 72.

27 In 1939, Ann used a student's accounting book to detail the movements of cattle, while another notebook shows cattle sold from various Burn's ranches in 1942 and 1943. Clifford/Alwood Papers, "Cattle, Bar U Ranch," and "P. Burns Ranches, 1942–1943."

28 Clifford, *Ann's Story*, pp. 181–82.

29 Author interview with Ann Clifford, 20 April 1994.

30 Sheppard, *Just About Nothing*, p. 82.

31 OHP, 11-4, Gordon Davis.

32 Parks Canada, Oral History Interview with Louise and Angus Mackenzie, April 1996.

33 Sheppard, *Just About Nothing*, p. 82.

34 Museum of the Highwood, videotaped interview with Raymond and Ann Clifford, 1989.

35 OHP, 11-4, Gordon Davis.

36 Ibid.

37 Clifford, *Ann's Story*, p. 82.

38 Stockmen's Foundation, "Heritage Voices," interview with Raymond Clifford, 13 January 1986.

39 Clifford, *Ann's Story*, p. 83.

40 OHP, 21-1, Jack Peach. This particular immigrant story of the connection between the ranches and their Chinese cooks would be interesting to follow up.

41 OHP, 11-3, Gordon Davis.

42 OHP, 11-11, Gordon Davis.

43 OHP, 21-4, Jack Peach.

44 OHP, 20-4, Hugh McLaughlin.

45 OHP, 16-3, Mary and Earl Wildman.

46 For fishing at the E.P. see Evans, *Prince Charming Goes West*, p. 91.

47 GA, M7771, f. 48, "Brief to Excess Profits Tax Act."

48 Clifford, *Ann's Story*, p. 80.

49 OHP, 24-5, Lee Alwood. The ration books and coupons were, however, an administrative nightmare.

50 OHP, 11-1, Gordon Davis.

51 Ibid.

52 Shorty Merino had been on the Alberta range before. In 1901 he was photographed with some other cowboys on the Bar U roundup wagon. His father was a Mexican mule driver, and Shorty had spent his childhood at a Roman Catholic Mission School in B.C. until he ran away to become a cowboy. In the late 1920s Merino had worked for the Gilchrists and had spent time on the range station at Manyberries, Alberta. In 1939, Shorty was working for the Bar S in the Dorothy area.

53 OHP, 11-12, Gordon Davis.

54 OHP, 05-1, George Read.

55 Clifford, *Ann's Story*, p. 76.

56 Ibid., 68.

57 OHP, 11-5, Gordon Davis; and 6-5, Tom Feist. Another specialist who has already been described was the harness man.

58 Parks Canada, interview with Floyd Erikson, October 1996.

59 He was given a fine funeral by the Danish Club in Calgary: Author Interview with Lee Alwood, 25 May 1994.

60 Parks Canada, interview with Floyd Erikson, October 1996.

61 See Russell H. Bennett, *The Compleat Rancher* (New York: Rhinehart and Co., 1946), p. 50.

62 GA, M160 f. 103, Information re ranches and farms, 1947.

63 Tom Feist explained that on smaller ranches a rider might be able to check cattle as many as eight times a day while at the Bar U it might only be once. OHP, 06-4, Tom Feist.

64 Museum of the Highwood, videotaped interview with Raymond and Ann Clifford, 1989.

65 Clifford/Alwood Collection, Cattle Book, 1939.

66 Wild hay was cut every second year.

67 OHP, 11-7, Gordon Davis.

68 Clifford/Alwood Collection, Cattle Book. Other farmers mentioned include: A . Husky, E. Thompson, W.A. Biggs, Mrs. A. Sleaman, Mr. Weiser, A. Murdock, and E. Hoch. A typical entry reads: "250 cows moved from the Biggs place to McMillan Colony, January 6th at 40 cents per head per month."

69 Museum of the Highwood, videotaped interview with Raymond and Ann Clifford, 1989.

70 GA, M7771, f. 48, "Brief to Excess Profits Tax Act."

71 J.W.G. MacEwan, *The Breeds of Farm Live-Stock in Canada* (Toronto: Thomas Nelson, 1941), pp. 316–37.

72 Jo Ann Jones-Hole, *Calgary Bull Sale, 1901–2000* (Calgary: Friesen, 2000).

73 See ibid., Appendix, for complete figures.

74 Pierre Berton, *The Great Depression, 1929–1939* (Toronto: McClelland and Stewart, 1990), p. 9.

75 R.D. Francis and H. Ganzevoort (eds.) *The Dirty Thirties in Prairie Canada* (Vancouver: Tantalus, 1973), p. 5.

76 Gerald Friesen, *The Canadian Prairies: A History* (Toronto: University of Toronto Press, 1984), p. 396.

77 James H. Gray, *The Winter Years* (Toronto: Macmillan, 1966); and Barry Broadfoot, *Ten Lost Years, 1929–1939: Memories of Canadians Who Survived the Depression* (Toronto: McClelland and Stewart, 1997).

78 This mass of specialized work is summarized and made accessible to non-specialists in an article by Alwynne B. Beaudoin, paleo-environmentalist with the Provincial Museum of Alberta in Edmonton. See Alwynne B. Beaudoin, "What they Saw: The Climatic and Environmental Context for Euro-Canadian Settlement in Alberta," *Prairie Forum* 24, no. 1 (Spring 1999): 1–40; and David J. Sauchyn and Alwynne B. Beaudoin, "Recent Environmental Change in the Southwestern Canadian Plains," *The Canadian Geographer* 42, no. 4 (Winter 1998): 337–53.

79 Wallace Stegner, *Where the Blue Bird Sings to the Lemonade Springs: Living and Writing in the West* (New York: Random House, 1992), p. 46; see also Wallace Stegner, *The American West as Living Space* (Ann Arbor: University of Michigan Press, 1987), essay "Living Dry."

80 Antony Joern and Kathleen H. Keeler (eds.), *The Changing Prairie: North American Grasslands* (New York: Oxford University Press, 1995).

81 Howard Lamar, "Comparing Depressions: The Great Plains and the Canadian Prairie Experience, 1929–1941," in Gerald D. Nash and Richard W. Etulain (eds.), *The Twentieth Century West: Historical Interpretations* (Albuquerque: University of New Mexico Press, 1989), 175–206; and Jones, *Empire of Dust*.

82 Beaudoin, "What they Saw," p. 38. Beaudoin's analysis of the post-glacial record leads her to suggest that we have not seen the extremes of precipitation that have been experienced in the past and which could occur in the future.

83 A.K. Chakravati, "Precipitation Deficiency Patterns in the Canadian Prairies, 1921–1970," *Prairie Forum* 1, no. 2 (November 1976): 95–110.

84 Alberta, Department of Agriculture, *Annual Report*, 1929, p. 15.

85 Alberta, Department of Agriculture, *Annual Report*, 1937, p. 63.

86 Alberta, Department of Agriculture, *Annual Report*, 1938, p. 80.

87 Berton, *Great Depression*, p. 15.

88 *High River Times*, 27 July 1933.

89 GA, M2398, f. 42, Carlyle to Halsey, 22 November 1934.

90 *High River Times*, 13 February 1936.

91 *High River Times*, 27 February 1936.

92 *High River Times*, 7 May 1936.

93 *High River Times*, 9 July 1936.

94 *High River Times*, 13 August 1936.

95 Alberta, Department of Agriculture, *Annual Report*, 1931, p. 15.

96 C.W. Vrooman, "Cattle Price Fluctuations," *Canadian Cattlemen* 4, no. 4 (March 1942).

97 Alberta, Department of Agriculture, *Annual Report*, 1933.

98 Max Foran, *Trails and Trials*, p. 161.

99 Broadfoot, *Ten Lost Years*, p. 2.

100 Alberta, Department of Agriculture, *Annual Report*, 1933, p. 15.

101 G.D. Chattaway, "Views on Cattle Ranching," *Canadian Cattlemen* 4, no. 4 (March 1942).

102 Alberta, Department of Agriculture, *Annual Report*, 1934, pp. 14–15.

103 These efforts did lead to an agreement which had beneficial effects on trade in wheat and pork, but which had little impact on trade in cattle. Appendix, Table 8.1.

104 Alberta, Department of Agriculture, *Annual Report*, 1930, p. 18.

105 Foran, *Trails and Trials*, p. 145.

106 Ibid., p. 155. Foran also quotes a pertinent comment in the *Brooks Bulletin* which illustrated the proportional share of meat prices taken by wholesalers and retailers. "A slice of cow is worth 5 cents in the cow, 14 cents in the hands of the packers, and $2.40 in a restaurant which specializes in atmosphere." "Essay on the Cow," *Brooks Bulletin*, 7 April 1938.

107 Berton, *Great Depression*, p. 507.

108 *Edmonton Bulletin*, 9 March 1931.

109 Jack Byers, "Recent Activities of the Western Stock Growers' Association," *Canadian Cattlemen* 1, no. 4 (December 1938).

110 Alberta, Department of Agriculture, *Annual Report*, 1937, pp. 29–31.

111 F.M. Baker, "Livestock Production Trends in the Western Provinces," *Canadian Cattlemen* 18, no. 4 (April 1955).

112 See Appendix, Figure 8.1.

113 Clifford/Alwood Collection, "Monthly Statements." These printed returns were completed by Ann Clifford and sent to the company office in Calgary. The returns give details of expenditures for board, general expenses, wages, and the total expenses for the period from 1934 to 1950. Appendix, Table 8.2.

114 GA, M160, f. 329, P. Burns Ranches, Income, Profit and Loss Accounts; Appendix, Table 8.3.

115 GA, M160, box 7, f. 271, "1937 Valuation."

116 GA, M160, f. 270, Burns to CP, 30 August 1929.

117 Ibid., Bark to Burns, 15 March 1932.

118 Ibid., Bark to Burns, 15 March 1932.

119 Ibid., "Circle Ranch Expenses," Appendix, Table 8.4.

120 Ibid., Burns to Charlesworth, 19 March 1945.

121 GA, M7771, box 2, f. 221.

122 GA, M7771, f. 21, Ledger Book.

123 GA, M160, f. 296, Burns to prospective purchaser, 1948.

124 GA, M7771, f. 25, Financial Statements. Appendix, Table 8.5.

125 GA, M160, f. 234, Abstract of Evaluation, 1937.

126 GA, M7771, f. 48, "Brief to Excess Profits Tax Act, 1940."

127 Appendix, Figure 8.2.

128 "Canada's Late Cattle King," *Canadian Cattlemen* 1, no. 1 (June 1938): 43.

Notes to Chapter 9

1 Jordan's concluding comment on the diffusion and mixing of various cultural traditions in the North American cattle industry. Terry G. Jordan, *North American Cattle Ranching Frontiers* (Albuquerque: University of New Mexico Press, 1993), p. 309.

2 GA, M7771, f. 59, "Memorandum re Company History," 1945.

3 GA, M160, f. 92, Newton to Wilson, 16 May 1941.

4 Clifford/Alwood Collection, "Cattle Book," 1938.

5 GA, M160, f. 92, Newton to Wilson, 13 June 1941.

6 See, for example, GA, M160, ff. 130 and 133, Farrell to Deschamps, 31 March 1942 and 3 May 1948.

7 GA, M160, f. 107, Stock return, December 1941, Walrond Ranch 1,588 head and Circle 1,955.

8 GA, M7771, f. 221, Harvey et al. to shareholders, 21 February 1940.

9 GA, M160, f. 364, Lease Indenture, 27 September 1943.

10 GA, M160, f. 272, Farrell to Ruthven, 19 August 1941; a similar letter went to Palethorpe at Medicine Hat concerning the corrals at Patricia.

11 M.M. Grimsen, "Hope Oliver retires," *Canadian Cattlemen* 16, no. 7 (July 1953): 7.

12 GA, M160, f. 108, RCMP to Burns, 25 April 1944.

13 "Cattle Drives Revisited," *Canadian Cattlemen* 11, no. 3 (December 1948).

14 GA, M160, f. 271, J. Burns to Newton, 10 February 1938.

15 GA, M160, f. 92, Newton to Wilson, 13 June 1941.

16 GA, M160, f. 273, Superintendent to Burns, 13 May 1943.

17 GA, M160, f. 273, MacEwan to Gray, 14 May 1943.

18 GA, M7771, f. 4, "Brief to Excess Profits Tax Act," p. 7.

19 GA, M160, f. 271, Farrell to Dominion Experimental Farms: and Experimental Farm, Fargo, ND, 25 February 1941.

20 GA, M160, f. 272, Superintendent to Farrell, 11 September 1942.

21 GA, M7771, f. 48, "Brief to Excess Profits Tax Act," p. 6.

22 GA, M160, f. 271, Hays to Farrell, 22 February 1941.

23 GA, M160, f. 274, Gray to Royal Bank, 16 December 1943.

24 GA, M7771, f. 48, "Brief to Excess Profits Tax Act," p. 7.

25 OHP, 12.03, Allan Baker.

26 Craig Brown (ed.), *The Illustrated History of Canada* (Toronto: Lester and Orpen Dennys, 1987), p. 457; and Pierre Berton, *Marching as to War: Canada's Turbulent Years, 1899–1953* (Toronto: Doubleday Canada, 2001).

27 For an overview see G.E. Britnell and V.C. Fowke, *Canadian Agriculture in War and Peace, 1935–50* (Stanford: Stanford University Press, 1962); and Max Foran, *Trails and Trials*, chapter 6, "Extraordinary Times."

28 Britnell and Fowke, *Canadian Agriculture*, Appendix, Table 9.1.

29 Annual Market Review of the Livestock and Meat Trade, 1944.

30 For summary statistics see *Canadian Cattlemen* 8, no. 4 (March 1946): 255.

31 Britnell and Fowke, *Canadian Agriculture*, p. 265.

32 In 1940, the Western Stock Growers Association passed a resolution assuring the government of "our earnest desire to assist and cooperate by every means in our power to prosecute to a successful conclusion of the war." GA, WSGA, box 2, f. 2, "Resolutions of the 44th Annual Convention," June 1940.

33 Price ceilings on live cattle would, it was felt, be impossible to administer in view of the wide variations in quality and type of animals marketed. This claim was borne out when attempts were made to impose ceilings on live cattle sales in the United States. Britnell and Fowke, *Canadian Agriculture*, p. 263.

34 Foran, *Trails and Trials*, p. 194; and *Lethbridge Herald*, 6 May 1948.

35 "The production of the Canadian cattle industry has thus expanded by approximately one million head [per year], yet over the period the cattle population has grown by approximately two and a half million head with most of this increase having taken place in the western provinces." "Cattle Outlook," *Canadian Cattlemen* 8, no. 4 (March 1946): 170.

36 The Dominion-Provincial Agricultural Conference advised against any increase in production in 1945 and advised a reduction in 1947. Britnell and Fowke, *Canadian Agriculture*, p. 270.

37 "With your Editor – Exports," *Canadian Cattlemen* 9, no. 2 (September 1946): 74.

38 "With your Editor – Canadian Cattle Embargo," *Canadian Cattlemen* 10, no. 2 (September 1947): 64.

39 Kenneth Coppock, "Record of Industry Controls, 1941–1944," *Canadian Cattlemen* 10, no. 4 (March 1948): 193.

40 Foran gives the context and a vivid account of the meeting. Foran, *Trails and Trials*, p. 201.

41 Kenneth R. Coppock, "The Beef Cattle Outlook," *Canadian Cattlemen* 12, no. 4 (November 1949): 35.

42 Exports of all cattle for 1948 were 446,000 head, a figure only exceeded in 1919. In addition to the live cattle exported to the United States, 80 million pounds of beef were also moved across the line, bringing the total export equivalent up to 611,900 head of cattle.

43 *Annual Market Review*, 1948, quoted in Britnell and Fowke, *Canadian Agriculture*, p. 272.

44 Britnell and Fowke, *Canadian Agriculture*, p. 272.

45 Kenneth R. Coppock, "The Beef Cattle Outlook," *Canadian Cattlemen* 12, no. 4 (November 1949): 35.

46 Appendix, Table 9.1.

47 Appendix, Table 9.2.

48 Appendix, Table 9.3.

49 Appendix, Table 9.4.

50 F.M. Baker, "Livestock Production Trends in the Western Prairies," *Canadian Cattlemen* 18, no. 4 (April 1955).

51 L. B. Thomson, "Costs of Beef Production," *Canadian Cattlemen* 1, no. 1 (June 1938).

52 G.D. Chattaway, "Views on Cattle Ranching," *Canadian Cattlemen* 4, no. 4 (March 1942).

53 Knud Elgaard, *Cattle Ranching in Southern Alberta* (Regina: Canada, Department of Agriculture, 1968), p. 23

54 GA, M160, f. 271, Farrell circular, 17 April 1939.

55 Ibid.,. Farrell to Rothwell, 30 June 1939.

56 Ibid., Dobson to Farrell, n.d.

57 GA, M160, f. 273, Kelly to Farrell, 15 April 1943; see also Farrell to Ranch Managers, 31 March 1942.

58 GA, M160, f. 107, Farrell to War Services, 14 December 1941; and on behalf of Alf Cote, 15 September 1942.

59 Ibid., Gray to War Services, 17 November 1942.

60 Ibid., Kelly to Oliver, 4 November 1942.

61 GA, M160, f. 272, Bull to Farrell, 16 March 1942.

62 Ibid., Wares to manager, 3 November 1942.

63 Appendix, Table 9.5.

64 Increase from 216 man-months in the mid-1930s, to 283 man-months in the late 1940s. Clifford/Alwood Collection, "Monthly Statements, P. Burns Ranches Ltd."

65 OHP, 09.02 Roy and Lenore McLean.

66 The Mount Sentinel Ranch is located about nine miles south of the Bar U headquarters and was established in 1898 by William C. Gardner on the former Charlie Knox place.

67 OHP, 017.3, Stan Wilson. Stan Wilson was the first president of The Friends of the Bar U Historic Ranch.

68 Author interview, Judith Baker Montano, 14 February 1994.

69 Author interview, Dick and Josephine Gardner, 4 November 1992.

70 The Clifford/Alwood Collection of Bar U Ranch documents (hereafter cited as CAC), is owned by Ann Clifford's niece, Jeanette Rousseau. I am most grateful for the chance to work with the documents. Among other material, the collection contains some thirty receipt books and seven monthly time books. This data covers a period from 1941 to 1950. The books were used by Raymond Clifford and Lee Alwood. They allow one to reconstruct a picture of the kinds of contracts awarded to Indians and the payments they received. However, there are gaps and it is impossible to know how much material is missing. Appendix, Table 9.6.

71 For a list of Stoneys employed at the Bar U Ranch, 1941–49, see Appendix, Table 9.7.

72 Museum of the Highwood and Bar U National Historic Park, Oral History Project, 1992–1993. David Finch, the interviewer for the project, produced an outline of the interviews and a cumulative index, page numbers refer to this publication. The project will be referred to as OHP followed by interview number and page number. For references to Stimson see OHP, 25.01, Johnny Lefthand; and 15.02, A.A. Leitch.

73 See OHP, and in chapter 3, p. 81, a photograph of Stimson dressed in Indian regalia and a discussion of his beadwork collection.

74 OHP, 25.1, Johnny Lefthand.

75 OHP, 13.1, Mark Lefthand; and 20.05, Hugh McLaughlin.

76 OHP, 016.1, Mary Wildman.

77 GA, M4562, f. 18, "Mr. and Mrs. George Lane," pp. 1–2; and Elizabeth Lane, "Memories," p. 20.

78 *The Albertan*, 29 September 1925.

79 CAC, Receipt Book, 15 April 1942.

80 CAC, Receipt Book, 17 June 1942.

81 CAC, Receipt Book, 2 December 1942.

82 Author interview with Lee Alwood, 25 May 1994. This was a representative figure for the number of Indians paid for branding. In 1942 fifteen men were employed and in 1947, thirteen.

83 See account and pictures of branding pp. 97–102

84 The horse occupies an important place in Native culture. See Morgan Baillargeon and Leslie Tepper (eds.), *Legends of our Time* (Vancouver: University of British Columbia Press, 1998).

85 CAC, Receipt Book, 14 November 1946.

86 OHP, 006.6, Tom Feist.

87 Author interview, Dick and Josephine Gardner, November 1992.

88 This was partially adjusted by the fact that Natives were paid $2 a day in 1941, while non-Natives got $50 a month, or $1.70 a day.

89 See Appendix, Table 9.7.

90 CAC, Receipt Book, October 1941. It includes a contract with Jonas Rider for two miles of new fence, another for picking potatoes and a third for cutting and pealing logs. Sometimes, too, money earned was paid directly to the Pekisko Trading Post in payment for bills that Indians had run up there.

91 CAC, Receipt Book, July 1948; and Monthly Time Book, 1949. Interestingly enough, in March and April 1947 eight Indians were paid as individuals for various lengths of time at a rate of $70 a month *with board*.

92 George Zarn, *Above the Forks* (Calgary: Zarn, 1978), not paginated.

93 This is mentioned by several respondents; see, for example, OHP, 13.01, Mark Lefthand.

94 OHP, 24.04, L. Alwood.

95 OHP, 16.01, Mary and Earl Wildman.

96 OHP, 24.01, Lee Alwood; and 09.02, Roy and Lenore McLean.

97 OHP, 20.05, Hugh McLaughlin; and Zarn, *Above the Forks*.

98 OHP, 25.01, Johnny Lefthand.

99 Zarn, *Above the Forks*.

100 OHP, 24.04, Lee Alwood.

101 See Sarah Carter, Postscript, "'He Country in Pants' No Longer – Diversifying Ranching History," in Evans, Carter and Yeo, *Cowboys, Ranchers and the Cattle Business*, pp. 161–65.

102 In 1942, 52 per cent of earnings were from fencing contracts; 32 per cent from working with stock; and 14 per cent from farm work. The remaining 2 per cent was not specified. CAC, Receipt Books, 1942.

103 A 1941 study suggested that an animal unit required 0.75 tons of hay to winter over. A herd of 5,000 head would need 3,750 tons. Canada, *Cattle Ranching in Western Canada, An Economic Study*, 1941, p. 15.

104 V.H. Lawrence, "Haying," *Alberta Historical Review* 2, no. 1 (Winter 1973): 18–20.

105 Verlyn Klinkenborg, *Making Hay* (New York: Nick Lyons Books, 1986), p. 23.

106 Lawrence also makes a distinction between their old mower and the new high-speed one available after the turn of the century, which could cut rough fescue grass. Lawrence, "Haying," p. 19.

107 Ings, *Before the Fences*, p. 10. Clearly, there was a considerable range in the capacity and efficiency of mowers.

108 See James A. Young, "Hay Making: The Mechanical Revolution on the Western Range," *Western Historical Quarterly* 14, no. 3 (July 1983): 320–24.

109 Terry G. Jordan, Jon T. Kilpinen, and Charles F. Gritzner, *The Mountain West: Interpreting the Folk Landscape* (Baltimore: Johns Hopkins University Press, 1997), p. 106. For context refer to chapter 6, "Material Culture of Haymaking."

110 OHP, 06.02, Tom Feist.

111 CAC, Receipt Book no. 4, 2 December 1942, "$212.50 for fencing stacks."

112 OHP, 06.02, Tom Feist.

113 Roy McLean started working with a horse sweep at the age of eight, and moved on to a dump rake at the age of ten. OHP, 09.01, Roy and Lenore McLean.

114 Young, "Hay Making," p. 321.

115 Holliday, *To Be a Cowboy*, pp. 132–33.

116 Rutherford, *Cattle Trade of Western Canada*, p. 19.

117 Chapter 7, "George Lane at Full Gallop."

118 Author interview with Lee Alwood, 25 May 1994.

119 GA, M160, f. 274, Burns Ranches to Canadian Wheat Board, December 1943.

120 OHP, 18.02, George Lawrence reporting on the winter of 1940.

121 OHP, 12.03. Alan Baker said that he experienced twenty-three years running during which he never had to feed any cattle. The present experience of the Cartwrights at the D Ranch, further up Pekisko Creek toward the mountains, supports this testimony.

122 OHP, 11.04, Gordon Davis.

123 OHP, 24. 08, Lee Alwood.

124 GA, M160, ff. 291 and 297, Bar U Ranch Inventories, 1940 and 1949.

125 For a review, see David Spector, *Agriculture on the Prairies, 1870–1940* (Hull: Minister of Supply and Services, 1983).

126 For a personal account of the progress of mechanization and the disappearance of draft horses, see James R. Dickenson, *Home on the Range: A Century on the High Plains* (New York: Scribner, 1995).

127 Alberta, Department of Agriculture, 1914, "Annual Report of Alberta Horse Breeders' Association," p. 258.

128 Alberta, Department of Agriculture, 1931, "Report of the Livestock Commissioner," p. 15.

129 F.M. Baker, "Livestock Production Trends in the Western Prairies," *Canadian Cattlemen* 18, no. 4 (April 1955).

130 G.C. Elliott and G.D. Chattaway, *Cattle Ranching in Western Canada* (Ottawa: Department of Agriculture, 1941), p. 21.

131 Knud Elgaard, *Cattle Ranching in Southern Alberta* (Regina: Canadian Department of Agriculture, 1968), p. 18

132 OHP, 11.03, Gordon Davis.

133 Clifford, *Ann's Story*, p. 73.

134 OHP, 01.03, Ray Fetterley.

135 Sheppard, *Just About Nothing*, p. 69.

136 Clifford, *Ann's Story*, p. 73; and OHP, 16.04, Mary Wildman.

137 A.B. McCullough, "The Seasonal Round in Ranching Country," unpublished research paper, Parks Canada, 1997, 11–19.

138 Clifford, *Ann's Story*, p. 73.

139 OHP, 01.06, Ray Fetterley.

140 Clifford, *Ann's Story*, p. 26.

141 Holliday, *To Be a Cowboy*, p. 105.

142 Clifford, *Ann's Story*, p. 69.

143 OHP, 06.03, Tom Feist.

144 See chapter 5, "Building an Internationally Famous Percheron Stud."

145 Clifford, *Ann's Story*, p. 69.

146 GA, M160, f. 273, Loree to Burns Ranches, May 1942.

147 Clifford, *Ann's Story*, p. 65.

148 GA, M160, ff. 271 and 272, Kasa to Burns Ranches, June 1939; Wyatt to Burns Ranches, December 1941; and Greenway to Farrell, May 1941.

149 GA, M160, f. 273, Hearn to Burns Ranches, May 1942.

150 GA, M160, f. 272, Hughes to Burns Ranches, January 1941.

151 GA, M160, f. 273, Kelly to Oliver, November 1942.

152 GA, M160, f. 329, Bar U Profit Statements, 1946, 1947, and 1948.

153 Raymond Clifford retold the story in two videotaped interviews, with the Stockmen's Association in 1986; and Museum of the Highwood, 1989.

154 The post-war period witnessed a determined effort throughout western Canada to reduce the number of horses, particularly the large number of idle animals running wild and eating valuable grass. See graph of the number of horses on farms; and MacEwan, *Heavy Horses*, p. 6.

155 Joseph L. Rosettis was born in Scotland in 1909 to Lithuanian parents. He grew up in Toronto and then worked for Associated Screen News in Montreal. He was sent to western Canada to take promotional pictures for the Canadian Pacific Railway. Joe spent the early years of the war working as a staff photographer for the *Calgary Herald*. In 1943 he opened his own studio. The assignment to photograph the Burns Ranches must have been a feather in his cap and a fillip to his business. GA, Joe and Kay Rosettis, biographical sketch.

156 Biographical information from author interview with Lee Alwood, 25 May 1994; and OHP, 24.1–16, Lee Alwood.

157 OHP, 18.06, George Lawrence.

158 Parks Canada, Oral History Interview with Floyd Erickson, 1 October 1996, p. 3.

159 Parks Canada, Oral History Interview with Oliver Christensen, 1994. p. 26.

160 OHP, 24.10, Lee Alwood.

161 Doug 'Shorty' Anderson was an exception. He had been raised around horses, as his family owned a livery stable in Innisfail. He also had particular expertise with mule teams, as he had used them on a road building contract with his uncle. He was hired during the summer of 1940 to rough break and drive a mule team on the Bar U farm, and spent the following winter as a rider on the ranch. Parks Canada. Oral History Interview with Doug "Shorty" Anderson, 23 August 1995.

162 Joining the crew at the Bar U as a "steady hand" or rider was rather like becoming part of an elite regiment with a long history. A young man was surrounded by stories and legends from the past, and expectations of new recruits were correspondingly high.

163 Parks Canada, Oral History Interview with Joe Hayes, 2 July 1996.

164 GA, M160, f. 315, Burns to National Selection Service, 17 May 1944.

165 Park Canada, Oral History Interview with Floyd Erickson, 1 October 1996.

166 For a complete biography see Holliday, *To Be a Cowboy*.

167 OHP, 24.14, Lee Alwood.

168 Parks Canada, Oral History Interview with Ed Peters, 1 October 1996.

169 The account of when and for how long various women cooked at the ranch and the farm was worked out from the time books kept by the foreman. CAC, Time Books, 1941–49.

170 OHP, 24.14, Lee Alwood.

171 Myrtle's family had already established a connection with the Bar U as her father, Everett Sylvester Farnham, had worked on the farm full-time during the summers of 1941 and 1942, and had obtained part-time work there on other occasions. CAC, Time Books, 1941–49.

172 Red Deer Museum and Archives, "Memoirs of Mrs. Myrtle Farnham Grierson."

173 Ibid., "First Happy Birthday, August 3, 1945."

174 OHP, 24.05, Lee Alwood.

175 Parks Canada, Oral History Interview, Oliver Christensen, 1994, p. 26.

176 Parks Canada, Oral History Interview, Floyd Erickson, 1 October 1996. This story is collaborated by Lee Alwood, OHP, 24.03; and Holliday, *To Be a Cowboy*, p. 99.

177 Parks Canada, Oral History Interview with Bert Sheppard by Jim Taylor and Robin Gyrogy, November 1989.

178 Holliday, *To Be a Cowboy*, p. 99.

179 The proximity of the ranch corrals and calf wintering sheds to Pekisko Creek does, however, offend our modern management sensibilities.

180 For discussion of this role, see Simon M. Evans, "Grazing the Grasslands: Exploring Conflicts, Relationships and Futures," *Prairie Forum* 26, no. 1 (Spring 2001): 67–84; and Marcy Houle, *The Prairie Keepers: Secrets of the Grasslands* (New York: Addison-Wesley, 1995).

181 *Calgary Herald*, 8 August 2002.

182 *Calgary Herald*, 2 October 2002.

183 There was concern lest "idle money" in banks might be confiscated or frozen by the government. GA, M160, f. 92, Dwyer to Newton, 26 February 1941.

184 GA, M160, f. 92, Burns to Dwyer, 23 January 1941.

185 Population of metropolitan Calgary in 1941 was 94,000; in 1951 it was 141,218, an increase of 33 per cent. See "Population Growth in Calgary Metropolitan Area, 1891–1971," in Max Foran, *Calgary: An Illustrated History* (Toronto: Lorimer, 1978), p. 176; and Hugh A. Dempsey, *Calgary: The Spirit of the West* (Saskatoon: Fifth House, 1994). Dempsey has this to say: "At the outbreak of war the population was 85,000. This had increased to 97,000 by 1944 and passed the magic 100,000 mark in 1946.... By early 1946, more than 2000 veterans and their families were on the waiting list for houses in the city." p. 131.

186 GA, M160, f. 92, Newton to Burns, 18 October 1947.

187 Ibid., Dwyer to Newton, 3 February 1941.

188 Ibid., Dwyer to Newton, 10 February 1941.

189 Ibid., Dwyer to Newton, 18 February 1941.

190 Ibid., Newton to Dwyer, 21 February 1941.

191 Ibid., Wilson to Burns, 23 April 1941.

192 Ibid., Newton to Wilson, 5 and 15 May 1941.

193 Ibid., Newton to Wilson, 5 May 1941.

194 Ibid., Wilson to Newton, 10 June 1941.

195 Ibid., Burns to Armitage, 5 April 1949.

196 Ibid., Tanner to Burns, 5 July 1946.

197 Ibid., Waldner to Burns, 3 July 1945.

198 This may well have been because the McMillan Colony had been established in 1937 and was located well within the proscribed limit of forty miles from the Bar U. The provisions of the Communal Property Act (1947) precluded the Hutterites from purchasing land within forty miles of an existing colony. For an analysis of legislation concerning Hutterites, and their response to it, see Government of Alberta, Select Committee of the Assembly, *Report on Communal Property, 1972* (Edmonton: Queen's Printer, 1973); and Simon M. Evans, "The Dispersal of Hutterite Colonies in Alberta, 1918–1971: The Spatial Expression of Cultural Identity," unpublished MA thesis, University of Calgary, 1973.

199 *Calgary Daily Herald*, 22 June 1929; and GA, M2269, C.P.R. Papers.

200 GA, M160, f. 92, Rempell to Tanner, 28 August 1949, and Tanner to Burns, 1 September 1949.

201 GA, M160, f. 92, Gooderham to Newton, 24 August 1948.

202 GA, M160, f. 92. There was a long correspondence between Hunt and Burns concerning the assignment of a lease. The final letter was Hunt to Newton, 20 July 1945. Sections 14 and 15 of township 17, Range 4, West of the Fifth Principal Meridian were also incorporated into the reserve. They made up part of the Kuck place purchased by Burns in March 1945, and sold to Peter Tjebbes in 1949. GA, M160, f. 92, Kuck to Burns, March 1945; and M7771, f. 53, "Real Estate," Kuck Place.

203 See, for example, a note from Cartwright which reads, "Following up our talk yesterday with Mr. Farrell and Mr. Newton, we are particularly interested in...." and then he goes on to detail various sections of land. GA, M160, f. 92, Cartwright to Burns, 20 April 1949.

204 *High River Times*, 25 May 1950.

205 *High River Times*, 3 August 1950.

206 The map was drawn using a variety of sources, some more specific and exact than others. There was a good deal of resale and shuffling of lands in the years immediately following the breakup. I am confident that the map conveys accurately a general picture of how the ranch was divided up. However, there may be cases where odd sections passed through other hands before reaching the designated owners shown. 1. See GA, M7771, f. 53, "Real Estate." This file includes "Sales of Lands: Fifth and Sixth Tenders." This paper gives the legal description of the lands both deeded and leased; the buyer and the total price. Nine buyers were involved in these tenders and the land was all in Twp. 18, Rge. 3, and Twp. 17, Rge. 2, along the northern margins of the ranch. Unfortunately the descriptions of tenders 1 through 4 have not survived. 2. GA, M160, f. 280, "Grazing Lease Assignments, 1950–51." This file includes details of fourteen assignments of leases which covered quite a lot of the leased acreage, but by no means all of it. For example there were no assignments to J. Allen Baker and only one to Cartwright. 3. M.D. Foothills, #31, Municipal Office. Calgary Regional Planning Commission, March 1964, Map Y2-5/54. This map shows landowners who were credited with road allowances. From this it was possible to isolate the area leased to Baker. There was also a map which detailed landowners by township in 1960, but did not specify sections. 4. PAA, Township Registers, 1979, Access. No. 74.32/449-450. Grazing Leases, Roll # 188. This microfilm of the grazing lease ledger devotes a page to each township. Typically, mention is made of the original lease in the 1880s, and the subsequent assignments. This provided a check on all the lease assignments of Bar U land. 5. PAA, Access. No., 78.133, 566 a) and b), Burns Ranches. These files contain information on taxes paid on

Burns ranches and includes legal descriptions of land and payments on leases by number. This information is available from other sources. 6. GA, Land Ownership Maps of Municipalities, M.D. Foothills (1964); M.D. Willow Creek (1964); and Improvement District no. 6. These maps provide a useful check which shows how the land was owned a few years after the break-up.

207 GA, M7771, f. 53, "Real Estate."

208 Nanton and District Historical Society, *Mosquito Creek Roundup* (Calgary: Friesen, 1975), p. 263; and interview with R.W. Gardner, June 1997.

209 GA, M771, f. 13, McLean to Burns, n.d.

210 Evans, *Prince Charming Goes West*, p. 196.

211 GA, M7771, f. 53, "Real Estate"; and M160, f. 92, Kuck to Burns, March 1945.

212 GA, M160, f. 128, Burns to De Paoli, 31 August 1949; and East Longview Historical Society, *Tales and Trails: A History of Longview and Surrounding Area* (Calgary: Northwest Publishing, 1973), p. 74.

213 GA, M160, f. 217, "Purchase of the Stampede Ranch."

214 GA, M160, f. 103, "P. Burns Ranches, Information on Ranches."

215 GA, M160, f. 92, Burns to Wilson, 5 May 1941.

216 GA, M160, f. 92, Burns to Durham, 16 December 1949.

217 See expiry dates for leases, GA, M160, f. 103, Information for prospective buyers, 1947.

218 GA, M160, f. 95, Stock returns, 1945.

219 GA, M160, f. 103, P. Burns Ranches Limited: Information re Farms and Ranches.

220 Appendix, Tables 9.8 and 9.9.

221 GA, M160, f. 109, "Stock Returns."

222 GA, M160, f. 92, "Memo. re Sale of Bar U Ranch"; Appendix, Table 9.10.

223 This behaviour was even more marked where there was a Royal connection: see Evans, *Prince Charming*, p. 196.

224 *High River Times*, 16 November 1950.

225 *High River Times*, 29 March 1951.

226 Appendix, Table 9.11.

227 See 1990 land ownership maps of the Municipalities of Foothills, Willow Creek and I.D. no. 6.

Notes to Chapter 10

1 Stegner, *Where the Blue Bird Sings to the Lemonade Springs*, p. xxii.

2 Appendix, Figure 2.1.

3 MacLachlan, *Kill and Chill*, p. 21.

4 MacLachlan, *Kill and Chill*, p. 77. Gradual increments in the freight rates on grain were begun in 1983 and the Crow rate was finally abolished in 1995.

5 Lakeside Packers in Brooks is the largest plant in Canada, while the Cargill operation in High River is also a large "state of the art" plant. For details see MacLachlan, *Kill and Chill*, pp. 260–65.

6 Estimated costs of shipping cattle and beef from western Canada to Ontario, 1993 (Dollars per head).

Fed cattle live	112.50
Beef carcasses	45.50
Boxed beef	29.91

Canada, International Trade Tribunal, *An Inquiry into the Competitiveness of Canadian Cattle and Beef Industries* (Ottawa: Supply and Services, 1993), p. 21.

7 Maurice Kraut, "Outlook for the Canadian Cattle Industry," (Calgary: Canada West Foundation, 1980), p. 18; and MacLachlan, *Kill and Chill*, pp. 17–19.

8 Jeremy Rifkin, *Beyond Beef: The Rise and Fall of the Cattle Culture* (New York: Dutton, 1992).

9 Canada, Trade Tribunal, *Inquiry*, p. 79.

10 For an overview of these pressures see, Simon Evans, "Grazing the Grasslands: Exploring Conflicts, Relationships and Futures," *Prairie Forum* 26, no. 1 (Spring 2001): 67–84.

11 Jacobs, *Cattle and Us*. Jacobs explains the whole revolution amusingly: see chapter 7, "Cattle Religions."

12 Sherm Ewing tells the whole story of the introduction of exotic breeds to North America in the words of the major players. He conveys the risks taken and the frustrations which developed with government regulations on both sides of the Atlantic. It surely was an exciting time. Sherm Ewing, *The Ranch: A Modern History of the North American Cattle Industry* (Missoula: Mountain Press Publishing Co., 1995).

13 Jacobs, *Cattle and Us*, pp. 144–47.

14 MacLachlan, *Kill and Chill*, p. 46.

15 Pat Burns was seventy when he bought the ranch.

16 There are obvious problems with writing "biography" while some of the protagonists are still alive. The passage of time provides a necessary perspective before decisions and their outcomes can be fairly evaluated. Moreover, events are too recent to be part of the public archival record.

17 He owned the ranch for twenty-seven years, from 1950 to 1977.

18 Elgaard found that there were only twenty ranches in the foothills over six thousand acres, and he had to omit these from his study because the class was too small to include. Knud Elgaard, *Cattle Ranching in Southern Alberta* (Regina: Canada Department of Agriculture, 1968), p. 4.

19 *Canadian Cattlemen* 22, no. 1 (January 1959): 3ff.

20 OHP, 12.03, Allen Baker.

21 Ewing, *The Ranch*, p. 309; and Jacobs, *Cattle and Us*, p. 156.

22 For an overview, see "The Ranching Industry in Canada: Report on Evaluation of Potential Sites for Commemoration," Report Prepared for the Historic Sites and Monuments Board of Canada, November 1989.

23 Harry Tatro, "Survey of Historic Ranches," Agenda Paper, Historic Sites and Monuments Board of Canada, June 1977.

24 Historic Sites and Monuments Board of Canada, "Ranching Industry," p. 5.

25 *Calgary Herald*, 24 March 1991.

26 Doug Nelson, "Preserving our heritage," *Canadian Cattlemen* 59, no. 5 (May 1996): 26.

27 Parks Canada, Bar U Ranch National Historic Site: Management Plan 1995, foreword.

28 "The Mighty Bar U," won an Iris award: Nelson, "Preserving our Heritage," p. 26.

29 For an overall appreciation by an eastern-based newspaper, see *Globe and Mail*, 13 March 1996.

30 For explanation of this term, see Simon Evans, "Tenderfoot to Rider: Learning 'Cowboying' on the Canadian Ranching Frontier during the 1800s," in Evans, Carter, and Yeo, *Cowboys, Ranchers and the Cattle Business*, 72.

31 Brian W. Dippie, "Charles M. Russell, Cowboy Culture, and the Canadian Connection," in Evans, Carter, and Yeo, *Cowboys, Ranchers and the Cattle Business*, p. 25.

32 R. Douglas Francis, *Images of the West: Reponses to the Canadian Prairies* (Saskatoon: Western Producer, 1989); see particularly "The West, the Nation and the Empire, 1845–1885."

33 Larraine Andrews, "Under Siege," *Alberta Views* 5, no. 3 (May/June 2002): 29.

34 Paul F. Starrs, *Let the Cowboy Ride*, p. 254.

35 "Big sky takes on big oil," *Fast Forward*, 12 March 2003, p. 14; and *Calgary Herald*, 2 October and 9 November 2002.

36 *Calgary Herald*, 15 January 2003.

37 See Evans, "Grazing the Grasslands," p. 79.

38 *Calgary Herald*, 25 March 2003.

39 Andrews, "Under Siege," p. 34.

Index

C

E

F

I

J

K

L

M

N

O

P

Q

R

S

T

U

V

W

Y

Z